TRANSDERMAL AND INTRADERMAL DELIVERY OF THERAPEUTIC AGENTS

APPLICATION OF PHYSICAL TECHNOLOGIES

TRANSDERMAL AND INTRADERMAL DELIVERY OF THERAPEUTIC AGENTS

APPLICATION OF PHYSICAL TECHNOLOGIES

AJAY K. BANGA

CRC Press
Taylor & Francis Group
Boca Raton London New York

CRC Press is an imprint of the
Taylor & Francis Group, an **informa** business

CRC Press
Taylor & Francis Group
6000 Broken Sound Parkway NW, Suite 300
Boca Raton, FL 33487-2742

First issued in paperback 2019

© 2011 by Taylor and Francis Group, LLC
CRC Press is an imprint of Taylor & Francis Group, an Informa business

No claim to original U.S. Government works

ISBN-13: 978-1-4398-0509-1 (hbk)
ISBN-13: 978-0-367-38278-0 (pbk)

Visit the Taylor & Francis Web site at
http://www.taylorandfrancis.com

and the CRC Press Web site at
http://www.crcpress.com

Dedication

This book is dedicated to my wife, Saveta, and my children, Manisha and Anshul, for their support and understanding during the preparation of this book.

Contents

Preface

About 10 years ago, I wrote a book with Taylor & Francis entitled *Electrically Assisted Transdermal and Topical Drug Delivery*. The book detailed two physical enhancement technologies—iontophoresis and electroporation—for delivering a range of drug molecules, especially macromolecules such as proteins (for example, insulin). However, since then, several developments have taken place in this field, such as the shift in electroporation applications from drug delivery to electrochemotherapy and DNA vaccination (Chapter 5). Recent advances in several fields, such as biomedical chemical engineering; material and pharmaceutical sciences; and collaboration among scientists has led to rapid progress in the development of physical enhancement strategies for intradermal/transdermal delivery. This resulted in the development of several other exciting skin enhancement technologies, such as skin microporation using microneedles, thermal ablation, radio-frequency ablation, and laser ablation, which results in micron-sized microchannels in the skin, thus enabling delivery of drugs. Some other technologies, such as sonophoresis, are also being actively investigated (Chapter 6). As a result, several products, based on physical enhancement strategies (see Chapter 10) which are the focus of this book, have been introduced in the marketplace, and several others are in the clinical development phase. Therefore, a decision was made to write a new book with an expanded scope, which will not only update the original two topics, but also add content relevant to several new developments. The devoted efforts resulted in this book and include the delivery of conventional drug molecules, peptides, proteins, and vaccines via physical enhancement methods for targeting into (intradermal) and through (transdermal) the skin.

This book is divided into 10 chapters, organized to introduce the various skin enhancement techniques and discuss their unique applications and commercialization. An introduction is provided in Chapter 1, followed by a discussion of experimental methods in Chapter 2 that will provide several useful tips to researchers. Microneedles and other skin microporation technologies are discussed in Chapter 3. The commercial development of these skin microporation technologies as well as the commercial development of iontophoresis are presented in Chapter 10. Fundamentals of iontophoretic delivery and its preclinical and clinical applications are addressed in Chapter 4. Chapter 7 provides a comparative evaluation of the various physical enhancement strategies and lists applications to specific drugs. This information will be useful to the industry in developing programs and making strategic deals with companies showcasing promising technologies. One of the most promising applications of microneedles, intradermal vaccination, is described in Chapter 9 along with other efforts for transcutaneous immunization. The delivery of proteins is discussed in Chapter 8.

I would like to thank my current and former graduate students and postdoctoral fellows for their direct or indirect help in putting this book together. My special thanks to Haripriya Kalluri, Theresa Montague, and Vishal Sachdeva for their assistance in the preparation of this book. Finally, I would like to thank the staff of Taylor & Francis for their efforts in bringing this book to the readers.

About the Author

Ajay K. Banga, Ph.D., is professor and department chair in the Department of Pharmaceutical Sciences at the College of Pharmacy and Health Sciences, Mercer University, Atlanta, Georgia. He also holds an Endowed Chair in transdermal delivery systems. His research expertise is in nontraditional approaches for transdermal drug delivery, especially for water-soluble drugs, including small conventional molecules and macromolecules. Dr. Banga has a Ph.D. in pharmaceutics from Rutgers University, New Jersey. He is currently major advisor to 10 Ph.D. students and has graduated 15 students under his guidance. Dr. Banga has more than 220 publications and scientific abstracts to his credit. His laboratory is funded by several start-up and leading pharmaceutical companies. He currently serves on the editorial boards of seven journals (as associate editor for one journal), and has served as the editor-in-chief (2002 to 2007) for a drug delivery journal. Dr. Banga has written two books in the areas of delivery of proteins and transdermal delivery. He has served on over 30 thesis/dissertation advisory committees and as a referee for more than 28 journals. He has given invited lectures in national and international conferences, pharmaceutical companies, federal agencies, and academic institutions. Dr. Banga is a Fellow of the American Association of Pharmaceutical Scientists.

1 Percutaneous Absorption and Enhancement Strategies

1.1 INTRODUCTION

Skin is the largest organ of the human body, with a surface area of about 2 m². Historically, skin was viewed as an impermeable barrier, as its primary purpose is to protect the body from the entry of foreign agents. However, we now recognize that intact skin can be used as a port for topical or continuous systemic administration of drugs. Local or topical dermatological delivery is useful when skin is the target site for the medication. Skin can also be used as a route of administration for systemic delivery of the drug via a transdermal patch. For drugs that have short half-lives, a transdermal route provides a continuous mode of administration, somewhat similar to that provided by an intravenous infusion. However, unlike an intravenous infusion, delivery is noninvasive, and no hospitalization is required. Once absorbed, hepatic circulation is bypassed, thus avoiding another major site of potential degradation (1–5). Drugs such as buprenorphine, capsaicin, clonidine, estradiol, fentanyl, granisetron, methylphenidate, rivastigmine, rotigotine, selegiline, nicotine, nitroglycerin, oxybutynin, scopolamine, and testosterone have been successfully marketed for transdermal delivery, with several brands available for many of these drugs. Patches with an estrogen and progestin combination for birth control and lidocaine patches for topical use are also available. A list of marketed products and those in clinical trials was compiled in Table 1.1. In addition, a salicylic acid patch is available for local treatment of plantar warts.

In the last 5 years, a new transdermal patch has been approved every 7.5 months (6), indicating the popularity and importance of the transdermal route of drug delivery. These drug patches are marketed in both the over-the-counter (OTC) and prescription (Rx) drug markets; furthermore, several cosmetic products are also available in the patch dosage form. The global market for all dermatological and transdermal products is in the billions (see Chapter 10). Nicotine patches have been successfully used by over a million smokers to quit smoking (7). Unlike a new drug entity that can cost over $500 million and take as long as 15 years to develop, a transdermal patch for an existing drug can be developed in 4 to 8 years at a fraction of the cost, around $10 to $15 million (8). Commercialization of transdermal drug delivery has required technology from many disciplines beyond pharmaceutical sciences, such as polymer chemistry, adhesion sciences, mass transport, web film coating, printing, and medical technology (9,10). For instance, while the use of electroporation to enhance penetration of chemotherapeutic agents into tumors is relatively new, the technique has been widely used to insert genes into cells for recombinant DNA work.

1

TABLE 1.1

Transdermal Products in Clinical Development or on the Market in the United States

Drug	Trade Name	Marketer	Technology	Developmental Status
Alprazolam	Alprazolam Patch	ZARS Pharma.	Amprivo™	Phase 1 completed
Buprenorphine	Butrans™	Purdue Pharma.	Passive patch	On market
	—	Zosano Pharma.	Drug coated microprojections/ Macroflux tech.	Phase 1
Capsaicin	Qutenza™	NeurogesX	Passive patch	On market
Clonidine	Catapres-TTS®	Boehringer Ingelheim	Passive patch	On market
	Clonidine Transdermal System	Mylan Tech.	Passive patch	On market
Cosmeceuticals/antiwrinkle patch	Isis Patch	Isis Biopolymer	Iontophoresis	On market
Diclofenac	Flector®	IBSA Institut Biochem	Passive patch	On market
Desmopressin		Zosano Pharma	Drug-coated microprojections/ Macroflux tech.	Phase 1
Estradiol	Vivelle®	Novartis	Passive patch	On market
	Estraderm®	Novartis	Passive patch	On market
	Climara®	Bayer Healthcare	Passive patch	On market
	Estradiol Transdermal System	Mylan Tech.	Passive patch	On market
	Alora®	Watson	Passive patch	On market
	Menostar®	Bayer Healthcare	Passive patch	On market
Estradiol Levonorgestrel	Climara Pro®	Bayer Healthcare	Passive patch	On market
Estradiol Norethindrone Acetate	CombiPatch®	Novartis	Passive patch	On market
Ethinyl Estradiol Norelgestromin	Ortho Evra®	Ortho McNeil Janssen	Passive patch	On market
Fentanyl	Duragesic®	Janssen (J&J)	Passive patch	On market
	Fentanyl Transdermal System	Mylan Tech.	Passive patch	On market

Drug	Product	Company	Technology	Status
	Fentanyl Skin Patch	Altea Therapeutics	Thermal ablation/Passport® technology	Phase 1
GLP-1 Analog	Vivaderm GLP-1 Analog	Trans Pharma Medical	Radio-frequency ablation/Vivaderm	Phase 1
Granisetron	Sancuso®	ProStrakan	Passive patch	On market
	ViaDerm-Granisetron	Trans Pharma Medical	Radio-frequency ablation/ViaDerm	Phase 1
Human Growth Hormone	ViaDerm-hGH	Trans Pharma Medical	Radio-frequency ablation/ViaDerm	Phase 2
Influenza Vaccine	—	Zosano Pharma	Drug coated microprojections/Macroflux tech.	Phase 1
	Intanza®/IDflu®	Sanofi Pasteur Ltd.	BD Soluvia microneedle technology	On market in Europe
Insulin	Insulin Skin Patch	Altea Therapeutics	Thermal ablation/Passport® technology	Phase 1
	—	Zosano Pharma	Drug coated microprojections/Macroflux tech.	Phase 1
	ViaDerm-Insulin	Trans Pharma Medical	Radio-frequency ablation/ViaDerm	Phase 1
Lidocaine	Lidoderm®	Endo Pharm.	Passive patch	On market
Lidocaine and Tetracaine	Synera®	ZARS Pharma	CHADD®	On market
Lidocaine HCl and Epinephrine	Lidosite®	Vyteris	Iontophoresis	On market
Methylphenidate	Daytrana™	Shire	Passive patch	On market
Nicotine	Nicoderm®	Ortho McNeil Janssen	Passive patch	On market
	Habitrol	Novartis	Passive patch	On market
	Nicotrol	McNeil Consumers	Passive patch	On market
	Prostep	Lederle Lab.	Passive patch	On market
	Nicoderm CQ®	GSK	Passive patch	On market

(Continued)

TABLE 1.1 (*Continued*)
Transdermal Products in Clinical Development or on the Market in the United States

Drug	Trade Name	Marketer	Technology	Developmental Status
Nitroglycerin	Nitrodisc	Roberts Pharm.	Passive patch	On market
	Transderm-Nitro	Novartis	Passive patch	On market
	Nitro-Dur	Schering/Key	Passive patch	On market
	Minitran™	Graceway Pharm.	Passive patch	On market
	Deponit	Schwarz Pharma	Passive patch	On market
	Nitroglycerin Transdermal System	Mylan Tech.	Passive patch	On market
Oxybutynin	Oxytrol®	Watson Lab.	Passive patch	On market
Parathyroid hormone	—	Zosano Pharma	Drug-coated microprojections/ Macroflux tech.	Phase 2
Rivastigmine	Exelon®	Novartis	Passive patch	On market
Selegiline	Emsam®	Mylan/Somerset	Passive patch	On market
Scopolamine	Transderm-Scop	Novartis	Passive patch	On market
Sumatriptan	Zelrix	Nupathe	Smart Relief (Iontophoresis)	Phase 3 completed
Testosterone	Testoderm	J&J	Passive patch	On market
	Androderm	Watson	Passive patch	On market
	Androderm Second Generation	Watson	Passive patch	Phase 3
Zolmitriptan	Vyteris Active Patch	Vyteris	Iontophoresis	Phase 1 completed

Transdermal delivery cannot and need not be used in every situation. A rationale to explore this route typically exists for drugs that are subject to an extensive first-pass metabolism when given orally or must be taken several times per day. Even then, only potent drugs can be administered through this route, because there are economical and cosmetic reasons that restrict the patch size. Though it is hard to make generalizations, the maximum patch size has been suggested to be about 50 cm^2, and the maximum possible dose that can be delivered may be around 50 mg per day (11). Furthermore, the drugs should be moderately lipophilic (typically, log P of about 1 to 3) to be able to have significant passive permeation into the skin and then be able to diffuse out from the skin into the aqueous systemic circulation. Lipophilic drugs with log P > 3, such as some antifungals and corticosteroids, tend to form a reservoir or depot in the skin. Passive permeation is also generally considered to be limited to drugs with a molecular weight (MW) less than 500 Da (12). Other factors that are important include solubility in the skin (13) and other parameters based on the physicochemical properties of the drug, including melting point, water solubility, number of atoms available for hydrogen bonding, as well as the previously mentioned MW and octanol-water partition coefficient (14). As the melting point decreases, solubility in skin increases, and therefore, flux increases. Because enantiomers may have different melting points, the lower melting form enantiomer may have higher flux than the other enantiomer or as compared to the racemic mixture. Similarly, eutectic mixtures have lower melting points and therefore may have higher flux (e.g., EMLA cream) (see Section 7.4.1.1). One of the factors limiting more molecules from being commercialized as skin patches is skin irritation, which is discussed in Chapter 10. Hydrophilic molecules and macromolecules do not normally permeate through the skin, and their delivery by physical enhancement technologies is the primary focus of this book.

1.2   STRUCTURE AND ENZYMATIC ACTIVITY OF THE SKIN

1.2.1   SKIN STRUCTURE

Skin is composed of an outer epidermis, an inner dermis, and the underlying subdermal tissue. A basement membrane separates the epidermis and dermis, whereas the dermis remains continuous with the subcutaneous fat layer (hypodermis) and adipose tissues (15,16). The dermis (or corium) is composed of a network of connective tissue consisting primarily of collagen fibrils and provides physiological support for the epidermis by supplying it with blood and lymphatic vessels. The epidermis includes several physiologically active epidermal tissues and the physiologically inactive stratum corneum. The physiologically active epidermis contains keratinocytes as the predominant cell type. Skin also has other cell types that represent the nonkeratinocytes which include melanocytes (pigment formation), Merkel cells, and Langerhans cells. Melanocytes produce the pigment melanin that can absorb ultraviolet (UV) radiation to protect the skin. Merkel cells reside just above the basement membrane, and these scarce cells are assumed to mediate touch. Langerhans cells mediate the immunological function and the immunology of skin as discussed in Chapter 9. Keratinocytes originate in a layer called stratum germinativum (basal layer) and undergo continuous differentiation and mitotic activity during the course of migration upward through the layers of spinosum, granulosum, and lucidum. The cells migrating from the stratum germinativum layer are slowly transforming and dying as they move upward away from their source of oxygen and nourishment. Upon reaching the stratum corneum, these cells are cornified and dead. The time required for the cells to proliferate from the stratum germinativum to the stratum corneum is about 28 days, of which 14 days are spent as corneocytes in the stratum corneum layer. Corneocytes are then sloughed off from the skin and into the environment (about one cell layer per day), a process called *desquamation*. The stratum corneum or the horny layer is the topmost layer of dead, flattened, keratin-filled cells (corneocytes). The stratum corneum is actually composed of about 10 to 15 layers of these flattened, cornified cells. The outermost layers of stratum corneum stay hydrated due to the presence of a natural moisturizing factor (NMF) that is a humectant composed of several components including a mixture of free amino acids. The entire epidermis is avascular and is supported by the underlying dermis. Skin also has several appendages, which include hair follicles, sebaceous glands, and sweat glands, occupying only about 0.1% of the total human skin surface (17). The hair follicle, hair shaft, and sebaceous gland form what is commonly termed the *pilosebaceous* unit and are discussed in more detail in the next section. Infants have a higher pore density because the pores of the skin are fixed at birth and do not regenerate. Resident flora in pores is usually not composed of pathogens. The pilosebaceous follicles have about 10% to 20% of the resident flora and cannot be decontaminated by scrubbing. Sebaceous glands secrete sebum, a mixture of triglycerides, phospholipids, and waxes. The exact purpose of sebum is unknown, but it helps to lubricate skin, and along with sweat, it helps to maintain skin surface pH around 5. Sebaceous glands are absent on the palms, soles, and nail beds. Sweat glands or eccrine glands respond to temperature via parasympathetic nerves, except on palms, soles, and axilla, where they respond to emotional stimuli via sympathetic nerves.

The total thickness of skin is about 2 to 3 mm, but the thickness of stratum corneum is only about 10 to 15 μm. However, most of the epidermal mass is concentrated in the stratum corneum, and this layer forms the principal barrier to the penetration of drugs. This rate-limiting barrier to transdermal permeation is packed with hexagonal cells, an arrangement that has provided a large surface area with the least mass. Species difference exists (e.g., the cells are stacked in vertical columns in mice but are distributed randomly in humans). In contrast to most other epithelia, the barrier function of epidermis is not based on tight junctions between the cells but rather on the lipid lamellae of the stratum corneum. Each corneocyte is bounded by a thick, proteinaceous envelope with the tough fibrous protein keratin as the main component. Earlier reports based on transmission electron microscopy suggested that the spaces between corneocytes are empty; however, this is now believed to be an artifact of sample preparation. In fact, an intact stratum corneum is a highly ordered structure. Freeze-fracture electron microscopy studies on hydrated human stratum corneum have confirmed that corneocytes are embedded in intercellular lipids, and are aligned parallel to the surface of the stratum corneum (18). As the epithelial cells migrate upward toward the stratum corneum, their plasma membrane seems to thicken due to a deposition of material on its inner and outer surface. This is the process of keratinization during which the polypeptide chains unfold and break down and then resynthesize into keratin, which is a tough, fibrous protein that forms the main component of the corneocyte. With increasing understanding of the skin, it is now being recognized that keratin represents a group of proteins that play a very important role as scaffolding filaments within epithelial cells (19). Corneocytes contain a compact arrangement of α-keratin filaments 60 to 80 Å in diameter and distributed in an amorphous matrix. The intercellular spaces of the stratum corneum are completely filled with broad, multiple lamellae. The lipids constituting these lamellae are composed of ceramides, cholesterol, and free fatty acids in approximately equal quantities (20). In contrast, the lipid composition in the viable epidermis is predominantly composed of phospholipids. These changes in the lipid composition have been demonstrated to occur during the keratinization process of the stratum corneum. The morphology of these lipids plays an important role in the barrier function of the stratum corneum (21). The dermis is composed of a loose connective tissue and contains collagen and elastin fibers. In the dermis, glycosaminoglycans or acid mucopolysaccharides are covalently linked to peptide chains to form proteoglycans, which is the ground substance that promotes the plasticity of the skin. In the ground substance, spindle-shaped cells called *fibroblasts* are interspersed between collagen bundles. In addition, mast cells are present in the ground substance. Nerves, blood vessels, and lymphatics are also present in the dermis. Lymphatics reach to the dermo-epidermal layer, and lymphatic uptake can be important for delivery of macromolecules via the skin (see Sections 9.2 and 10.5). Epidermal appendages such as hair follicles and sweat glands are also embedded in the dermis (22).

1.2.1.1 Hair Follicles in the Skin

Recently, there has been considerable interest in transfollicular delivery of drugs and particulates; therefore, an understanding of the pilosebaceous unit composed of the hair follicle, hair shaft, and sebaceous glands is important. The hair follicle is an invagination of the epidermis deep into the dermis. It is associated with a capillary network and therefore provides an effective area of absorption if drug molecules can enter the follicles. The follicular openings typically occupy only about 0.1% of skin surface but can be as high as 10% on the face and scalp. There are no hair follicles on the lips, palms, and soles of the feet (23). Follicle density is typically around 14 to 32 follicles/cm^2 in the back, thorax, arm, thigh, or calf regions but can be as high as 292 follicles/cm^2 on the forehead. Follicle diameters comparatively have less variation but still may vary from about 30 to 80 μm (24). Terminal hairs extend about 3 mm into the hypodermis, while the shorter vellus hairs such as on the face extend only about 1 mm into the dermis. The follicular shaft consists of sebum, which is a bacteriostatic and fungistatic mix of short-chain fatty acids. The composition varies among species, but human sebum is rich in free fatty acids, squalene, wax esters, triglycerides, cholesterol, and cholesterol esters (25). Sebum may facilitate the transport of lipophilic molecules over hydrophilic molecules due to its fatty acid content. Accumulation of drug molecules in follicles has been reported to increase with increasing dye lipophilicity and with increasing content of propylene glycol in formulations. In contrast, follicular passive transport of calcein, a hydrophilic charged dye was negligible but was enabled by iontophoresis (26). It may also be possible to modify drug delivery by using a solvent such as ethanol to dissolve and draw out sebum from the follicles. Uptake of particles into hair follicles and skin microchannels is discussed in Section 3.2.6.

1.2.2 ENZYMATIC ACTIVITY OF THE SKIN

Skin has considerable enzymatic activity that includes cytochrome P450 isozymes that may be localized to specific cell types, especially in the epidermis and pilosebaceous system. Enzymes identified in the stratum corneum include lipase, protease, phosphatase, sulfatase, and glycosidase activity (27). The total skin blood flow is only about 6.25% of the total liver blood flow. Thus, the metabolism is relatively lower in skin though the spectrum of reactions in the skin is similar to those observed in the liver (28). The distribution of hydrolytic enzymes in the skin that metabolize β-estradiol 17-acetate to β-estradiol has been reported to be species dependent but was the highest in the basement layer of the epidermis for human skin (29). Xenobiotic substances are first chemically activated by oxidation with the involvement of cytochrome P450 isozymes. These enzymes are localized mainly in the endoplasmic reticulum, and the activity is highest in the microsomal fraction of skin homogenates. The catalytic activity of the enzymes in hair follicles is particularly high. While epidermal activities of cytochrome P450 in skin are only about 1% to 5% of those in the liver, the transferase activity in skin can be as high as 10% of hepatic values (30). The enzymatic activity of the skin varies with the anatomical site. For instance, hydrocortisone 5α-reductase activity was detected only in the human foreskin, while high levels of testosterone 5α-reductase were found in the scrotal skin. The distribution of enzymatic activity within the various skin layers is not well known due to difficulties with the experimental methodologies. Because blood capillaries lay just under the epidermis–dermis junction, drugs may only have minimal contact with dermal enzymes before they are taken up by the general circulation. Thus, the enzymatic activity of the epidermis may be more important as a barrier to drug absorption (22). Enzymatic hydrolysis of drug in skin has been reported (31) and may differ between *in vivo* and *in vitro* conditions, with *in vitro* results sometimes overestimating metabolism due to increased enzymatic activity and lack of removal by capillaries (32,33). The proteolytic activity of the skin is discussed separately in Section 8.3.1.

1.3 PERCUTANEOUS ABSORPTION

1.3.1 MECHANISMS OF PERCUTANEOUS ABSORPTION

The stratum corneum is a predominantly lipophilic barrier that minimizes transepidermal water loss. It has several defensive functions but is best known as the principal barrier to percutaneous absorption (34). Transdermal delivery of small molecules has thus been considered as a process of interfacial partitioning and molecular diffusion through this barrier. A typical mathematical model treats stratum corneum as a two-phase protein-lipid heterogeneous membrane having the lipid matrix as the continuous phase (35–37). Several theoretical skin-permeation models have been proposed which predict the transdermal flux of a drug based on a few physicochemical properties of the drug (38–42). These models often make some assumptions about the barrier properties of skin and predict the transdermal flux of a drug from a saturated aqueous solution, given the knowledge of water solubility and molecular weight of the drug and its lipid-protein partition coefficient. An analogy of "bricks and mortar" is often given for this model. Based on this model, drugs can diffuse through the stratum corneum via a transepidermal or a transappendageal route. Transepidermal drug penetration through the stratum corneum can take place between the cells (intercellular) or through the protein-filled cells (transcellular route). The relative contribution of these routes depends on the solubility, partition coefficient, and diffusivity of the drug within these protein or lipid phases. The transappendageal route has generally been considered to contribute only to a very limited extent to the overall kinetic profile of transdermal drug delivery. It has been shown that the penetration of retinoic acid was greater through hairless guinea pig as compared to haired guinea pig skin, again suggesting that the structure and composition of the stratum corneum are more important than follicular density for passive permeation (43). However, several factors play a role in determining which transport pathways will dominate. For example, during iontophoretic delivery of ionic molecules, hair follicles and sweat ducts can act as diffusion shunts for permeation. A detailed discussion of the mechanisms of iontophoretic delivery can be found in Chapter 4.

1.3.2 THEORETICAL BASIS OF PERCUTANEOUS ABSORPTION

Passive diffusion of a nonelectrolyte in the absence of any bulk flow is expressed by Fick's first law of diffusion as:

$$J = -D \, dC/dx \tag{1.1}$$

where J is the flux, D is the diffusion coefficient, and dC/dx is the concentration gradient over a distance x. Fick's first law of diffusion can be used to describe skin permeation of drugs; however, the concentration gradient across skin tissue cannot be easily measured but can be approximated by the product of permeability coefficient (P_s) and concentration difference across the skin (C_s). The steady-state transdermal flux, J_S, through the skin barrier is thus given as:

$$J_S = P_S \, C_S \tag{1.2}$$

where P_S, the permeability coefficient, is defined by

$$P_S = K \cdot D/h \tag{1.3}$$

where K is the partition coefficient, and h is the thickness of skin. The cumulative amount of drug permeating through the skin (Q_t) is given by

$$Q_t = K \cdot D \cdot C_S/h \, (t - h^2/6D) \tag{1.4}$$

where C_S is the saturated reservoir concentration when a sink condition is maintained in the receptor solution. Differentiation of Equation 1.4 with respect to time will yield Equation 1.2, which describes the steady-state transdermal flux. When the steady-state line is extrapolated to the time axis, the value of lag time, t_L, is obtained by the intercept at $Q = 0$.

$$t_L = h^2/6D \qquad (1.5)$$

The intercept t_L is a measure of the time it takes for the penetrant to achieve a constant concentration gradient across the skin. The lag time method is commonly used for analysis of permeation data from *in vitro* experiments with an infinite dosing technique (i.e., where skin is separated by an infinite reservoir of drug on the donor side and a perfect sink as receptor). The steady-state flux is calculated from the slope of the linear permeation profile, and the x-intercept provides the lag time of diffusion. The diffusion coefficient can then be calculated from Equation 1.4 using the known donor phase concentration, thickness of the barrier, and the measured partition coefficient. However, the lag time method could be subjective, as it requires judgment to determine the linearity of the permeation profile (44). Because the diffusion coefficients of commonly used drugs across skin range from 10^{-6} to 10^{-13} cm²/s, lag times across skin can range from a few minutes to several days. Low lag times are required for most therapeutic indications. Physical enhancement strategies discussed in this book such as microporation and iontophoresis can provide the means to lower lag times.

Considering the complexity of skin, these equations do a reasonably good job in analyzing and predicting the data from skin flux experiments (16,45). Using equations and physicochemical properties of drug molecules, several modeling studies and mathematical models have been used to predict percutaneous absorption across stratum corneum for both hydrophilic and hydrophobic drugs based on mechanistic and empirical approaches (5,46–50). Nevertheless, it should be recognized that percutaneous absorption can vary from site to site and person to person. Variability of 30% to 40% is not unusual and may confound studies designed to find out how gender, age, and race may affect transdermal delivery.

1.4 PASSIVE PERMEATION ENHANCEMENT STRATEGIES

This section will briefly discuss the formulation-type approaches to enhance percutaneous absorption of drugs. These include chemical enhancers, prodrugs, nanocarriers such as microemulsions, liposomes, or similar approaches that do not breach the skin by a physical force and do not apply an active driving force on the drug. Several other variations of such approaches may fit here (e.g., charged molecules will normally not penetrate the skin but can be made to form a lipophilic ion pair with an oppositely charged species, and the ion pair can have some passive permeability) (51,52). However, a detailed discussion of these approaches is outside the scope of this book.

1.4.1 Chemical Penetration Enhancers

Chemical penetration enhancers have been widely investigated by many researchers over the last 20 years. However, most of this research has not translated into products in the marketplace. During this same time period, several products were introduced in the market based on physical enhancement strategies (see Chapter 10), which are the focus of this book. The use of chemical penetration enhancers to enhance percutaneous absorption of drugs was reviewed (53–58). A penetration enhancer is usually a small molecule, and mechanisms of enhancement include modification of intercellular lipid matrix with increased membrane fluidity. Over 360 chemicals have been tested as enhancers, and examples of commonly investigated chemical enhancers include 1-dodecylazacycloheptan-2-one (Azone or laurocapram), alkanols, alkanoic acids and their esters, amides (including urea), bile salts, chelating agents, cyclodextrins, dimethylsulfoxide (DMSO), essential oils, ethanol (59), fatty acid derivatives, iminosulfuranes, isopropyl myristate (60), octyl salicylate, oleic acid, surfactants, and terpenes (57,61–63). Many of these surfactants work better when the formulation also contains either ethanol or propylene glycol. In addition, water is an effective penetration enhancer (51). Azone, with a log P of around 6.2, was especially designed as a chemical enhancer and is effective at low concentrations. Nonionic surfactants are generally safer for the skin but are less effective than ionic surfactants as enhancers. Based on several high-throughput screening studies, it was suggested that a combination of some enhancers may act synergistically and may therefore be more effective. These screening methods are typically based on the effect of enhancer combinations on the electrical conductivity of the skin. A plate having several wells (donors) is superimposed on another plate having matching wells (receivers) with the skin clamped in between the plates. This allows high-throughput screening of several wells quickly with simple resistance measurements after treating with enhancers, and the procedure uses very little skin. Results have been validated using model drugs and Franz diffusion cells (64–67). Chemicals such as ethanol, oleic acid, or oleyl alcohol have been used in some marketed estradiol or combination patches. Ethanol has also been used in some marketed fentanyl and testosterone patches. Chemicals used in sunscreens are being developed for use in commercial products (see Section 7.4.1). Some products that have chemical enhancers do not claim them as enhancers (e.g., a product may contain propylene glycol as a solvent, with the drug solubilized by a surfactant and terpene added as fragrance to the formulation) (16).

Penetration enhancers may be promising for some smaller molecules, but their use is unlikely to be successful for delivery of macromolecular drugs in therapeutic levels. Also, because enhancers modify the structure of skin (68,69), reversibility of effect and safety upon long-term usage will have to be demonstrated to get regulatory approval. In addition to increasing absorption of the drug, enhancers may nonspecifically increase their own absorption or that of formulation excipients. Thus, long-term local and systemic toxicity of chemical enhancers needs to be evaluated once they have been formulated into the final transdermal dosage form. An ideal enhancer must be nontoxic, nonirritating, nonallergenic, pharmacologically inert, and compatible with most drugs and excipients. However, no single agent meets all the desirable attributes of an enhancer. A combination of enhancers may thus be required. In addition to these chemical enhancers, many of the generally recognized as safe (GRAS) parenteral vehicles can also enhance percutaneous drug absorption. Many of the transdermal and dermal delivery products on the market are formulated with cosolvents (70). These cosolvents include propylene glycol, polyethylene glycol 400, isopropyl myristate, isopropyl palmitate, ethanol, water, and mineral oil. Propylene glycol has been reported to act synergistically with several enhancers. It has been reported to dramatically increase the flux of highly lipophilic drugs in combination with lauric acid, presumably because it enhances the penetration of the enhancer into the skin and due to their synergistic lipid fluidization effects on the stratum corneum (71). Propylene glycol permeates from the formulation into the skin, so it is possible that it may get depleted from the formulation applied to skin in clinically relevant doses, possibly leading to a time-dependent permeation of the drug due to gradual reduction of the enhancer effect (72). Similarly, any volatile components or other solvents may be lost over time; therefore, the composition of the formulation may change over time (73). Ethanol can permeate rapidly through human skin at a flux of around 1 mg/cm²/hour, and as ethanol is depleted, drug concentration in the formulation can increase to a supersaturated solution, allowing effective drug delivery from some spray formulations. Transcutol® (Gattefosse, France), a monoethyl ether of diethylene glycol, is believed to similarly partition into the stratum corneum and increase the solubility of both hydrophilic and lipophilic drugs in the skin (61,74). It has been reported to reduce the flux of hydrophobic drugs across the skin, but a recent study suggested that the same is not true for hydrophilic drugs (75). NexMed (San Diego, California) is developing an enhancer technology (NexACT), and several products are in clinical trials. Because an enhancer is delivered to the skin, its pharmacokinetics must be determined so as to know its half-life in skin, degree of absorption, mechanism of elimination, and metabolism. Also, the reversibility of skin barrier properties should be determined, as any breach of the barrier properties of stratum corneum could result in infection (76). Use of a transdermal patch creates occlusive conditions on the skin, and this leads to increased hydration and irritation of skin underneath the patch. As discussed, increased skin hydration may increase the permeation of the enhancer which may cause even more irritation or toxicity. Further, the enhancer may increase the permeation of the formulation components along with the drug. Therefore, a careful evaluation of the long-term local and systemic toxicity of the chemical enhancers in the final transdermal dosage form is required (53).

1.4.2 PRODRUGS

Prodrugs have also been used to enhance transdermal delivery. Typically, they are designed by attaching lipophilic moieties to the parent compound to enable percutaneous absorption. The parent compound is then generated in the skin by the enzymatic activity of skin. In one study, several prodrugs of naltrexone were reported to hydrolyze to naltrexone on passing through skin. Methyl-3-O-carbonate-naltrexone was found to provide the highest *in vitro* flux across hairless guinea pig skin and also provided a mean steady-state plasma concentration of 7.1 ng/ml for *in vivo* studies (77). However, this chemical modification of the parent compound creates a new chemical entity, making this approach less practical.

1.4.3 NANOCARRIERS FOR INTRADERMAL OR TRANSDERMAL DELIVERY

Nanotechnology advances in recent years have created many colloidal drug carriers that have been investigated for delivery into skin (78). These nanocarriers include microemulsions, dendrimers, micelles, quantum dots or solid lipid nanoparticles (see Section 3.2.6), and liposomes. Dendrimers have been investigated as carriers for topical drug delivery and offer the advantages of ease of synthesis and monodispersity (79,80). Nanocarrier approaches have been widely used for cosmetic applications (81) and can also be useful to enable the delivery of highly lipophilic drugs (82). Niosomes, vesicles made of nonionic surfactants, have also been widely used in the cosmetic industry (83). Other approaches include carbon nanotubes that involve an electrosensitive transdermal delivery system (84).

1.4.4 MICROEMULSIONS

Microemulsions were discovered several decades ago and are receiving increasing attention as potential vehicles for delivery of pharmaceuticals and cosmeceuticals into skin (60,85–90). Microemulsions are thermodynamically stable dispersions that form spontaneously when the ratios of oil, water, and surfactant (and usually a cosurfactant) are in the microemulsion region of a ternary phase diagram. They are optically transparent as the dispersed phase droplet size is small, typically in the 10 to 140 nm range. They may be oil-in-water or water-in-oil type, and if the amount of water and oil is equal, they may form a bicontinuous microemulsion with a constantly changing interface (87). They may increase intradermal or transdermal delivery of the dissolved drug due to the penetration enhancer effect of the oils used or due to the surfactant, or due to increased concentrations by solubilization of the drug (86). They have been reported to increase the permeation of both hydrophilic and lipophilic drugs (85,90) and have been tested in human subjects (89).

1.4.5 LIPOSOMES

The use of liposomes as drug carriers for topical or transdermal delivery has been reviewed (91–93). Liposomes have been widely used in the cosmetic industry with several products already on the market (94,95). Potential advantages include enhancement of drug delivery, solubilization of poorly soluble drugs, local depot for the sustained release of topically effective drugs, reduction of side effects or incompatibilities, or as rate-limiting barriers for the modulation of systemic absorption (96). It has also been suggested that topically applied liposomes can enhance the delivery of drug into sebaceous glands (97). The potential use of liposomes to enhance the transport of conventional drugs such as estradiol, triamcinolone acetonide, hydrocortisone, progesterone, betamethasone, methotrexate, and econazole, has been widely investigated (91). Liposome formulations with various lipid composition, size, charge, and type, all resulted in a significantly higher flux and permeability of triamcinolone acetonide through rat skin than a commercially available ointment (98). Dyphylline liposomes have also been delivered to the skin for potential use as a topical treatment for psoriasis (99). Econazole, an imidazole derivative useful for treatment of dermatomycosis, became the first approved dermatic product (Pevaryl® Lipogel, Janssen-Cilag AG, Switzerland) available in some European countries (100). It is generally believed that intact liposomes do not traverse the skin (92,101). It has been suggested that liposomes adsorb and fuse with the surfaces of the skin, and its constituents may induce changes in the ultrastructure of the intercellular lipid regions in the deeper layers of the stratum corneum, and thus produce a penetration-enhancing effect (101,102). The ability of liposomes to fuse with skin depends on the liposome composition and appears to be a prerequisite for skin penetration (103). The size of the liposomes also appears to be important for their penetration into skin and location of the depot in the skin (104,105). Though liposomes are mostly prepared from phospholipids, some have been prepared from stratum corneum lipids (106,107) and have been used to enhance topical and transdermal delivery (98,108). The use of liposomes as a carrier for peptide/protein drugs is less common but is now being exploited for various drug delivery applications (109). Liposomes present an opportunity to deliver the hydrophilic peptides/proteins to the skin as lipophilic vesicles, but alternatives such as microneedles (Chapter 3) perhaps offer a better way to deliver hydrophilic macromolecules into skin. A topical formulation of liposomally encapsulated interferon was found to be effective to reduce lesion scores in the cutaneous herpes simplex virus guinea pig model, while application of interferon formulated as a solution or as an emulsion was ineffective (110). When lipids similar in composition to stratum corneum were used instead of phospholipids, the amount of interferon deposited in the deeper skin layers was doubled (108).

Another vesicle that appears to be related to liposomes but is considered to be different has been termed *transfersome*, though the term is actually a trademark of IDEA AG (Munich, Germany). Transfersomes are much more deformable (sometimes termed *ultradeformable*) and adaptable and can apparently bring large molecules into the skin through intact permeability barrier. They are claimed to penetrate the skin barrier spontaneously and distribute throughout the body, possibly via the lymphatic system (111,112). It has been shown that a water gradient across the skin can play a role in drug transport (113). Transfersomes have been believed to penetrate into the skin under a water gradient as the driving force (52). However, in a study on human volunteers, it was reported that elastic liposomes (100 to 120 nm) did not pass through the stratum corneum, though they were seen in the stratum corneum based on tape stripping studies and freeze fracture electron microscopy. They were seen up to the 9th tape strip after 1-hour nonocclusive treatment or up to the 15th tape strip after 4-hour treatment. However, extensive vesicle fusion both at surface and deeper layers of stratum corneum was reported. They were believed to pass into stratum corneum through channel-like regions that represent imperfections in the intercellular lipid lamellae. Occlusive conditions further reduced penetration of vesicles into the stratum corneum (114). A recent study used calcein as a model hydrophilic molecule and used different experimental approaches to suggest that ultradeformable vesicles do not carry hydrophilic drugs across the skin. The authors actually suggested that the vesicles reduced the permeation of calcein relative to a buffer control, most likely due to controlling the release of calcein from the formulation into the skin (115). Another type of elastic vesicle that contains ethanol was described as an *ethosome* (116,117).

1.4.6 CYCLODEXTRINS

The natural cyclodextrins, α-, β-, and γ-, are cyclic oligosaccharides of 6, 7, and 8 glucopyranose units, respectively. The ring structure resembles a truncated core, and the fundamental basis of their pharmaceutical applications is the capability to form inclusion complexes due to the hydrophobic property of the cavity. The cavity size is the smallest (.5 D) for α-cyclodextrin, followed by that of β-cyclodextrin (.6 D) and γ-cyclodextrin (.8 D) (118,119). Cyclodextrin derivatives (as compared to natural cyclodextrins) exhibit higher solubility and lower toxicity by the parenteral route, while retaining their efficacy of molecular encapsulation. Hydroxypropyl-β-cyclodextrin (HPβ-CD) is a derivative with the most solubility (>50%) and the least toxicity. The necessary toxicological and human clinical data on HPβ-CD are available, and its approval as an excipient is expected. Cyclodextrin–drug complexes have been marketed in several other countries (120–122). Cyclodextrins have also been investigated as transdermal penetration enhancers (123–125). For example, cyclodextrins have also been used to enhance the skin penetration of hydrocortisone (126,127). Because cyclodextrins form inclusion complexes with drugs, it has been suggested that they may act as enhancers by interacting with components of the skin (128,129). Cyclodextrins have also been used to improve the skin delivery of highly lipophilic drugs (82). Cyclodextrins have high MW (typically, >1 kDa) and are therefore expected to have indirect effects on the skin permeation of drugs rather than directly entering the skin. Cyclodextrins can be used to bring the drug in aqueous solution to be delivered by iontophoresis. We have used hydroxypropyl β-cyclodextrin (HPβ-CD) to bring hydrocortisone into aqueous solution in relatively high concentrations and then deliver it iontophoretically (by electroosmosis) across human cadaver skin from this cyclodextrin solution (130).

1.5 ACTIVE ENHANCEMENT STRATEGIES FOR PERCUTANEOUS ABSORPTION

Skin is impermeable to macromolecules such as peptides, proteins, oligonucleotides, and DNA. The physical and enzymatic barriers together create a formidable barrier for any permeation under normal circumstances. Various physical (83,131,132), chemical, and biochemical techniques have been attempted to surmount these barriers. These techniques include iontophoresis, electroporation, phonophoresis, microneedles, and microdermabrasion (Figure 1.1). In this chapter, an overview of skin enhancement techniques is provided, and the physical enhancement techniques (i.e., iontophoresis, electroporation, sonophoresis, and microporation) will be discussed in detail in subsequent chapters of this book. A novel technique that uses a transdermal delivery system with jet injection was also reported, where a jet injector containing physiological saline first makes a pore in the skin prior to application of the drug (see Chapter 9). Alternatively, the drug can be directly delivered to the skin in dry powder form using a supersonic flow of helium gas to accelerate the particles to a high velocity (131,133).

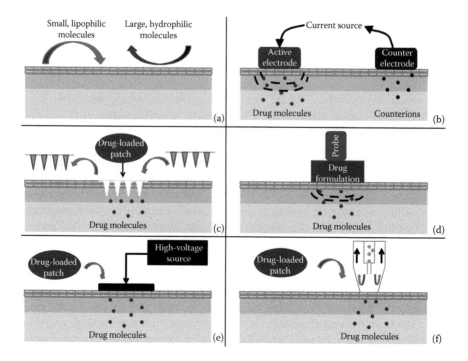

FIGURE 1.1   Skin is not permeable to hydrophilic molecules (a), but such delivery can be enabled by enhancement technologies such as iontophoresis (b), microneedles (c), sonophoresis (d), electroporation (e), or microdermabrasion (f). **(See color insert.)**

Invasive techniques such as subcutaneous injections and implants are not being discussed here. Patents issued on these physical enhancement techniques as well as on other transdermal enhancement techniques were reviewed recently (95) but are not discussed here. Enhancement techniques that were not widely investigated, such as pressure wave (134) or magnetophoresis, are not discussed here. Magnetophoresis involves the application of magnetic field and has been reported to enhance the permeation of diamagnetic molecules into the skin (51,95).

1.5.1   MICROPORATION

Skin microporation creates microscopic temporary holes in the skin to allow transport of hydrophilic molecules and macromolecules. Skin microporation can be achieved by several technologies including microneedles, thermal ablation, radiofrequency ablation, or laser ablation. A detailed discussion of these technologies is provided in Chapter 3, and aspects relating to commercial development of microporation technologies are discussed in Chapter 10.

1.5.2   IONTOPHORESIS

Transdermal and intradermal delivery of drugs can be assisted by electrical energy. Enhancement via electrical energy can involve iontophoresis, electroosmosis, or electroporation (135). Electroosmosis is a phenomenon that accompanies ionto-phoresis. Iontophoresis and electroosmosis are discussed in detail in Chapter 4. Electroporation, on the other hand, involves a different mechanism and is discussed in detail in Chapter 5. Briefly, iontophoresis implies the use of small amounts of physiologically acceptable electric current to drive ionic (charged) and neutral drugs into the body (136–141). By using an electrode of the same polarity as the charge on the drug, the drug is driven into the skin mainly by electrostatic repulsion, though electroosmosis (see Section 4.7) also plays a role. This technique is not new and has been used clinically for delivering medication to surface tissues for several decades. However, its potential is now being rediscovered for transdermal systemic delivery of ionic drugs, including peptides that are normally difficult to administer except by parenteral route. Iontophoresis has been observed to enhance the transdermal per-meation of ionic drugs severalfold, and this can expand the horizon of transdermal controlled drug delivery for systemic medication. In addition to the usual benefits of transdermal delivery, iontophoresis presents a unique opportunity to provide pro-grammed drug delivery. This is because the drug is delivered usually in proportion to the current applied, which can be readily adjusted. Such dependence on current may also make drug absorption via iontophoresis less dependent on biological variables, unlike most other drug delivery systems. Also, patient compliance will improve, and because the dosage form includes electronics, means to remind patients to replace the dose can be built into the system. The dose can also be titrated for individual patients by adjusting the current. Several new developments have taken place in the field of iontophoretic drug delivery since the topic was last reviewed by the author over 20 years ago (142). The iontophoretic delivery of over 50 drugs has been inves-tigated in the last 30 years and over 500 papers have been published just since 2001 (143). Several products have been marketed, with stories of success and failure (see Chapter 10).

1.5.3 SONOPHORESIS

Sonophoresis or phonophoresis implies the transport of drug molecules under the influence of ultrasound (144–146) and is discussed in detail in Chapter 6. The drug is delivered from a coupling (contact) agent that transfers ultrasonic energy from the ultrasonic device to the skin. The exact mechanism involved is not known, but enhancement presumably results from thermal, mechanical, and chemical alterations in the skin induced by ultrasonic waves. It has been shown that low-frequency ultrasound can deliver therapeutic doses of proteins such as insulin, interferon, and erythropoietin (147) or even particles up to 25 μm into skin (148).

1.5.4 MICRODERMABRASION

Microdermabrasion is a skin resurfacing or rejuvenation technique widely used worldwide as a cosmetic office procedure. Also known as body polishing, it involves blowing aluminum oxide crystals (typically 100 to 300 μm) or other abra-sive crystals onto the face to partially remove the stratum corneum (149–151). The body then replaces the lost skin with new and healthy cells to rejuvenate the skin. This procedure is also used for treatment of scars, wrinkles, hyperpigmentation, and photodamage. The procedure is relatively simple and can be accomplished in 15 to 30 minutes. A recent study used histology techniques to characterize the skin of rhesus macaques and human volunteers following microdermabrasion, and it was shown that the technique could be controlled to remove the stratum corneum without causing damage to deeper tissues. Several passes were required to completely remove the stratum corneum (152). Microdermabrasion devices were originally classified by the U.S. Food and Drug Administration (FDA) as Class 1 devices rendering a noninvasive procedure; therefore, they have not undergone rigorous clinical trials and were later classified as exempt. Though microdermabrasion is widely used in cosmetic clinics, rigorous scientific studies to validate the efficacy of microdermabrasion in a large group of patients is therefore lacking (153). In a study with eight patients, some disruption of skin barrier was shown as the transepidermal water loss (TEWL) values were found to be high at 24 hours after microdermabrasion of the face but then recovered by the next measurement at 7 days (149). Because microdermabrasion causes disruption of the stratum corneum, it can be expected that it may increase the permeation of hydrophilic drugs. Lee et al. reported that microdermabrasion enhanced the permeation of a hydrophilic drug by 8- to 24-fold across skin but did not enhance the permeation of a lipophilic drug (154). Delivery of vitamin C across skin by microdermabrasion was also reported (155).

1.6   APPLICATIONS IN VETERINARY MEDICINE

Topical route of delivery is attractive to the farming industry because it is less labor intensive to apply drugs to an animal's skin than to administer more conventional dosage forms such as drenches, injections, and inoculations. These systems are less likely to cause trauma and tissue damage and are less likely to interfere with nutritional status. Several studies have shown that chemicals can be transported across sheep and cattle skins in therapeutic amounts for topical or systemic effect, and these have been reviewed (156). The route of passage of drugs across the skin of farm animals is of interest and was investigated (157). Topical dosage forms used in veterinary medicine include dusting powders, suspensions, lotions, liniments, and creams. Products such as dips, sprays, pour-on, or spot-on formulations are usually meant for systemic absorption and may include a penetration enhancer (158). Delivery of dexamethasone phosphate into the equine tibiotarsal joint by iontophoresis was investigated (159). In certain skin disorders in animals, the infecting organism resides in the pilosebaceous unit, from which it is not easily eradicated by topical medication. In these cases, iontophoresis could be especially useful due to the appendageal pathways of drug transport during iontophoresis. Iontophoresis of methylene blue under positive electrode was reported to eradicate heavy infections of *Demodex folliculorum* from canine skin, while iontophoresis of potassium iodide under negative electrode alleviated *Trichophyton verrucosum* infection of bovine skin (160).

REFERENCES

1. L. Brown and R. Langer. Transdermal delivery of drugs, *Ann. Rev. Med.*, 39:221–229 (1988).
2. V. V. Ranade. Drug delivery systems. 6. Transdermal drug delivery, *J. Clin. Pharmacol.*, 31:401–418 (1991).
3. B. J. Thomas and B. C. Finnin. The transdermal revolution, *Drug Discov. Today*, 9:697–703 (2004).
4. R. Langer. Transdermal drug delivery: Past progress, current status, and future prospects, *Adv. Drug Deliv. Rev.*, 56:557–558 (2004).
5. J. Hadgraft. Skin deep, *Eur. J. Pharm. Biopharm.*, 58:291–299 (2004).
6. M. R. Prausnitz and R. Langer. Transdermal drug delivery, *Nat. Biotechnol.*, 26:1261–1268 (2008).
7. M. R. Prausnitz, S. Mitragotri, and R. Langer. Current status and future potential of transdermal drug delivery, *Nat. Rev. Drug Discov.*, 3:115–124 (2004).
8. G. W. Cleary. Transdermal and transdermal-like delivery system opportunities: Today and the future, *Drug Del. Technol.*, 3:34–40 (2003).
9. R. O. Potts and G. W. Cleary. Transdermal drug delivery: Useful paradigms, *J. Drug Target.*, 3:247–251 (1995).
10. G. Levin. Advances in radio-frequency transdermal drug delivery, *Pharm. Technol.* s12–s19 (2008).
11. R. H. Guy. Current status and future prospects of transdermal drug delivery, *Pharm. Res.*, 13:1765–1769 (1996).
12. J. D. Bos and M. M. Meinardi. The 500 Dalton rule for the skin penetration of chemical compounds and drugs, *Exp. Dermatol.*, 9:165–169 (2000).
13. Q. Zhang, J. E. Grice, P. Li, O. G. Jepps, G. J. Wang, and M. S. Roberts. Skin solubility determines maximum transepidermal flux for similar size molecules, *Pharm Res.*, 26:1974–1985 (2009).
14. B. M. Magnusson, W. J. Pugh, and M. S. Roberts. Simple rules defining the potential of compounds for transdermal delivery or toxicity, *Pharm Res.*, 21:1047–1054 (2004).
15. J. J. Berti and J. J. Lipsky. Transcutaneous drug delivery: A practical review, *Mayo Clin. Proc.*, 70:581–586 (1995).
16. A. C. Williams. *Transdermal and topical drug delivery*, Pharmaceutical Press, London, 2003.
17. A. K. Banga and Y. W. Chien. Dermal absorption of peptides and proteins. In K. L. Audus and T. J. Raub (eds), *Pharmaceutical biotechnology. Vol. 4. Biological barriers to protein delivery*, Plenum Press, New York, 1993, pp. 179–197.
18. D. A. vanHal, E. Jeremiasse, H. E. Junginger, F. Spies, and J. A. Bouwstra. Structure of fully hydrated human stratum corneum: A freeze fracture electron microscopy study, *J. Invest. Dermatol.*, 106:89–95 (1996).
19. S. M. Morley. Keratin and the skin: Past, present and future, *Q. J. Med.*, 90:433–435 (1997).
20. M. Fartasch. The nature of the epidermal barrier: Structural aspects, *Adv. Drug Del. Rev.*, 18:273–282 (1996).
21. M. Denda, J. Koyama, R. Namba, and I. Horii. Stratum-corneum lipid morphology and transepidermal water loss in normal skin and surfactant-induced scaly skin, *Arch. Dermatol. Res.*, 286:41–46 (1994).
22. I. Steinstrasser and H. P. Merkle. Dermal metabolism of topically applied drugs: Pathways and models reconsidered, *Pharmaceutical Acta Helvetiae*, 70:3–24 (1995).
23. V. M. Meidan, M. C. Bonner, and B. B. Michniak. Transfollicular drug delivery—Is it a reality? *Int. J. Pharm.*, 306:1–14 (2005).

24. N. Otberg, H. Richter, H. Schaefer, U. Blume-Peytavi, W. Sterry, and J. Lademann. Variations of hair follicle size and distribution in different body sites, *J. Invest. Dermatol.*, 122:14–19 (2004).
25. A. C. Lauer, Percutaneous drug delivery to the hair follicle, *Cut. Ocular Toxicol.*, 20:475–495 (2001).
26. R. Alvarez-Roman, A. Naik, Y. N. Kalia, H. Fessi, and R. H. Guy. Visualization of skin penetration using confocal laser scanning microscopy, *Eur. J. Pharm. Biopharm.*, 58:301–316 (2004).
27. D. Howes, R. Guy, J. Hadgraft, J. Heylings, U. Hoeck, F. Kemper, H. Maibach, J. P. Marty, H. Merk, J. Parra, D. Rekkas, I. Rondelli, H. Schaefer, U. Tauber, and N. Verbiese. Methods for assessing percutaneous absorption—The report and recommendations of ECVAM workshop 13, *ATLA. Altern. Lab. Anim.*, 24:81–106 (1996).
28. U. Tauber. Drug metabolism in the skin: Advantages and disadvantages. In J. Hadgraft and R. H. Guy (eds), *Transdermal drug delivery: Developmental issues and research initiatives*, Marcel Dekker Inc., New York, 1989, pp. 99–134.
29. T. Hikima, K. Yamada, T. Kimura, H. I. Maibach, and K. Tojo. Comparison of skin distribution of hydrolytic activity for bioconversion of beta-estradiol 17-acetate between man and several animals *in vitro*, *Eur. J. Pharm. Biopharm.*, 54:155–160 (2002).
30. H. F. Merk, F. K. Jugert, and S. Frankenberg. Biotransformations in the skin. In F. N. Marzulli and H. I. Maibach (eds), *Dermatotoxicology*, Taylor and Francis, London, 1996, pp. 61–73.
31. K. H. Valia, K. Tojo, and Y. W. Chien. Long-term permeation kinetics of estradiol: (III) Kinetic analysis of the simultaneous skin permeation and bioconversion of estradiol esters, *Drug Dev. Ind. Pharm.*, 11:1133–1173 (1985).
32. D. B. Guzek, A. H. Kennedy, S. C. McNeill, E. Wakshull, and R. O. Potts. Transdermal drug transport and metabolism. I. Comparison of *in vitro* and *in vivo* results, *Pharm. Res.*, 6:33–39 (1989).
33. R. O. Potts, S. C. McNeill, C. R. Desbonnet, and E. Wakshull. Transdermal drug transport and metabolism. II. The role of competing kinetic events, *Pharm. Res.*, 6:119–124 (1989).
34. P. M. Elias. Stratum corneum defensive functions: An integrated view, *J. Invest. Dermatol.*, 125:183–200 (2005).
35. A. S. Michaels, S. K. Chandrasekaran, and J. E. Shaw. Drug permeation through human skin: Theory and *in vitro* experimental measurement, *AICHE J.*, 21:985–996 (1975).
36. P. M. Elias. Epidermal lipids, barrier function, and desquamation, *J. Invest. Dermatol.*, 80:44s–49s (1983).
37. P. M. Elias. Structure and function of the stratum corneum permeability barrier, *Drug Dev. Res.*, 13:97–105 (1988).
38. K. Tojo. Random brick model for drug transport across stratum corneum, *J. Pharm. Sci.*, 76:889–891 (1987).
39. R. O. Potts and R. H. Guy. Predicting skin permeability, *Pharm. Res.*, 9:663–669 (1992).
40. W. J. Pugh and J. Hadgraft. Ab initio prediction of human skin permeability coefficients, *Int. J. Pharm.*, 103:163–178 (1994).
41. A. J. Lee, J. R. King, and D. A. Barrett. Percutaneous absorption: A multiple pathway model, *J. Control. Release*, 45:141–151 (1997).
42. L. A. Kirchner, R. P. Moody, E. Doyle, R. Bose, J. Jeffery, and I. Chu. The prediction of skin permeability by using physicochemical data, *ATLA*, 25:359–370 (1997).
43. G. Hisoire and D. Bucks. An unexpected finding in percutaneous absorption observed between haired and hairless guinea pig skin, *J. Pharm. Sci.*, 86:398–400 (1997).
44. J. C. Shah. Analysis of permeation data—Evaluation of the lag time method, *Int. J. Pharm.*, 90:161–169 (1993).

45. B. W. Barry. Reflections on transdermal drug delivery, *Pharm. Sci. Technol. Today*, 2:41–43 (1999).
46. Y. N. Kalia and R. H. Guy. Modeling transdermal drug release, *Adv. Drug Deliv. Rev.*, 48:159–172 (2001).
47. S. Mitragotri. Modeling skin permeability to hydrophilic and hydrophobic solutes based on four permeation pathways, *J. Control. Release*, 86:69–92 (2003).
48. F. Yamashita and M. Hashida. Mechanistic and empirical modeling of skin permeation of drugs, *Adv. Drug Deliv. Rev.*, 55:1185–1199 (2003).
49. I. T. Degim. New tools and approaches for predicting skin permeability, *Drug Discov. Today*, 11:517–523 (2006).
50. S. Hansen, A. Naegel, M. Heisig, G. Wittum, D. Neumann, K. H. Kostka, P. Meiers, C. M. Lehr, and U. F. Schaefer. The role of corneocytes in skin transport revised—A combined computational and experimental approach, *Pharm. Res.*, 26:1379–1397 (2009).
51. B. W. Barry. Novel mechanisms and devices to enable successful transdermal drug delivery, *Eur. J. Pharm Sci.*, 14:101–114 (2001).
52. K. Higaki, C. Amnuaikit, and T. Kimura. Strategies for overcoming the stratum corneum, *Am. J. Drug Del.*, 1:187–214 (2003).
53. T. K. Ghosh and A. K. Banga. Methods of enhancement of transdermal drug delivery: Part IIB, Chemical permeation enhancers, *Pharm. Technol.*, 17(5):68–76 (1993).
54. T. K. Ghosh and A. K. Banga. Methods of enhancement of transdermal drug delivery: Part IIA. Chemical permeation enhancers, *Pharm. Technol.*, 17(4):62–90 (1993).
55. R. B. Walker and E. W. Smith. The role of percutaneous penetration enhancers, *Adv. Drug Del. Rev.*, 18:295–301 (1996).
56. H. Y. Thong, H. Zhai, and H. I. Maibach. Percutaneous penetration enhancers: An overview, *Skin Pharmacol. Physiol.*, 20:272–282 (2007).
57. S. A. Ibrahim and S. K. Li. Effects of chemical enhancers on human epidermal membrane: Structure-enhancement relationship based on maximum enhancement (E(max)), *J. Pharm. Sci.*, 98:926–944 (2009).
58. I. B. Pathan and C. M. Setty. Chemical penetration enhancers for transdermal drug delivery systems. *Tropical J. Pharm. Res.*, 8:173–179, 2009.
59. Y. S. Krishnaiah, V. Satyanarayana, and R. S. Karthikeyan. Effect of the solvent system on the *in vitro* permeability of nicardipine hydrochloride through excised rat epidermis, *J. Pharm. Pharmaceut. Sci.*, 5:123–130 (2002).
60. M. Rangarajan and J. L. Zatz. Effect of formulation on the topical delivery of alpha-tocopherol, *J. Cosmet. Sci.*, 54:161–174 (2003).
61. H. A. Ayala-Bravo, D. Quintanar-Guerrero, A. Naik, Y. N. Kalia, J. M. Cornejo-Bravo, and A. Ganem-Quintanar. Effects of sucrose oleate and sucrose laureate on *in vivo* human stratum corneum permeability, *Pharm. Res.*, 20:1267–1273 (2003).
62. Y. Song, C. Xiao, R. Mendelsohn, T. Zheng, L. Strekowski, and B. Michniak. Investigation of iminosulfuranes as novel transdermal penetration enhancers: Enhancement activity and cytotoxicity, *Pharm. Res.*, 22:1918–1925 (2005).
63. M. B. Pierre, E. Ricci Jr., A. C. Tedesco, and M. V. Bentley. Oleic acid as optimizer of the skin delivery of 5-aminolevulinic acid in photodynamic therapy, *Pharm. Res.*, 23:360–366 (2006).
64. P. Karande and S. Mitragotri. High throughput screening of transdermal formulations, *Pharm. Res.*, 19:655–660 (2002).
65. P. Karande, A. Jain, and S. Mitragotri. Discovery of transdermal penetration enhancers by high-throughput screening, *Nat. Biotechnol.*, 22:192–197 (2004).
66. P. Karande, A. Jain, and S. Mitragotri. Insights into synergistic interactions in binary mixtures of chemical permeation enhancers for transdermal drug delivery, *J. Control. Release*, 115:85–93 (2006).

67. V. K. Rachakonda, K. M. Yerramsetty, S. V. Madihally, R. L. Robinson, Jr., and K. A. Gasem. Screening of chemical penetration enhancers for transdermal drug delivery using electrical resistance of skin, *Pharm. Res.*, 25:2697–2704 (2008).

68. B. W. Barry. Mode of action of penetration enhancers in human skin, *J. Control. Release*, 6:85–97 (1987).

69. B. W. Barry. Lipid-protein-partitioning theory of skin penetration enhancement, *J. Control. Release*, 15:237–248 (1991).

70. W. R. Pfister and D. S. T. Hsieh. Permeation enhancers compatible with transdermal drug delivery systems. Part I: Selection and formulation considerations, *Pharm. Technol.*, 14(9):132–140 (1990).

71. A. P. Funke, R. Schiller, H. W. Motzkus, C. Gunther, R. H. Muller, and R. Lipp. Transdermal delivery of highly lipophilic drugs: *In vitro* fluxes of antiestrogens, permeation enhancers, and solvents from liquid formulations, *Pharm. Res.*, 19:661–668 (2002).

72. L. Trottet, C. Merly, M. Mirza, J. Hadgraft, and A. F. Davis. Effect of finite doses of propylene glycol on enhancement of *in vitro* percutaneous permeation of loperamide hydrochloride, *Int. J. Pharm.*, 274:213–219 (2004).

73. J. Hadgraft. Skin, the final frontier, *Int. J. Pharm.*, 224:1–18 (2001).

74. D. R. de Araujo, C. Padula, C. M. Cereda, G. R. Tofoli, R. B. Brito, Jr., E. de Paula, S. Nicoli, and P. Santi. Bioadhesive films containing benzocaine: Correlation between *in vitro* permeation and *in vivo* local anesthetic effect, *Pharm. Res.*, 27:1677–1686 (2010).

75. S. Mutalik, H. S. Parekh, N. M. Davies, and N. Udupa. A combined approach of chemical enhancers and sonophoresis for the transdermal delivery of tizanidine hydrochloride, *Drug Deliv.*, 16:82–91 (2009).

76. W. R. Pfister and D. S. T. Hsieh. Permeation enhancers compatible with transdermal drug delivery systems: Part II: System design considerations, *Pharm. Technol.*, 14(10):54–60 (1990).

77. S. Valiveti, K. S. Paudel, D. C. Hammell, M. O. Hamad, J. Chen, P. A. Crooks, and A. L. Stinchcomb. *In vitro/in vivo* correlation of transdermal naltrexone prodrugs in hairless guinea pigs, *Pharm. Res.*, 22:981–989 (2005).

78. G. Cevc and U. Vierl. Nanotechnology and the transdermal route: A state of the art review and critical appraisal, *J. Control. Release*, 141:277–299 (2010).

79. A. S. Chauhan, S. Sridevi, K. B. Chalasani, A. K. Jain, S. K. Jain, N. K. Jain, and P. V. Diwan. Dendrimer-mediated transdermal delivery: Enhanced bioavailability of indomethacin, *J. Control. Release*, 90:335–343 (2003).

80. G. T. Tolia and H. H. Choi. The role of dendrimers in topical drug delivery, *Pharm. Technol.*, Nov:88–98 (2008).

81. V. B. Patravale and S. D. Mandawgade. Novel cosmetic delivery systems: An application update, *Int. J. Cosmet. Sci.*, 30:19–33 (2008).

82. L. Trichard, M. B. Delgado-Charro, R. H. Guy, E. Fattal, and A. Bochot. Novel beads made of alpha-cyclodextrin and oil for topical delivery of a lipophilic drug, *Pharm. Res.*, 25:435–440 (2008).

83. B. W. Barry. Is transdermal drug delivery research still important today? *Drug Discov. Today*, 6:967–971 (2001).

84. J. S. Im, B. C. Bai, and Y. S. Lee. The effect of carbon nanotubes on drug delivery in an electro-sensitive transdermal drug delivery system, *Biomaterials*, 31:1414–1419 (2010).

85. P. J. Lee, R. Langer, and V. P. Shastri. Novel microemulsion enhancer formulation for simultaneous transdermal delivery of hydrophilic and hydrophobic drugs, *Pharm. Res.*, 20:264–269 (2003).

86. A. Kogan and N. Garti. Microemulsions as transdermal drug delivery vehicles, *Adv. Colloid Interface Sci.*, 123–126:369–385 (2006).

87. P. Boonme. Applications of microemulsions in cosmetics, *J. Cosmet. Dermatol.*, 6:223–228 (2007).
88. A. Azeem, M. Rizwan, F. J. Ahmad, Z. I. Khan, R. K. Khar, M. Aqil, and S. Talegaonkar. Emerging role of microemulsions in cosmetics, *Recent Pat. Drug Deliv. Formul.*, 2:275–289 (2008).
89. A. H. Elshafeey, A. O. Kamel, and M. M. Fathallah. Utility of nanosized microemulsion for transdermal delivery of tolterodine tartrate: Ex-vivo permeation and in-vivo pharmacokinetic studies, *Pharm Res.*, 26:2446–2453 (2009).
90. J. Hosmer, R. Reed, M. V. Bentley, A. Nornoo, and L. B. Lopes. Microemulsions containing medium-chain glycerides as transdermal delivery systems for hydrophilic and hydrophobic drugs, *AAPS. PharmSciTech.*, 10:589–596 (2009).
91. A. Rolland. Particulate carriers in dermal and transdermal drug delivery: Myth or reality? In A. Rolland (ed), *Pharmaceutical particulate carriers: Therapeutic applications*, Marcel Dekker, New York, 1993, pp. 367–421.
92. H. Schreier and J. Bouwstra. Liposomes and niosomes as topical drug carriers—Dermal and transdermal drug delivery, *J. Cont. Rel.*, 30:1–15 (1994).
93. G. M. El Maghraby, B. W. Barry, and A. C. Williams. Liposomes and skin: From drug delivery to model membranes, *Eur. J. Pharm. Sci.*, 34:203–222 (2008).
94. G. Betz, A. Aeppli, N. Menshutina, and H. Leuenberger. *In vivo* comparison of various liposome formulations for cosmetic application, *Int. J. Pharm.*, 296:44–54 (2005).
95. M. Rizwan, M. Aqil, S. Talegaonkar, A. Azeem, Y. Sultana, and A. Ali. Enhanced transdermal drug delivery techniques: An extensive review of patents, *Recent Pat. Drug Deliv. Formul.*, 3:105–124 (2009).
96. N. Weiner, L. Lieb, S. Niemiec, C. Ramachandran, Z. Hu, and K. Egbaria. Liposomes: A novel topical delivery system for pharmaceutical and cosmetic applications, *J. Drug Target.*, 2:405–410 (1994).
97. T. Tschan, H. Steffen, and A. Supersaxo. Sebaceous-gland deposition of isotretinoin after topical application: An *in vitro* study using human facial skin, *Skin Pharmacol.*, 10:126–134 (1997).
98. H. Y. Yu and H. M. Liao. Triamcinolone permeation from different liposome formulations through rat skin *in vitro*, *Int. J. Pharm.*, 127:1–7 (1996).
99. E. Touitou, N. Shaco-Ezra, N. Dayan, M. Jushynski, R. Rafaeloff, and R. Azoury. Dyphylline liposomes for delivery to the skin, *J. Pharm. Sci.*, 81:131–134 (1992).
100. R. Naeff. Feasibility of topical liposome drugs produced on an industrial scale, *Adv. Drug Del. Rev.*, 18:343–347 (1996).
101. H. E. J. Hofland, J. A. Bouwstra, H. E. Bodde, F. Spies, and H. E. Junginger. Interactions between liposomes and human stratum corneum *in vitro*: Freeze fracture electron microscopical visualization and small angle X-ray scattering studies, *Br. J. Dermatol.*, 132:853–856 (1995).
102. A. Bhatia, R. Kumar, and O. P. Katare. Tamoxifen in topical liposomes: Development, characterization and in-vitro evaluation, *J. Pharm. Pharmaceut. Sci.*, 7:252–259 (2004).
103. M. Kirjavainen, A. Urtti, I. Jaaskelainen, T. M. Suhonen, P. Paronen, R. ValjakkaKoskela, J. Kiesvaara, and J. Monkkonen. Interaction of liposomes with human skin *in vitro*—The influence of lipid composition and structure, *Biochim. Biophys. Acta*, 1304:179–189 (1996).
104. J. Duplessis, C. Ramachandran, N. Weiner, and D. G. Muller. The influence of particle size of liposomes on the deposition of drug into skin, *Int. J. Pharm.*, 103:277–282 (1994).
105. R. Natsuki, Y. Morita, S. Osawa, and Y. Takeda. Effects of liposome size on penetration of *dl*-tocopherol acetate into skin, *Biol. Pharm. Bull.*, 19:758–761 (1996).
106. P. W. Wertz, W. Abraham, L. Landmann, and D. T. Downing. Preparation of liposomes from stratum corneum lipids, *J. Invest. Dermatol.*, 87:582–584 (1986).

107. S. Kitagawa, N. Yokochi, and N. Murooka. pH-dependence of phase transition of the lipid bilayer of liposomes of stratum corneum lipids, *Int. J. Pharm.*, 126:49–56X (1995).
108. K. Egbaria, C. Ramachandran, D. Kittayanond, and N. Weiner. Topical delivery of liposomally encapsulated interferon evaluated by *in vitro* diffusion studies, *Antimicrob. Agents Chemother.*, 34:107–110 (1990).
109. G. Storm, F. Koppenhagen, A. Heeremans, M. Vingerhoeds, M. C. Woodle, and D. J. A. Crommelin. Novel developments in liposomal delivery of peptides and proteins, *J. Control. Release*, 36:19–24 (1995).
110. N. Weiner, N. Williams, G. Birch, C. Ramachandran, C. Shipman, and G. Flynn. Topical delivery of liposomally encapsulated interferon evaluated in a cutaneous herpes guinea pig model, *Antimicrob. Agents Chemother.*, 33:1217–1221 (1989).
111. G. Cevc, A. Schatzlein, and G. Blume. Transdermal drug carriers: Basic properties, optimization and transfer efficiency in the case of epicutaneously applied peptides, *J. Control. Release*, 36:3–16 (1995).
112. A. Paul, G. Cevc, and B. K. Bachhawat. Transdermal immunization with large proteins by means of ultradeformable drug carriers, *Eur. J. Immunol.*, 25:3521–3524 (1995).
113. S. Bjorklund, J. Engblom, K. Thuresson, and E. Sparr. A water gradient can be used to regulate drug transport across skin, *J. Control. Release*, 143:191–200 (2010).
114. P. L. Honeywell-Nguyen, H. W. Wouter Groenink, A. M. de Graaff, and J. A. Bouwstra. The *in vivo* transport of elastic vesicles into human skin: Effects of occlusion, volume and duration of application, *J. Control. Release*, 90:243–255 (2003).
115. A. P. Bahia, E. G. Azevedo, L. A. Ferreira, and F. Frezard. New insights into the mode of action of ultradeformable vesicles using calcein as hydrophilic fluorescent marker, *Eur. J. Pharm. Sci.*, 39:90–96 (2010).
116. E. Esposito, E. Menegatti, and R. Cortesi. Ethosomes and liposomes as topical vehicles for azelaic acid: A preformulation study, *J. Cosmet. Sci.*, 55:253–264 (2004).
117. D. Paolino, G. Lucania, D. Mardente, F. Alhaique, and M. Fresta. Ethosomes for skin delivery of ammonium glycyrrhizinate: *In vitro* percutaneous permeation through human skin and *in vivo* anti-inflammatory activity on human volunteers, *J. Control. Release*, 106:99–110 (2005).
118. J. Szejtli. Cyclodextrins in drug formulations: Part I, *Pharm. Technol.*, 15:36–44 (1991).
119. D. Duchene and D. Wouessidjewe. Pharmaceutical uses of cyclodextrins and derivatives, *Drug Dev. Ind. Pharm.*, 16:2487–2499 (1990).
120. J. Szejtli. Cyclodextrins in drug formulations: Part II, *Pharm. Technol.*, 15:24–38 (1991).
121. C. E. Strattan. 2-Hydroxypropy1-beta-cyclodextrin: Part I, Patents and regulatory issues, *Pharm. Technol.*, 16(1):69–74 (1992).
122. C. E. Strattan. 2-Hydroxypropyl-beta-cyclodextrin, Part II: Safety and manufacturing issues, *Pharm. Technol.*, 16(2):52–58 (1992).
123. H. Arima, H. Adachi, T. Irie, and K. Uekama. Improved drug delivery through the skin by hydrophilic β-Cyclodextrins: Enhancement of anti-inflammatory effect of 4-biphenylacetic acid in rats, *Drug Invest.*, 2:155–161 (1990).
124. U. Vollmer, B. W. Muller, J. Peeters, J. Mesens, B. Wilffert, and T. Peters. A study of the percutaneous absorption-enhancing effects of cyclodextrin derivatives in rats, *J. Pharm. Pharmacol.*, 46:19–22 (1994).
125. N. Kaur, R. Puri, and S. K. Jain. Drug-cyclodextrin-vesicles dual carrier approach for skin targeting of anti-acne agent, *AAPS. PharmSciTech.*, 11:528–537 (2010).
126. T. Loftsson, G. Frioriksdottir, G. Ingvarsdottir, B. Jonsdottir, and A. M. Siguroardottir. The Influence of 2-Hydroxypropyl-beta-cyclodextrin on diffusion rates and transdermal delivery of hydrocortisone, *Drug Dev. Ind. Pharm.*, 20:1699–1708 (1994).

127. A. Preiss, W. Mehnert, and K. H. Fromming. Penetration of hydrocortisone into excised human skin under the influence of cyclodextrins, *Pharmazie.*, 50:121–126 (1995).
128. J. Szejtli. Medicinal applications of cyclodextrins, *Med. Res. Rev.*, 14:353–386 (1994).
129. M. Vitoria, L. B. Bentley, R. F. Vianna, S. Wilson, and J. H. Collett. Characterization of the influence of some cyclodextrins on the stratum corneum from the hairless mouse, *J. Pharm. Pharmacol.*, 49:397–402 (1997).
130. S. Chang and A. K. Banga. Transdermal iontophoretic delivery of hydrocortisone from cyclodextrin solutions, *J. Pharm. Pharmacol.*, 50:635–640 (1998).
131. S. E. Cross and M. S. Roberts. Physical enhancement of transdermal drug application: Is delivery technology keeping up with pharmaceutical development? *Curr. Drug Deliv.*, 1:81–92 (2004).
132. A. Nanda, S. Nanda, and N. M. Ghilzai. Current developments using emerging transdermal technologies in physical enhancement methods, *Curr. Drug Deliv.*, 3:233–242 (2006).
133. N. Inoue, D. Kobayashi, M. Kimura, M. Toyama, I. Sugawara, S. Itoyama, M. Ogihara, K. Sugibayashi, and Y. Morimoto. Fundamental investigation of a novel drug delivery system, a transdermal delivery system with jet injection, *Int. J. Pharm.*, 137:75–84 (1996).
134. A. G. Doukas and N. Kollias. Transdermal drug delivery with a pressure wave, *Adv. Drug Deliv. Rev.*, 56:559–579 (2004).
135. J. E. Riviere and M. C. Heit. Electrically-assisted transdermal drug delivery, *Pharm. Res.*, 14:687–697 (1997).
136. J. Singh and M. S. Roberts. Transdermal delivery of drugs by iontophoresis: A review, *Drug Design Del.*, 4:1–12 (1989).
137. P. G. Green, M. Flanagan, B. Shroot, and R. H. Guy. Iontophoretic drug delivery. In K. A. Walters and J. Hadgraft (eds), *Pharmaceutical skin penetration enhancement*, Marcel Dekker, New York, 1993, pp. 311–333.
138. P. Singh and H. I. Maibach. Transdermal iontophoresis: Pharmacokinetic considerations, *Clin. Pharmacokinet.*, 26(5):327–334 (1994).
139. P. Singh and H. I. Maibach. Iontophoresis in drug delivery: Basic principles and applications, *Crit. Rev. Ther. Drug Carr. Syst.*, 11:161–213 (1994).
140. J. Singh and K. S. Bhatia. Topical iontophoretic drug delivery: Pathways, principles, factors, and skin irritation, *Med. Res. Rev.*, 16:285–296 (1996).
141. A. K. Banga. New technologies to allow transdermal delivery of therapeutic proteins and small water-soluble drugs, *Am. J. Drug Del.*, 4:221–230 (2006).
142. A. K. Banga and Y. W. Chien. Iontophoretic delivery of drugs: Fundamentals, developments and biomedical applications, *J. Control. Release*, 7:1–24 (1988).
143. A. Sieg and V. Wascotte. Diagnostic and therapeutic applications of iontophoresis, *J. Drug Target.*, 17:690–700 (2009).
144. S. Mitragotri and J. Kost. Low-frequency sonophoresis: A review, *Adv. Drug Deliv. Rev.*, 56:589–601 (2004).
145. N. B. Smith. Perspectives on transdermal ultrasound mediated drug delivery, *Int. J. Nanomedicine*, 2:585–594 (2007).
146. J. Kushner, D. Blankschtein, and R. Langer. Heterogeneity in skin treated with low-frequency ultrasound, *J. Pharm. Sci.*, 97:4119–4128 (2008).
147. S. Mitragotri, D. Blankschtein, and R. Langer. Ultrasound-mediated transdermal protein delivery, *Science*, 269:850–853 (1995).
148. L. J. Weimann and J. Wu. Transdermal delivery of poly-*l*-lysine by sonomacroporation, *Ultrasound Med. Biol.*, 28:1173–1180 (2002).
149. P. Rajan and P. E. Grimes. Skin barrier changes induced by aluminum oxide and sodium chloride microdermabrasion, *Dermatol. Surg.*, 28:390–393 (2002).

150. T. Fujimoto, K. Shirakami, and K. Tojo. Effect of microdermabrasion on barrier capacity of stratum corneum, *Chem. Pharm. Bull. (Tokyo)*, 53:1014–1016 (2005).

151. P. Savardekar. Microdermabrasion, *Indian J. Dermatol. Venereol. Leprol.*, 73:277–279 (2007).

152. H. S. Gill, S. N. Andrews, S. K. Sakthivel, A. Fedanov, I. R. Williams, D. A. Garber, F. H. Priddy, S. Yellin, M. B. Feinberg, S. I. Staprans, and M. R. Prausnitz. Selective removal of stratum corneum by microdermabrasion to increase skin permeability, *Eur. J. Pharm. Sci.*, 38:95–103 (2009).

153. P. E. Grimes. Microdermabrasion, *Dermatol. Surg.*, 31:1160–1165 (2005).

154. W. R. Lee, R. Y. Tsai, C. L. Fang, C. J. Liu, C. H. Hu, and J. Y. Fang. Microdermabrasion as a novel tool to enhance drug delivery via the skin: An animal study, *Dermatol. Surg.*, 32:1013–1022 (2006).

155. W. R. Lee, S. C. Shen, W. Kuo-Hsien, C. H. Hu, and J. Y. Fang. Lasers and microdermabrasion enhance and control topical delivery of vitamin C, *J. Invest. Dermatol.*, 121:1118–1125 (2003).

156. I. H. Pitman and S. J. Rostas. Topical drug delivery to cattle and sheep, *J. Pharm. Sci.*, 70:1181–1194 (1981).

157. D. M. Jenkinson, G. Hutchison, D. Jackson, and L. McQueen. Route of passage of cypermethrin across the surface of sheep skin, *Res. Vet. Sci.*, 41:237–241 (1986).

158. K. A. Walters and M. S. Roberts. Veterinary applications of skin penetration enhancers. In K. A. Walters and J. Hadgraft (eds), *Pharmaceutical skin penetration enhancement*, Marcel Dekker, New York, 1993, pp. 345–364.

159. J. Blackford, T. J. Doherty, K. E. Ferslew, and P. C. Panus. Iontophoresis of dexamethosone-phosphate into the equine tibiotarsal joint, *J. Vet. Pharmacol. Ther.*, 23:229–236 (2000).

160. D. M. Jenkinson and G. S. Walton. The potential use of iontophoresis in the treatment of skin disorders, *Vet. Rec.*, 94:8 (1974).

2 Experimental Methods and Tools for Transdermal Delivery by Physical Enhancement Methods

2.1 INTRODUCTION

To ensure reliable results, transdermal experiments must be carefully designed considering several factors that might influence *in vitro* and *in vivo* studies. For *in vitro* studies, these include considerations of the diffusion cells, skin source and type, solvents, buffer and other inactive excipients in formulation, cell design, and analysis, among others. Factors affecting delivery mediated by microporation, iontophoresis, electroporation, or sonophoresis are discussed in individual chapters in this book and are not further discussed here. For *in vivo* studies, additional considerations involve assessment of skin damage or irritation, sensitivity of the analytical method, and patch design considerations. Problems relating to assay sensitivity arise because the amount of drug absorbed from the skin is small and is often below assay sensitivity following dilution in the body fluids. Tracers are often used in animal studies, but the appearance of radioactivity in urine does not account for metabolism of the drug in skin. Thus, plasma concentrations need to be directly measured, and biological response can be measured in some cases. Radioimmunoassay or enzyme-linked immunosorbent assay (ELISA) kits can be used, as these are very sensitive and are rapidly becoming available for many drugs. For a good correlation of *in vitro* and *in vivo* studies, the principal barrier to transport should be the same.

2.2 CONSIDERATIONS IN SELECTING THE SKIN MODEL

2.2.1 ANIMAL MODELS FOR TRANSDERMAL DELIVERY

Human studies would be ideal for investigating drug delivery. However, this is not always feasible, especially in the early stages of development. *In vivo* animal models are therefore often used. Because the use of primates is restricted, porcine skin is often used as it has a stratum corneum thickness similar to that of human skin, unlike hairless rodents (1). Furthermore, the hair follicle density of pig (about 11 hair follicles/cm²) and human skin (about 6 hair follicles/cm²) is similar as opposed to rats (289/cm²) or mice (658/cm²), which have significantly higher hair follicle densities. The term *hairless* used with several animal models in the literature may be a misnomer, as these animals are not really hairless and have rudimentary follicles with missing hair shafts. The hairless rat has a follicle density ($75/cm^2$) much lower than regular hairy rats and has been widely used as an animal model in transdermal studies. Despite several differences mentioned above, we have found good correlation of the full-thickness hairless rat skin with dermatomed human skin for drug transport (2). Also, the physiological properties of hairless rat skin, as reflected by pore closure in microneedle-treated skin, correlated well with what others found with human skin (3).

Animal skins, especially mice skin, are typically more permeable than human skin and may thus overestimate drug delivery. Hairless mouse skin was not found to be a good *in vitro* substitute for human skin when studying drug penetration (4) and is also more susceptible to hydration damage as compared to human skin (5). The permeability coefficients of morphine, fentanyl, and sufentanil across full-thickness hairless mouse skin were found to be one order of magnitude higher than those for human epidermis (6). These differences can be even more in the presence of chemical enhancers. For absorption of leuprolide, *in vitro* permeability through nude mouse skin was 10 or 100 times higher than that obtained for cadaver skin, depending on the type of enhancer used in the formulation. An exception is shed snake skin, which was found to be at least 10 times less permeable than cadaver skin (7). It should be noted that shed snake skin differs from human stratum corneum in that it is devoid of appendageal structures (8) and is anion selective, unlike human skin that is cation selective (9). Another skin type used is rabbit inner pinna skin (10). Also, skin from animal species often has a higher proteolytic activity than human skin. Proteolytic activity of skin is discussed in Section 8.3.1. Skin from different species can also differ in other enzymatic activity (e.g., the esterase activity of rat skin was much higher than that in human or mini pig's skin) (11). Human skin has thermoregulatory eccrine glands all over the body, while other mammals including primates only have them on palmar and plantar surfaces and in the perianal region. The skin of hairless rodents and pigs lacks eccrine glands.

Using a series of compounds, it was shown that skin of miniature swine has the closest permeability characteristics to that of human skin, especially for lipophilic drug molecules. *In vitro* studies with human skin have also been shown to correlate well with *in vivo* studies in pigs. Using histological and echogenic techniques, pig skin was reported to be similar to human skin. However, pig skin has a greater density of elastin and collagen bundles, leading to reduced elasticity relative to human skin (12). Also, the diameter of hair follicles is larger for pigs (177 μm) compared to that for humans (70 μm) (13–17). In a study that compared topical drug delivery in isolated pig ear skin and human volunteers, pig ear skin was found to be a good quantitative model for dermatopharmacokinetic studies to compare topical bioavailability of different formulations. Because isolated and excised skin lacks a functioning blood supply, such studies are best avoided where sink conditions may be needed (e.g., for long contact time with a drug of high lipophilicity) (18). Pig ear skin has also been reported to be a good model for investigating delivery of topically applied formulations into hair follicles (17). Hairless guinea pig skin was reported to be closer to human skin as compared to skin of normal-haired guinea pigs or other rodents (19); therefore, this animal model can also be used for transdermal studies, especially for irritation or sensitization studies. In another study, skin permeability of nicorandil was determined across excised skin samples from hairless mouse, hairless rat, guinea pig, dog, pig, and human. Permeability was found to be the highest in hairless mice among the six species tested, but a good agreement was found only between pig and human skin (20). However, it should be noted that the extra body fat on the pig may change the drug distribution relative to man and confound the results from *in vivo* studies (21). The skin site is also important, and care should be taken to make comparisons. For instance, the systemic bioavailability of parathion in weanling pigs was higher from the back than from the abdomen, and the absorption as well as cutaneous biotransformation were altered by occlusion (22). Thus, the site and dosing method should be controlled and specified. The age of skin may also be important for drug delivery. In a study on the passive permeability of water and mannitol across rat skin, age was not a significant factor for permeability, but changes in dermal thickness and hair follicle depth were found to be influenced by age (23). These factors may affect the delivery of a drug across skin and need to be taken into consideration.

2.2.2 USE OF HUMAN SKIN

The European center for validation of alternative methods recommended that there should be a concerted effort toward using human skin as the primary *in vitro* model for skin permeability studies. However, because the supply of human skin is limited, animal skin will continue to be used. Pig skin and hairless guinea pig skin may be acceptable alternatives (24). However, human skin should be used where possible. Human cadaver skin or human skin from surgery (tummy tuck or breast removal) can be obtained from a local tissue bank, hospital, or national or international tissue source. Serological testing should be performed on the skin to make sure that it tests negative for HIV and hepatitis B. Skin should preferably be frozen within a few hours of death or excision. Because the rate-limiting barrier to drug absorption, the stratum corneum, is a dead layer of cells, the viability of the skin need not be preserved; therefore, it can be frozen for future use. Slow programmed freezing in the presence of a protectant such as glycerol may be required (25). Frozen human skin has been widely used in transdermal studies. The resistivity of frozen (and thawed) skin is not as high as that of human skin *in vivo*, but it is still satisfactory. The skin should be prepared carefully and electrically prescreened for defects. In a study to evaluate the electrical properties of frozen skin, the majority of tissue samples had specific resistances at 10 µA of ≥ 35 kΩ cm^2 and sodium ion permeability coefficients comparable to human *in vivo* values. The skin bank should be instructed to prepare the skin carefully to minimize the effect of freezing. In order to ensure an intact skin barrier, other studies have used human skin samples that had at least 100 kΩ-cm^2 resistance (26) or in the 20 to 60 kΩ-cm^2 range (27). If skin viability is needed for some metabolism-type studies, freshly obtained skin can be stored in a physiological buffer with preservative and refrigerated for use within hours or even within a few days at the most, depending on the nature of the study. The source of fresh skin will be surgical procedures such as biopsies, breast skin from mastectomy or reduction, or abdominal skin from tummy-tuck surgery. Similarly, when studies are needed to be done to monitor a physiological process (e.g., pore closure following skin micro-poration), a live animal model should be used or human studies should be conducted (28). Human skin can be supplied by tissue banks, dermatomed to a specific thickness (typically 250 µm to 800 µm). Alternatively, epidermis can be separated from the full-thickness skin. Full-thickness skin should generally not be used for *in vitro* studies because it does not mimic the *in vivo* situations in the sense that the drug will have to cross the entire length of the skin to reach the receptor compartment of the diffusion cells. In contrast, under actual *in vivo* conditions, the drug just needs to cross the epidermis to reach the blood supply under the epidermal layer from where it becomes available to the systemic circulation. When physical enhancement techniques are used, an appropriate skin model should be used. For example, because microneedles may typically penetrate the entire epidermis to reach the top layers of the dermis, use of heat-separated epidermal sheets in such studies may not be advisable (28). Some special considerations may exist when using enhancement technologies. For example, the electrical properties of fresh, excised human skin are also similar, so the use of frozen tissue in iontophoresis studies seems justified. However, fresh skin was found to be less conductive than frozen skin at low current levels, so that the driving voltage required for *in vivo* delivery devices may be somewhat higher than that expected based on *in vitro* studies with frozen skin. Also, fresh skin showed a trend toward lower sodium ion permeability, though the difference was not statistically significant. This may suggest that studies with frozen skin may not predict the electroosmotic flow as well as those with fresh skin (29).

2.2.3 ISOLATED PERFUSED PORCINE SKIN FLAP MODEL

The isolated perfused porcine skin flap (IPPSF) model was developed as a novel alternative animal model for dermatology and cutaneous toxicology (30). It has been used in several transdermal delivery and iontophoresis investigations (31–34). Transdermal delivery of a drug involves movement from the delivery system into the skin, then into the vasculature, and finally systemic disposition and pharmacodynamic effects. Although most *in vitro* models can only investigate the first step, the IPPSF model can go a step further and investigate the movement into vasculature as well. Furthermore, it provides a large dosing surface area and the convenience of continuously sampling venous perfusate. The abdominal skin of weanling pigs is typically used, and a single-pedicle axial pattern tubed skin flap is created following a surgical procedure. The sole vascular supply of the tube is cannulated and perfused *ex vivo* with Krebs-Ringer bicarbonate buffer containing albumin and glucose. The skin flap is then maintained in a Plexiglas chamber with controlled temperature and humidity. Viability is maintained for about 24 hours (35). Because the IPPSF model possesses a viable epidermis and intact vasculature, it is also useful for studies of cutaneous metabolism of drugs (34). A mathematical model to predict percutaneous absorption and subsequent disposition based on the biophysical parameters measured with the *ex vivo* perfused skin preparations was described (36). The model predicted *in vivo* serum concentrations of an iontophoretically delivered peptide (31). In another study using arbutamine, a novel catecholamine, the concentration-time profiles were predicted on the basis of an IPPSF study. For two different sets of iontophoretic dosing conditions, the profiles predicted by IPPSF studies compared well with those seen in humans (37). The use of perfused pig flap is a very attractive and useful model, but it is surgically and technically demanding and expensive, and the preparations are not usually viable for extended periods of time (21,24).

2.2.4 ARTIFICIAL SKIN EQUIVALENTS

A cultured or living skin equivalent of human origin contains human dermal fibroblasts in a collagen matrix (38). This artificial skin closely resembles human skin, as it has an epidermis and dermis, the former with a well-differentiated stratum corneum. However, it lacks appendages such as hair follicles and sweat glands, and it obviously lacks the vasculature as well. Iontophoretic transport of pindolol hydrochloride, salmon calcitonin, and benzyl alcohol across living skin equivalent was found to correlate well with that across guinea pig skin *in vitro* (39). However, more typically, artificial skin equivalents may be as much as 10 times more permeable to drugs as compared to excised human skin (40). Furthermore, the cost of artificial skin is generally high. However, they are sometimes used by industry, especially smaller companies that may not have an in-house vivarium and may not want to go through the biosafety approvals required for use of human skin. Use of these skin equivalents may still be better than using synthetic membranes, depending on the intended application. For example, use of synthetic membranes would be inappropriate for a study of chemical penetration enhancers, as these membranes do not have the lipid content and other histological attributes of skin which are targeted by these chemical enhancers. Use of artificial skin for skin irritation studies may be a good use of these cultures, because the procedures have been well developed and alternatives to animal testing are highly desirable, especially for cosmeceuticals (see Chapter 10). Human skin or living skin equivalent xenografted onto immunodeficient nude mice has also been used in transdermal or basic research studies (41,42). In one study, human skin xenograft from the thigh of a 24-year-old donor was removed and stored in regular medium with antibiotics. A piece of skin 2 × 1.5 cm was removed from the back of the mice, and human skin of the same size was trimmed and transplanted the same day onto anesthetized mice by suturing into place. Mice were then used after 9 weeks when the graft was well healed (43).

2.2.5 SKIN TREATMENT

Subcutaneous fat should be removed from skin before use. As mentioned earlier, use of full-thickness human skin can result in underestimation of *in vivo* delivery, because blood circulation under the epidermis will normally pick up the drug. However, drug delivery may vary according to the type of skin layer employed for the study (i.e., full thickness versus epidermis only versus stratum corneum only). For example, *in vitro* iontophoretic transport of nafarelin across human cadaver skin varied with the type and layer of skin used. The cumulative amount (nmol/cm^2) delivered in 24 hours was 3.97 for whole skin, 28 for epidermis, and 125 for dermis (44). In another study with full-thickness human skin, an applied voltage of 0.5 V did not result in any measurable flux for two polypeptides, leuprolide and a cholecystokinin-8 analogue.

However, pretreatment of skin with ethanol, followed by iontophoresis was found to be effective for delivery (45). Therefore, freshly separated epidermis can serve as a substitute for transdermal studies, unless full-thickness skin is desired. In a typical procedure, the whole skin is immersed in water at 60°C for 45 seconds, at which time the epidermis can be peeled off from the dermis. Alternate methods for separation of epidermis are also available, such as the use of ethylenediaminetetraacetic (EDTA) acid or microwave. For the former procedure, skin can be placed on a filter paper saturated with 0.75% EDTA and kept at 37°C for 2 hours after which epidermis can be separated from the rest of the skin. If desired, epidermis can then be placed on a filter paper saturated with 0.0001% trypsin solution overnight at 37°C to digest the dermis and obtain stratum corneum (46).

2.2.6   USE OF SYNTHETIC MEMBRANES

Skin is a complex biological tissue; thus, permeability measurements across skin tend to have a high variation. In order to avoid these problems or sometimes just for convenience, various synthetic membranes have been tried in transdermal research. These membranes, however, may not be predictive of what to expect with skin, because skin is a complex biological tissue, and these membranes are not. The ideal membrane should be hydrophobic to mimic the lipophilic skin barrier and prevent excessive flow of water. The membrane should also be conductive if its use in iontophoresis is being considered. This combination of characteristics is hard to find. Nevertheless, the use of synthetic membranes may be desired in some situations, such as initial mechanistic preliminary studies to narrow down the number of experiments to be performed with skin or as a routine quality control for batch-to-batch variations in commercial production. Synthetic membranes have been widely used in iontophoresis studies. The use of two hydrophilic and two hydrophobic membranes on the passive and iontophoretic transport of salbutamol sulfate has been investigated. Slower transport was observed with the hydrophobic membranes (47). Nucleopore™ (Whatman, Inc., Piscataway, New Jersey) is a synthetic membrane with essentially cylindrical, aqueous-filled pores (pore radius 75 Å; porosity 0.001). It has a polyvinylpyrrolidone-coated polycarbonate backbone with a net negative charge and a nominal thickness of 6 μm. It has been used in iontophoresis research by stacking 50 such membranes together to form a net negatively charged, random pore network for diffusion with a resistance of about 1.5 kΩ, which is of the same order of magnitude as skin (48–51). The enhanced transport of cations and anions across nucleopore porous membranes in the presence of an applied electric field was found to be asymmetric, possibly due to the direct effect of the field and convective solvent flow (48). Thus, it seems the membrane is somewhat representative of what to expect with skin, because iontophoresis is accompanied by electroosmosis. Drug release rates from hydrogels through cellophane membranes have also been investigated. In the absence of current, release was matrix controlled with a linear relationship between the square root of time and amount of drug released. As current was applied, this relationship changed to a linear relationship with time for the amount of drug released (52,53). A microporous polyolefin membrane with hydrophilic urethane polymer–filled pores has also been used for iontophoretic delivery of the ionized drugs dexamethasone sodium phosphate, hydrocortisone sodium phosphate, prednisolone sodium succinate, and one nonionized drug, cortisone acetate. As expected, the electric fields interacted more efficiently with the charged molecules (54). Synthetic membranes may also be used as an integral part of the patch to be used on skin. Ion-exchange membranes have been investigated in this respect. In this case, the membrane should inhibit passive delivery but not delivery under an electric field. By eliminating passive release, release of the drug would be turned on and off simply by turning the current on and off. The membrane would also protect against unintended passive absorption of the drug from abraded or compromised skin. Ion-exchange membranes can also be used in an iontophoretic patch system to inhibit transport of competing counterions (55). Heterogeneous cation-exchange membranes have also been prepared by mixing conductive sulfonated polystyrene beads into a nonconductive silicone rubber matrix. They have found use in modulating iontophoretic delivery from implantable devices (56). A perfluorosulfonic acid cation-exchange membrane was also used for iontophoretic delivery of acetate ions and was stated to be a good model for skin (57). Iontophoretic transport of diclofenac sodium across a cellophane membrane has also been investigated (58).

2.3 TRANSDERMAL STUDIES

Horizontal or vertical configuration diffusion cells are typically used for *in vitro* transdermal studies. In many cases, skin can be pretreated with the enhancement technology being used, such as phonophoresis or microneedles, and then mounted on the diffusion cells. However, in other cases, such as iontophoresis which acts primarily on the drug molecule rather than on the skin, skin is first mounted and then the electrodes are applied. For *in vivo* studies, various animal models have been used as discussed earlier. For patch design used in such *in vivo* studies, the reader is referred to Chapter 10. Final testing may be done in human subjects, which is also discussed in Chapter 10, along with a discussion of skin irritation and safety and regulatory issues.

2.3.1 Diffusion Cells: Vertical Configuration

Vertical diffusion cells (Figure 2.1) are often used for *in vitro* permeation studies. They typically represent the finite dosing technique, as small amounts of drug can be placed on the skin. A commonly used vertical cell is the Franz diffusion cell, developed by Dr. Thomas Franz. It has been in use for over 30 years (59). Other cells similar in design are also available. The general cell design includes a donor half that is exposed to room temperature (25°C) and a receptor half maintained at 37°C, thus simulating conditions that closely approximate the *in vivo* situation. Excised human cadaver, animal skin, or membranes can be mounted onto these vertical diffusion cells. Because the donor side is exposed to the environment, this cell is sometimes called the one-chamber cell. Also, skin on the donor side is not necessarily hydrated in this setup because ointments or other semisolid preparations can be used in some studies (60).

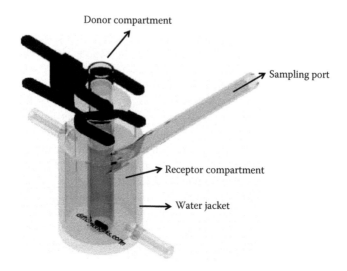

FIGURE 2.1 Vertical diffusion cell showing the donor and receptor compartments as well as the sampling arm and water jacket of the receptor compartment. (Reproduced with permission from DiffusionCells.com, http://diffusioncells.com.)

The Franz cell can also be used when skin enhancement technologies are utilized. When using the Franz cells for sonophoresis, the sonicator probe can be placed in the donor solution, though in a more common configuration of sonophoresis devices, the skin will be treated prior to mounting on the Franz cells. Similarly, skin can be treated with microneedles, electroporation, or microdermabrasion before mounting. When using Franz cells for iontophoresis, placement of the counterelectrode in the sampling arm of the receptor compartment may be considered as being "inside" the skin. Any actual device, on the other hand, will place both electrodes "on" the skin. Thus, it may be preferable to design cells that place both electrodes on the same side of the skin to simulate the *in vivo* situation. Feasibility of such a cell design was demonstrated by delivering morphine and clonidine across full-thickness hairless mouse skin. It was shown that significant lateral transport does not take place in this cell design (61). For evaluation of commercially available iontophoresis electrodes in clinical use, use of a donor chamber is not required. For such studies, a cell design that uses two pieces of skin placed at either end of a central receptor compartment was used. The reservoir-type electrode filled with drug can then be placed on one side and a dispersive electrode on the other side, and samples can be taken from the central compartment (62,63). Commercially available horizontal cells can also be modified for a somewhat similar single-compartment design that is clinically relevant (64). In the case of a simple ion (benzoate), it was observed that changes in diffusion cell configuration and placement of electrodes relative to skin had little effect on transport (62), perhaps suggesting that the simple diffusion cells still have a role for *in vitro* investigations, especially for simple molecules. Flow-through systems are commercially available for either design, to facilitate sampling and maintaining sink conditions throughout the duration of the study. Development of automated diffusion apparatus has also been discussed in the literature (65), and some instruments are now commercially available. The automated instruments are especially useful for *in vitro* release testing of semisolids. These tests look at drug transport across synthetic membranes from semisolid formulations and are designed as a test for batch-to-batch reproducibility or to get regulatory approval for small changes in formulations during scale-up or for postapproval changes after marketing (66–68).

2.3.2 DIFFUSION CELLS: HORIZONTAL CONFIGURATION

Valia-Chien or various other types of horizontal cells have also been used for transdermal studies. This setup, sometimes called the two-chambered cell, almost always represents infinite dose technique, because the amount of drug permeating into receptor is small relative to the total amount present in the donor. Thus, it is used for studying the mechanisms of transport and can also be used to measure absorption if the drug delivery device is intended for application to skin to produce steady-state levels. These cells should be hydrodynamically calibrated so that the studies give an intrinsic permeation rate that is independent of the hydrodynamic conditions of the cell (69). Each half-cell may have one or two ports. If two ports are present, one port serves as the sampling port while the other port can be used to insert the electrode. Excised human cadaver, animal skin, or a membrane can be mounted between the two half-cells. The exposed surface area of the skin should be known for calculations, a typical value being 0.64 cm^2. An external water bath maintains the temperature of the circulating water in the jackets at 32°C for horizontal configuration cells (or 37°C for vertically oriented cells). Magnetic stirrers are used to stir the solutions in both compartments continuously.

2.3.3 DEVICES USED WITH SKIN ENHANCEMENT TECHNIQUES

Iontophoresis power supplies, which are commercially available for topical application of drugs in physical therapy clinics, are discussed separately in Chapters 4 and 10. These can also be used for *in vivo* studies in experimental animal models. The wearable patches in development also have their own miniaturized power supply, as discussed in Chapter 10. Other vendors, such as Keithley Instruments Inc. (Cleveland, Ohio), provide power supplies that can also be used for *in vitro* and *in vivo* studies. For *in vitro* studies, they can typically power several transdermal diffusion cells at one time, allowing several triplicate experiments to be carried out simultaneously. Many researchers also use custom-built power supplies. A constant (DC) current of 0.5 mA/cm^2 or less should typically be used, as higher currents are usually not acceptable for human studies. In constant current iontophoresis, the voltage drop across skin adjusts to keep the current constant. When there is an increase in resistance of the system, the power supply will respond by increasing the voltage to drive the same current through the skin. Devices used for electroporation and sonophoresis studies are discussed in Chapters 5 and 6, respectively. Skin microporation can be accomplished by microneedles or with other technologies or devices based on thermal, radio-frequency, and laser ablation; the devices used to accomplish this are discussed in Chapter 3. For application of microneedles, applicators have been considered for uniform application forces.

Several factors are involved in the selection of an applicator—a simple laboratory technique involves the back of a syringe (70). More details on applicators are discussed in Chapter 3.

2.3.4 TYPICAL EXPERIMENTAL SETUP

Skin should be held against light and observed for the absence of pinholes prior to initiation of experiments. This is especially important when epidermal membrane or isolated stratum corneum is being used. Skin should then be mounted between the donor and receptor cells. The integrity of skin samples should preferably be checked by measuring its electrical conductivity with an ohmmeter or by alternative techniques such as transport of tritiated water or transepidermal water loss (TEWL) measurements. In a study with porcine epidermis, the authors used the skin only if the resistance was greater than 30 kΩ-cm^2 (71). Similarly, in another study, rat skin was used if the resistance was greater than 30 kΩ-cm^2 (72). In a study that used microneedles on human epidermal membrane, the epidermal membrane was used if the resistance was higher than 15 kΩ-cm^2. Skin resistance can vary among human subjects between 20 and 400 kΩ, depending on body site, and is lowered to around 7 kΩ by a short application of current (73). This is the basis of the Hybresis iontophoresis patch (see Section 10.6.2).

Due to variability inherent in transdermal delivery, 6 to 12 replicates may be needed, especially when human skin is used to finalize a formulation after the initial screening studies have been completed. If the drug has high intrinsic permeability through skin such that a significant percentage of the drug is being delivered, the receptor should be replaced periodically or flow-through diffusion cells can be used. In order to maintain sink conditions for adequate diffusion gradient and prevent back diffusion, the thermodynamic activity in the receptor should never exceed 10% of that in the donor formulation. The receptor fluid must be stirred continuously and thermostated so as to maintain a temperature of 32°C for the skin (24). Ethanol, up to 25%, has been commonly used in the receptor to maintain sink conditions. When using ethanol in the donor or receptor formulations, *in vitro* studies should be designed carefully because permeation of ethanol across skin from donor to receptor compartments may cause precipitation of the drug in the donor compartment. Alternatively, its transport from receptor to donor may modify the formulation (40). Also, ethanol will have an enhancer effect on the skin.

A typical *in vitro* experiment should last less than 7 days, with a common time period being 24 to 72 hours. The stratum corneum obtained from human skin and mounted on a diffusion cell at 37°C was shown to lose its barrier function at around 200 hours. Immediately after separation, the passive electrical properties of skin were similar to those *in vivo*, but they changed within hours. At about 300 hours, the magnitude of skin impedance was only about 1% to 2% of the value observed immediately after separation (74). The duration of study will, of course, also be dictated by the therapeutic rationale for the compound being investigated. It was suggested that protecting human epidermal membrane from physical stress is an essential element in maintaining its permeability and electrical resistance for extended periods. When supported by a porous synthetic membrane, the epidermal membrane could be used for successive passive permeability experiments over 5 days with extensive washing between the experiments. The advantage of successive experiments is to investigate experimental variables without factoring in skin-to-skin variability (75). For long-duration experiments, suitable preservatives may have to be included in the solutions. The effect of these preservatives on the permeability of skin or the lack of their effectiveness (which may result in microbial growth) should be carefully evaluated (76). If the viability of skin has to be maintained (e.g., for investigating metabolism of drugs in skin) during the *in vitro* study, fresh skin should be used with a growth medium such as Eagle's minimal essential medium or Dulbecco's modified phosphate-buffered saline supplemented with glucose. Skin viability can then be maintained for about 24 hours in a flow-through diffusion cell, as measured by aerobic ^{14}C-labeled glucose utilization (77). The calculations should take into account the amount of drug lost with each sampling period, as the sample is typically replaced with fresh buffer. Results can be plotted as cumulative amount versus time or flux versus time. For comparing several experiments, the cumulative amount versus time curve can be used to calculate the slope during the delivery period.

2.3.5 VARIABILITY OF PERCUTANEOUS ABSORPTION

It should be realized that skin can be quite variable between subjects or even at different sites on the same subject (78). For example, the percutaneous absorption of ketoprofen in human subjects was similar when applied to the back or arm but lower when applied to the knee (79). The effect of other variables such as age, race, and gender on drug delivery through skin is less clear (80). There are known physiological and structural differences in skin based on sex and racial differences, but these differences have not been clearly correlated with differences in percutaneous absorption (81). This could perhaps be because the variability inherent in percutaneous absorption overshadows the comparatively smaller contributions resulting from these differences in male versus female skin or in various ethnic groups. For example, the permeability coefficient of narcotic analgesics through human cadaver skin sites as diverse as the sole of the foot, chest, thigh, and abdomen were reported to be remarkably similar. In another study, permeability through skin sections obtained from different cadavers varied four- to fivefold, but no trends in permeation as a function of age or gender were seen (82). In general, ionic permeants produce more variable flux data than neutral permeants during passive transport through skin. This variability seems to increase as the hydrophilicity of the drug increases (83,84). As mentioned, ethnic differences may also influence absorption. In a comparison between American and Taiwanese smokers, transdermal delivery of nicotine was higher for Taiwanese smokers. The differences between the ethnic groups were statistically significant for all patch sizes (85). However, when the contraceptive patch, Ortho Evra™ was tested in special populations, there was no significant difference in absorption with respect to Caucasians, Hispanics, and blacks. Only a small percentage (10% to 25%) of the overall variability in pharmacokinetics was associated with demographic parameters of age, body weight, body surface area, and race (86). Skin permeability may also vary by up to threefold between cadavers of the same species, resulting in considerable interexperimental variations (14). Therefore, when human cadaver skin is used for experiments, a large number of replicates are recommended. In an iontophoresis study with four ethnic groups, the effect on the skin barrier function and irritation resulting from a 4-hour application of current (0.2 mA/cm^2) on a 6.5-cm^2 area was evaluated. The ethnic groups were Caucasians, African Americans, Hispanics, and Asians, with 10 subjects in each group. Bioengineering skin instrumentation was used to monitor TEWL, capacitance, skin temperature, and skin color. Skin barrier function was not dramatically affected, and racial differences observed were subtle, not major (87). Racial differences in skin function in physiologic and pathologic conditions were reviewed (88,89). An understanding of these differences can help to explain any observed differences in absorption, irritation, sensitization, or erythematous reactions (89). The incidence of skin irritation (see Section 10.4) may also depend on the site of application (90).

2.4 ANALYSIS

For *in vitro* studies, samples taken from the receptor are analyzed by one or more of several possible means. For studies with radiolabeled compounds, ^{14}C label in a metabolically stable position is preferable. If ^{3}H label is used, the results should be validated by an high-performance liquid chromatography (HPLC) assay with ultraviolet (UV), radiochemical detector, or any other suitable stability-indicating assay. It may also be possible to lyophilize the receptor solution samples and resuspend in buffer to remove any ^{3}H label that may have exchanged with water (26). In general, care should be taken when using isotopes, because only very small amounts typically permeate through the skin and any permeation of the free tracer will disproportionately confound the data. HPLC or ELISA analysis is more commonly employed to get reliable data. When cumulative amount permeated is calculated, the amount lost in each sample should be taken into consideration. This can be easily set up in any commercially available spreadsheet program. The cumulative amounts can then be divided by the surface area of the skin to express results in terms of drug permeated per square centimeter of the skin, which allows more meaningful comparisons of work done in different laboratories. Cumulative amount can be plotted as a function of time, and steady-state flux (slope) can be calculated by linear regression. Analysis of the drug inside the skin is more involved compared to analysis of the drug permeating across the skin into the receiver compartment. Different methods to quantify drug inside the skin include tape stripping, microdialysis, skin biopsy, suction blister, follicle removal, and confocal Raman spectroscopy (91).

For *in vitro* studies, one reason why measuring the drug inside the skin is difficult is partly because the skin is in direct contact with the rather high drug concentration in the donor solution, which can easily confound the results. The donor compartment should be emptied and rinsed with a suitable buffer several times before the skin is dismounted from the Franz cell setup. Skin should then be removed and rinsed thoroughly, preferably alternating rinsing and drying instead of continuously rinsing. Following rinsing, tape stripping studies may be carried out for quantification of drug in the stratum corneum (see Section 2.4.1). The underlying skin can also be further assayed for drug level in the epidermal and dermal layers. Autoradiography or immunohistochemical staining studies can be performed to study the localization of drug within skin (92,93). However, skin sectioning cannot normally be performed on humans unless limited to punch biopsies (21). An indirect measurement of the amount of drug in skin can be done by measuring the amount lost from the donor. However, loss due to reasons other than absorption needs to be considered, and it may be hard to quantify a very small difference between two very large values. It should be realized that the amount in skin will vary as a function of time due to several factors and will most likely also vary between *in vitro* and *in vivo* studies as well (94).

2.4.1 TAPE STRIPPING

As many drugs tend to accumulate in the stratum corneum, a widely used method to quantify the amount of drug in skin involves stripping off the stratum corneum by consecutive applications of adhesive tapes, followed by extraction of drug from the tapes and into a suitable buffer, and finally analysis of the buffer samples. The amount of drug in the remaining or underlying skin (rest of epidermis and dermis) can then be quantified by developing a method for drug extraction from the skin and calculating the recovery coefficient. Alternatively, the underlying skin can be sectioned parallel to the skin surface with a freezing microtome to quantify the amount of drug in different layers of skin. In 1998, the U.S. Food and Drug Administration (FDA) issued a draft guidance document to propose tape stripping as a method to evaluate dermatopharmacokinetics (DPK) of drugs, to compare bioequivalence of topical products applied to skin. However, when two independent laboratories compared tretinoin gel formulations, they came to opposing conclusions on which formulation is more bioavailable. The draft guidance was then withdrawn in 2002. It is believed that differences in tape stripping protocols (e.g., type of tape used, number of strippings, and so forth) used in different laboratories were a reason for the observed discrepancy. Efforts are now underway to standardize protocols and possibly revive the guidance document for a more limited use for drugs that have their site of action within the skin, such as antifungals (95–97). It was suggested that the dermatopharmacokinetic method can be more sensitive and discriminating than efficacy-based clinical trials (98). Some of the changes proposed to standardize protocols include removal of the complete stratum corneum as observed by a sharp increase in TEWL values and measurement of the weight of each tape strip poststripping. A roller can be used to press the adhesive tape on the skin, to standardize the pressure used, and to minimize the influence of skin furrows and wrinkles that may confound the data. Differential weighing of tape strips as they are removed allows for better control of the methodology (though this can be tedious and time consuming) by knowing the amount removed for reproducible results. Also, knowing the area of the tape and using a value of 1 g/cm^3 for the density of stratum corneum, the thickness of the stratum corneum removed can be calculated from these weightings. As the amount of drug in tape strips can be analyzed separately, the results can be plotted as a function of distance penetrated by the drug in the skin (18,99,100). The amount of stratum corneum removed in each tape strip depends on the type of formulation and other factors; therefore, the amount of different formulations on a certain tape strip (e.g., tape strip number 4) should not be directly compared (101).

Most investigators reject the first one or two tape strips to remove the contribution of any formulation that was not washed or wiped away before tape stripping, but there are no clear guidelines on this at this time. It should be noted that in some cases, the amount of drug in the hair follicles may be equally important or even more important, such as in skin conditions like acne, alopecia, and some skin tumors. Also, in iontophoresis-assisted delivery, drug molecules are preferably pushed into skin via appendages such as hair follicles. In these cases, the follicles can be removed by a cyanoacrylate adhesive on a glass slide as described in the literature (102).

2.4.2 MICRODIALYSIS

Microdialysis probes can be inserted into tissue to measure unbound drug levels in the interstitial fluid, thereby allowing continuous *in vivo* monitoring of drug delivery (103). Microdialysis can also provide a drug concentration-time profile inside the skin, typically in the dermis, and therefore can be very useful for dermatological formulations that have their site of action within the skin. Unlike tape stripping that gives information about the drug levels in the stratum corneum, microdialysis provides information about drug levels in the deeper layers of skin. However, the technique is technically challenging and somewhat invasive compared to skin stripping studies. A very fine hypodermic needle is used to insert a linear probe into the skin and then the needle is withdrawn, leaving the probe in the skin. At least 1 hour is needed to let the skin recover from the trauma of insertion. A concentric probe is also available which can be inserted under the skin if desired. The probe has a semipermeable dialysis membrane (Figure 2.2) that is positioned under the drug patch, and a perfusion fluid is then passed through the probe. Some of the drug

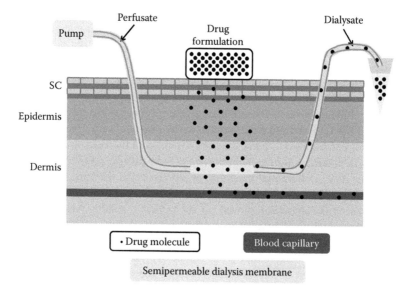

FIGURE 2.2 Linear microdialysis probe inserted into the dermis with the dialysis membrane of the probe located under the patch having the drug formulation. **(See color insert.)**

entering the skin also enters the probe, and a recovery factor is used to normalize the data to plot a concentration-time profile of the drug in the skin. In a study with 15 healthy volunteers, microdialysis probes were inserted into defined skin layers directly under the transdermal delivery system, and complete concentration versus time profiles were obtained for nicotine (104). Microdialysis can also be used when active delivery systems are employed (e.g., linear microdialysis probes inserted into subject's forearm skin have been used to investigate the kinetics of propranolol delivery mediated by iontophoresis) (105). Using linear microdialysis probes, we found that the depth of insertion in the hairless rat model was 0.54 mm as measured by a 20-MHz probe on an ultrasonic echography instrument, suggesting that the probe was in the dermis. Methotrexate was then delivered across skin by iontophoresis or microneedles, and the drug levels were measured in the dialysate fluid (106). Use of microdialysis probes for *in vitro* studies in excised skin has also been investigated (107). The method has several limitations and may not apply to large (>20 kDa) molecules or those with high lipophilicity. However, efforts are underway to enable microdialysis sampling of proteins and other macromolecules by using large-pore membranes, slower infusion rates, and osmotic balancing (108–111).

2.5 INSTRUMENTAL TOOLS USED TO CHARACTERIZE TRANSDERMAL TRANSPORT

In recent years, many noninvasive bioengineering instruments were introduced into cosmetology, and some of these instruments can also be used to study transdermal delivery of drugs, especially changes in skin, for example, formation of microchannels created in skin following microporation (see Chapter 3) or to monitor the inserted microdialysis probes in skin and the resulting reactions (112). One of the biophysical parameters of the skin which becomes especially meaningful when microchannels are created in skin is the TEWL from skin. A passive diffusion of water constantly takes place from the deeper layers of skin to the outside. When the skin barrier is damaged, TEWL increases and can be monitored by an open chamber diffusion technique where the water vapor pressure gradient is measured as per Fick's law. An example of a commercially available instrument to achieve this is the Tewameter®. TEWL measurements are typically made for *in vivo* studies and have only limited use for *in vitro* studies (113). More recently, a closed-chamber instrument, the VapoMeter device (Delfin Technologies Ltd., Kuopio, Finland), became available which uses a humidity sensor in a closed chamber that is created when the probe is in touch with skin. A study with 16 human subjects reported that lower TEWL readings were recorded with the closed-chamber system and that the open-chamber system can detect small changes in TEWL. However, the VapoMeter is faster, more portable, and offers more flexibility in measuring different locations on the skin (114). Skin microporation induces significant changes in TEWL; therefore, the VapoMeter should suffice in most cases. TEWL is usually expressed in units of $g/h.m^2$. We used both the VapoMeter (115) and the Cyberderm evaporimeter (Cortex Technology, Denmark) (116) to show that a TEWL baseline value of around 6 to 8 $g/h.m^2$ is observed for intact hairless rat skin, which then increases significantly when treated with maltose or metal microneedles. A positive control of tape-stripped skin was also performed and was found to have the highest TEWL values indicating complete removal of the stratum corneum barrier (115,116). A correlation between TEWL and percutaneous absorption was reported in the past, except for highly lipophilic drugs (117).

Impedance spectroscopy (IS) measures the impedance of tissue to the flow of a small alternating current as a function of frequency and fits the resulting spectrum to equivalent circuit models of skin to deduce its resistive and capacitive components (see Section 4.2). In addition to TEWL, IS provides information about the skin barrier function. When combined with attenuated total reflectance Fourier transform infrared (ATR-FTIR) spectroscopy, these three noninvasive biophysical techniques allow tracking of the interaction of a formulation with the stratum corneum (118). IS can be used to monitor the effect of iontophoresis on skin as ions are driven into the skin under the electric field. However, because iontophoresis does not induce major changes in skin barrier function, these and other instrumental tools may not be as beneficial as found for techniques that cause physical changes in the skin (119). Confocal laser scanning microscopy (CLSM) has been widely used to study transport of fluorescent or fluorescent-tagged molecules and particles into skin appendages (such as hair follicles) and through microchannels in the skin. CLSM allows imaging of the fluorescent probe within tissues without much preparation or destruction of skin. However, only a limited range of fluorophores and lasers are available, and typically only qualitative or semiquantitative information is obtained. Care has to be taken to avoid artifacts from autofluorescence of tissue. Nevertheless, the technique is very useful and has been widely used for skin imaging. Calcein, a charged hydrophilic dye, or Nile red, a lipophilic neutral dye, is commonly used for CLSM imaging (120). Using rhodamine B dye, microchannels created by 150-µm-long microneedles (applied by an applicator) were scanned in z-axis at 10-µm increments by CLSM. It was shown that the fluorescence intensity decreased with depth. Penetration of the dye was seen until 80 µm by CLSM imaging, and this was considered to be the depth of the microchannels (121). In our studies, we inserted 500-µm-long microneedles into skin to create microchannels, following which calcein dye and 2-µm fluorescent particles were applied to the site. Both were monitored simultaneously by dual dye imaging for their permeation profiles. Green fluorescence indicated transport of calcein, red fluorescence indicated transport of the particles, and areas where both the dyes were colocalized were indicated by yellow (122). When using tagged molecules or vesicles, it should be realized that CLSM is providing information about the fluorescent label and not necessarily about penetration of the tagged molecule or vesicle if the label may have separated (123). Another technique, confocal Raman spectroscopy, can enable the analysis of the molecular composition of skin as a function of distance from the surface. For example, it has been used to quantitate the water concentration in the stratum corneum with a depth resolution of 5 µm and also to semiquantitate concentration profiles of the major components of the natural moisturizing factor (NMF), a collection of water-soluble compounds found in the stratum corneum (124).

High-frequency ultrasound (>10 MHz) can be used for dermatological scanning in pulsed mode by rapidly switching between emission of ultrasound and the echo of the sound coming back to the same transducer. Instruments designed for such measurements take advantage of the fact that epidermis and dermis reflect ultrasound differently. Ultrasonic echography using Dermascan (Cortex, Denmark), a commercially available instrument, has been performed to measure skin thickness in pigs and humans using a 20-MHz two-dimensional (2D) probe. In this study, the authors also injected hexafluoride sulfate, an ultrasound contrast media, by hollow microneedles to detect the location of fluid accumulation in the dermis using three-dimensional ultrasound echography (12). We have used a 20-MHz probe that has a resolution of 60 × 200 microns and a penetration capability of 10 to 15 mm with an ultrasound velocity of 1580 m s^{-1} when placed perpendicular to the skin with a thin layer of ultrasound gel. This procedure was carried out to measure the depth of a microdialysis probe inserted into skin (106). A literature review of studies with ultrasonic echogenicity was carried out to find the changes in skin as a function of age, though no clear trends were identified (125). These instruments typically have different scanning modes. In A-mode scanning, these instruments display a line with peaks on the monitor, with each peak representing echoes from different layers of the skin. In B-mode scanning, this information is processed electronically to display cross-sectional images of the skin. A B-scan measurement with 20-MHz probe was reported to measure dermal thickness (126). Optical coherence tomography (OCT) is another instrumental tool that uses near-infrared light to provide high-resolution three-dimensional imaging of skin in real time without touching the skin and has a resolution of about 15 μm (127). Because OCT is based on light, it can provide images that have a higher resolution than magnetic resonance imaging (MRI) or ultrasound, and can image deeper (1 to 2 mm below the surface) than the typical CLSM. OCT was recently used to monitor changes in skin following microporation (128). Near-infrared spectroscopy (NIRS) uses the near-infrared region of the electromagnetic spectrum from about 800 to 2500 nm and can have tissue penetration depths of micrometers to centimeters, depending on the frequency. It has been investigated as an analytical method to directly quantitate the amount of drug penetrating into the skin, as an alternative to tape stripping, followed by extraction and HPLC analysis. The drug powders first need to be scanned to make sure that they have near-infrared (NIR) chromophores. Medendorp et al. used econazole nitrate and estradiol to demonstrate that NIR spectrometry can be used to quantitate penetration of these drugs *in vitro* into human skin. The results paralleled those obtained by HPLC analysis, and they suggested that this all-optical method can potentially be used to measure topical bioavailability/bioequivalence for different formulations of these drugs (129). Laser Doppler flowmetry is an optical technique that uses a laser beam to measure microcirculation, especially cutaneous blood flow based on the shifted (Doppler) backscattering of laser light by moving red blood cells. The cutaneous blood flow provides an indirect measure of the extent of irritation. It thus provides a simple noninvasive test that can provide continuous monitoring. A laser Doppler perfusion monitor has been used to show that in human subjects, the maximum blood flow under skin was achieved much faster following a microneedle injection of a vasodilator methyl nicotinate as compared to its topical application (130). Using laser Doppler perfusion imaging, it has been shown that some nonspecific vasodilatation may occur during iontophoresis, especially for cathodal iontophoresis (131). We also used Chromameter to measure skin erythema and laser Doppler velocimetry to measure blood flow under the skin (106). Chromameter has been used for many years to measure skin blanching induced by application of corticosteroid formulations to skin (132). Scanning electron microscopy (SEM) is a type of electron microscopy that images the surface of a sample by scanning it with a high-energy beam of electrons to produce high-resolution images. SEM has been widely used to characterize microneedles. We used field-emission SEM to characterize maltose microneedles (93,116).

Use of some of these bioengineering instruments to monitor skin irritation is discussed in Section 10.4. In addition to the sophisticated tools discussed here, simple techniques have been very helpful in transdermal studies. Staining techniques have been widely used to show creation of microchannels on the surface of skin following skin microporation. These techniques rely on the hydrophilic dye being taken up by the areas of the skin surface breached by microporation. Intact skin with hydrophobic stratum corneum does not take up the hydrophilic dye, such as methylene blue. On the other hand, a lipophilic dye coated on microneedles has been used to stain the skin when the microneedles penetrate the skin. The dye marks the track of the projections, as it does not diffuse. The depth of penetration can then be determined by subsequent histology studies (133). A video microscope can also be used to facilitate skin imaging (93). Simple histology studies have provided very useful information about changes in skin structure. Preparation of skin samples for histological analysis involves embedding frozen nonfixed skin in a water-based media that is then snap frozen in liquid nitrogen and cryosectioned, followed by hematoxylin and eosin (H&E) staining for microscopic observation (93). We used an image analysis software developed by Altea Therapeutics (Atlanta, Georgia) for investigating the uniformity of microchannels created after microporation. Calcein dye is applied on the site, and a fluorescent image is taken by a camera fitted with a long-pass filter. The fluorescent image is then processed based on the bloom around each pore which is indicative of the volumetric distribution of calcein in each pore relative to other pores. A pore permeability index (PPI) is then computed for each pore, and a histogram of all PPI values is generated. Using hairless rat skin treated with maltose microneedles, we have shown that this histogram shows a relatively narrow distribution, indicating uniformly created microchannels (116). Several other techniques such as tape stripping and microdialysis which have been discussed earlier in this chapter are also available for quantification of drug levels in skin layers.

REFERENCES

1. C. Cullander. What are the pathways of iontophoretic current flow through mammalian skin? *Adv. Drug Del. Rev.*, 9:119–135 (1992).
2. R. S. Upasani and A. K. Banga. Response surface methodology to investigate the iontophoretic delivery of tacrine hydrochloride, *Pharm. Res.*, 21:2293–2299 (2004).
3. H. Kalluri and A. K. Banga. Formation and closure of microchannels in skin following microporation, *Pharm. Res.*, March 31 [Epub] (2010).
4. G. P. Kushla and J. L. Zatz. Influence of pH on lidocaine penetration through human and hairless mouse skin *in vitro*, *Int. J. Pharm.*, 71:167–173 (1991).

5. J. R. Bond and B. W. Barry. Limitations of hairless mouse skin as a model for *in vitro* permeation studies through human skin: Hydration damage, *J. Invest. Dermatol.*, 90:486–489 (1988).
6. S. D. Roy, S. Y. E. Hou, S. L. Witham, and G. L. Flynn. Transdermal delivery of narcotic analgesics: Comparative metabolism and permeability of human cadaver skin and hairless mouse skin, *J. Pharm. Sci.*, 83:1723–1728 (1994).
7. M. F. Lu, D. Lee, and G. S. Rao. Percutaneous absorption enhancement of leuprolide, *Pharm. Res.*, 9:1575–1579 (1992).
8. W. H. M. C. Hinsberg, J. C. Verhoef, L. J. Bax, H. E. Junginger, and H. E. Bodde. Role of appendages in skin resistance and iontophoretic peptide flux: Human versus snake skin, *Pharm. Res.*, 12:1506–1512 (1995).
9. J. Hirvonen, K. Kontturi, L. Murtomaki, P. Paronen, and A. Urtti. Transdermal iontophoresis of sotalol and salicylate—The effect of skin charge and penetration enhancers, *J. Control. Release*, 26:109–117 (1993).
10. Y. Y. Huang, S. M. Wu, and C. Y. Wang. Response surface method: A novel strategy to optimize iontophoretic transdermal delivery of thyrotropin-releasing hormone, *Pharm. Res.*, 13:547–552 (1996).
11. J. J. Prusakiewicz, C. Ackermann, and R. Voorman. Comparison of skin esterase activities from different species, *Pharm. Res.*, 23:1517–1524 (2006).
12. P. E. Laurent, S. Bonnet, P. Alchas, P. Regolini, J. A. Mikszta, R. Pettis, and N. G. Harvey. Evaluation of the clinical performance of a new intradermal vaccine administration technique and associated delivery system, *Vaccine*, 25:8833–8842 (2007).
13. M. J. Bartek, J. A. LaBudde, and H. I. Maibach. Skin permeability *in vivo*: Comparison in rat, rabbit, pig and man, *J. Invest. Dermatol.*, 58:114–123 (1972).
14. I. P. Dick and R. C. Scott. Pig ear skin as an in-vitro model for human skin permeability, *J. Pharm. Pharmacol.*, 44:640–645 (1992).
15. N. A. Monteiro-Riviere, A. O. Inman, and J. E. Riviere. Identification of the pathway of iontophoretic drug delivery: Light and ultrastructural studies using mercuric chloride in pigs, *Pharm. Res.*, 11(2):251–256 (1994).
16. C. L. Slough, M. J. Spinelli, and G. B. Kasting. Transdermal delivery of etidronate (EHDP) in the pig via iontophoresis, *J. Memb. Sci.*, 35:161–165 (1988).
17. J. Lademann, H. Richter, M. Meinke, W. Sterry, and A. Patzelt. Which skin model is the most appropriate for the investigation of topically applied substances into the hair follicles? *Skin Pharmacol. Physiol.*, 23:47–52 (2010).
18. C. Herkenne, A. Naik, Y. N. Kalia, J. Hadgraft, and R. H. Guy. Pig ear skin *ex vivo* as a model for *in vivo* dermatopharmacokinetic studies in man, *Pharm. Res.*, 23:1850–1856 (2006).
19. H. Sueki, C. Gammal, K. Kudoh, and A. M. Kligman. Hairless guinea pig skin: Anatomical basis for studies of cutaneous biology, *Eur. J. Dermatol.*, 10:357–364 (2000).
20. K. Sato, K. Sugibayashi, and Y. Morimoto. Species differences in percutaneous absorption of nicorandil, *J. Pharm. Sci.*, 80:104–107 (1991).
21. V. P. Shah, G. L. Flynn, R. H. Guy, H. I. Maibach, H. Schaefer, J. P. Skelly, R. C. Wester, and A. Yacobi. *In vivo* percutaneous penetration/absorption, *Int. J. Pharm.*, 74:1–8 (1991).
22. G. L. Qiao and J. E. Riviere. Significant effects of application site and occlusion on the pharmacokinetics of cutaneous penetration and biotransformation of parathion *in vivo* in swine, *J. Pharm. Sci.*, 84:425–432 (1995).
23. I. P. Dick and R. C. Scott. The influence of different strains and age on *in vitro* rat skin permeability to water and mannitol, *Pharm. Res.*, 9:884–887 (1992).

24. D. Howes, R. Guy, J. Hadgraft, J. Heylings, U. Hoeck, F. Kemper, H. Maibach, J. P. Marty, H. Merk, J. Parra, D. Rekkas, I. Rondelli, H. Schaefer, U. Tauber, and N. Verbiese. Methods for assessing percutaneous absorption—The report and recommendations of ECVAM workshop 13, *ATLA. Altern. Lab. Anim.*, 24:81–106 (1996).

25. G. B. Kasting and L. A. Bowman. DC electrical properties of frozen, excised human skin, *Pharm. Res.*, 7:134–143 (1990).

26. M. R. Prausnitz, E. R. Edelman, J. A. Gimm, R. Langer, and J. C. Weaver. Transdermal delivery of heparin by skin electroporation, *Biotechnology*, 13:1205–1209 (1995).

27. W. H. M. Craanevanhinsberg, L. Bax, N. H. M. Flinterman, J. Verhoef, H. E. Junginger, and H. E. Bodde. Iontophoresis of a model peptide across human skin *in vitro*: Effects of iontophoresis protocol, pH, and ionic strength on peptide flux and skin impedance, *Pharm. Res.*, 11:1296–1300 (1994).

28. A. K. Banga. Microneedle mediated transdermal delivery: How to contribute to meaningful research to advance this growing field? *TransDermal*, 1:8–13 (2009).

29. G. B. Kasting and L. A. Bowman. Electrical analysis of fresh, excised human skin: A comparison with frozen skin, *Pharm. Res.*, 7:1141–1146 (1990).

30. J. E. Riviere, K. F. Bowman, N. A. Monteiro-Riviere, L. P. Dix, and M. P. Carver. The isolated perfused porcine skin flap (IPPSF): I. A novel *in vitro* model for percutaneous absorption and cutaneous toxicology studies, *Fundam. Appl. Toxicol.*, 7:444–453 (1986).

31. M. C. Heit, P. L. Williams, F. L. Jayes, S. K. Chang, and J. E. Riviere. Transdermal iontophoretic peptide delivery—*In vitro* and *in vivo* studies with luteinizing hormone releasing hormone, *J. Pharm. Sci.*, 82:240–243 (1993).

32. J. E. Riviere, B. Sage, and P. L. Williams. Effects of vasoactive drugs on transdermal lidocaine iontophoresis, *J. Pharm. Sci.*, 80:615–620 (1991).

33. M. C. Heit, N. A. Monteiroriviere, F. L. Jayes, and J. E. Riviere. Transdermal iontophoretic delivery of luteinizing hormone releasing hormone (LHRH): Effect of repeated administration, *Pharm. Res.*, 11:1000–1003 (1994).

34. J. E. Riviere, J. D. Brooks, P. L. Williams, E. McGown, and M. L. Francoeur. Cutaneous metabolism of isosorbide dinitrate after transdermal administration in isolated perfused porcine skin, *Int. J. Pharm.*, 127:213–217 (1996).

35. J. E. Riviere. Isolated perfused porcine skin flap. In F. N. Marzulli and H. I. Maibach (eds), *Dermatotoxicology*, Taylor and Francis, London, 1996, pp. 337–351.

36. P. L. Williams and J. E. Riviere. A biophysically based dermatopharmacokinetic compartment model for quantifying percutaneous penetration and absorption of topically applied agents. 1. Theory, *J. Pharm. Sci.*, 84:599–608 (1995).

37. J. E. Riviere, P. L. Williams, R. S. Hillman, and L. M. Mishky. Quantitative prediction of transdermal iontophoretic delivery of arbutamine in humans with the *in vitro* isolated perfused porcine skin flap, *J. Pharm. Sci.*, 81:504–507 (1992).

38. C. Augustin, V. Frei, E. Perrier, A. Huc, and O. Damour. A skin equivalent model for cosmetological trials: An *in vitro* efficacy study of a new biopeptide, *Skin Pharmacol.*, 10:63–70 (1997).

39. D. F. Hager, F. A. Mancuso, J. P. Nazareno, J. W. Sharkey, and J. R. Siverly. Evaluation of a cultured skin equivalent as a model membrane for iontophoretic transport, *J. Control. Release*, 30:117–123 (1994).

40. A. C. Williams. *Transdermal and topical drug delivery*, Pharmaceutical Press, London, 2003.

41. I. Higounenc, M. Demarchez, M. Regnier, R. Schmidt, M. Ponec, and B. Shroot. Improvement of epidermal differentiation and barrier function in reconstructed human skin after grafting onto athymic nude mice, *Arch. Dermatol. Res.*, 286:107–114 (1994).

42. C. A. L. Valle, L. Germain, M. Rouabhia, W. Xu, R. Guignard, F. Goulet, and F. A. Auger. Grafting on nude mice of living skin equivalents produced using human collagens, *Transplantation*, 62:317–323 (1996).
43. L. Zhang, L. N. Li, Z. L. An, R. M. Hoffman, and G. A. Hofmann. *In vivo* transdermal delivery of large molecules by pressure-mediated electroincorporation and electroporation: A novel method for drug and gene delivery, *Bioelectrochem. Bioenerg.*, 42:283–292 (1997).
44. A. M. R. Bayon and R. H. Guy. Iontophoresis of nafarelin across human skin *in vitro*, *Pharm. Res.*, 13:798–800 (1996).
45. V. Srinivasan, M. Su, W. I. Higuchi, and C. R. Behl. Iontophoresis of polypeptides: Effect of ethanol pretreatment of human skin, *J. Pharm. Sci.*, 79:588–591 (1990).
46. P. V. Raykar, M. Fung, and B. D. Anderson. The role of protein and lipid domains in the uptake of solutes by human stratum corneum, *Pharm. Res.*, 5:140–150 (1988).
47. A. M. R. Bayon, J. Corish, and O. I. Corrigan. *In vitro* passive and iontophoretically assisted transport of salbutamol sulphate across synthetic membranes, *Drug Develop. Ind. Pharm.*, 19:1169–1181 (1993).
48. S. M. Sims, W. I. Higuchi, and V. Srinivasan. Interaction of electric field and electro-osmotic effects in determining iontophoretic enhancement of anions and cations, *Int. J. Pharm.*, 77:107–118 (1991).
49. A. J. Hoogstraate, V. Srinivasan, S. M. Sims, and W. I. Higuchi. Iontophoretic enhancement of peptides: Behaviour of leuprolide versus model permeants, *J. Control. Release*, 31:41–47 (1994).
50. K. D. Peck, V. Srinivasan, S. K. Li, W. I. Higuchi, and A. H. Ghanem. Quantitative description of the effect of molecular size upon electroosmotic flux enhancement during iontophoresis for a synthetic membrane and human epidermal membrane, *J. Pharm. Sci.*, 85:781–788 (1996).
51. S. K. Li, A. H. Ghanem, K. D. Peck, and W. I. Higuchi. Iontophoretic transport across a synthetic membrane and human epidermal membrane: A study of the effects of permeant charge, *J. Pharm. Sci.*, 86:680–689 (1997).
52. Y. B. Bannon, J. Corish, and O. I. Corrigan. Iontophoretic transport of model compounds from a gel matrix across a cellophane membrane, *Drug Develop. Ind. Pharm.*, 13:2617–2630 (1987).
53. Y. B. Bannon, J. Corish, O. I. Corrigan, and J. G. Masterson. Iontophoretically induced transdermal delivery of salbutamol, *Drug Develop. Ind. Pharm.*, 14:2151–2166 (1988).
54. Y. H. Tu and L. V. Allen. *In vitro* iontophoretic studies using synthetic membranes, *J. Pharm. Sci.*, 78:211–213 (1989).
55. Q. Xu, S. A. Ibrahim, W. I. Higuchi, and S. K. Li. Ion-exchange membrane assisted transdermal iontophoretic delivery of salicylate and acyclovir, *Int. J. Pharm.*, 369:105–113 (2009).
56. S. P. Schwendeman, G. L. Amidon, V. Labhasetwar, and R. J. Levy. Modulated drug release using iontophoresis through heterogeneous cation-exchange membranes. 2. Influence of cation-exchanger content on membrane resistance and characteristic times, *J. Pharm. Sci.*, 83:1482–1494 (1994).
57. L. L. Miller and G. A. Smith. Iontophoretic transport of acetate and carboxylate ions through hairless mouse skin. A cation exchange membrane model, *Int. J. Pharm.*, 49:15–22 (1989).
58. S. Nakhare, N. K. Jain, and H. V. Verma. Iontophoretic cellophane membrane delivery of diclofenac sodium, *Pharmazie*, 49:672–675 (1994).
59. S. Raney, P. Lehman, and T. Franz. 30th Anniversary of the Franz cell finite dose model: The crystal ball of topical drug development, *Drug Del. Technol.*, 8:32–37 (2008).
60. T. J. Franz. The finite dose technique as a valid *in vitro* model for the study of percutaneous absorption in man, *Curr. Probl. Dermatol.*, 7:58–68 (1978).

61. P. Glikfeld, C. Cullander, R. S. Hinz, and R. H. Guy. A new system for *in vitro* studies of iontophoresis, *Pharm. Res.*, 5:443–446 (1988).
62. N. H. Bellantone, S. Rim, M. L. Francoeur, and B. Rasadi. Enhanced percutaneous absorption via iontophoresis I. Evaluation of an *in vitro* system and transport of model compounds, *Int. J. Pharm.*, 30:63–72 (1986).
63. T. J. Petelenz, J. A. Buttke, C. Bonds, L. B. Lloyd, J. E. Beck, R. L. Stephen, S. C. Jacobsen, and P. Rodriguez. Iontophoresis of dexamethasone: Laboratory studies, *J. Control. Release*, 20:55–66 (1992).
64. L. H. Chen and Y. W. Chien. Development of a skin permeation cell to simulate clinical study of iontophoretic transdermal delivery, *Drug Develop. Ind. Pharm.*, 20:935–945 (1994).
65. S. A. Akhter, S. L. Bennett, I. L. Waller, and B. W. Barry. An automated diffusion apparatus for studying skin penetration, *Int. J. Pharm.*, 21:17–26 (1984).
66. G. L. Flynn, V. P. Shah, S. N. Tenjarla, M. Corbo, D. DeMagistris, T. G. Feldman, T. J. Franz, D. R. Miran, D. M. Pearce, J. A. Sequeira, J. Swarbrick, J. C. Wang, A. Yacobi, and J. L. Zatz. Assessment of value and applications of *in vitro* testing of topical dermatological drug products, *Pharm Res.*, 16:1325–1330 (1999).
67. M. Rapedius and J. Blanchard. Comparison of the Hanson microette and the Van Kel apparatus for *in vitro* release testing of topical semisolid formulations, *Pharm Res.*, 18:1440–1447 (2001).
68. W. W. Hauck, V. P. Shah, S. W. Shaw, and C. T. Ueda. Reliability and reproducibility of vertical diffusion cells for determining release rates from semisolid dosage forms, *Pharm Res.*, 24:2018–2024 (2007).
69. K. Tojo, Y. Sun, M. M. Ghannam, and Y. W. Chien. Characterization of a membrane permeation system for controlled drug delivery studies, *AICHE J.*, 31:741–746 (1985).
70. G. Yan, K. S. Warner, J. Zhang, S. Sharma, and B. K. Gale. Evaluation needle length and density of microneedle arrays in the pretreatment of skin for transdermal drug delivery, *Int. J Pharm.*, 391:7–12 (2010).
71. N. S. Murthy, V. A. Boguda, and K. Payasada. Electret enhances transdermal drug permeation, *Biol. Pharm. Bull.*, 31:99–102 (2008).
72. S. N. Murthy, Y. L. Zhao, S. W. Hui, and A. Sen. Electroporation and transcutaneous extraction (ETE) for pharmacokinetic studies of drugs, *J Control. Release*, 105:132–141 (2005).
73. T. M. Parkinson, M. A. Szlek, and J. D. Isaacson. Hybresis: The hybridization of traditional with low-voltage iontophoresis, *Drug Del. Technol.*, 7:54–60 (2007).
74. F. Pliquett and U. Pliquett. Passive electrical properties of human stratum corneum *in vitro* depending on time after separation, *Biophys. Chem.*, 58:205–210 (1996).
75. K. D. Peck, A. H. Ghanem, W. I. Higuchi, and V. Srinivasan. Improved stability of the human epidermal membrane during successive permeability experiments, *Int. J. Pharm.*, 98:141–147 (1993).
76. K. B. Sloan, H. D. Beall, W. R. Weimar, and R. Villanueva. The effect of receptor phase composition on the permeability of hairless mouse skin in diffusion cell experiments, *Int. J. Pharm.*, 73:97–104 (1991).
77. I. Steinstrasser and H. P. Merkle. Dermal metabolism of topically applied drugs: Pathways and models reconsidered, *Pharmaceutical Acta Helvetiae*, 70:3–24 (1995).
78. R. H. Guy. Current status and future prospects of transdermal drug delivery, *Pharm. Res.*, 13:1765–1769 (1996).
79. A. K. Shah, G. Wei, R. C. Lanman, V. O. Bhargava, and S. J. Weir. Percutaneous absorption of ketoprofen from different anatomical sites in man, *Pharm. Res.*, 13:168–172 (1996).
80. J. J. Berti and J. J. Lipsky. Transcutaneous drug delivery: A practical review, *Mayo Clin. Proc.*, 70:581–586 (1995).

81. M. D. Donovan. Sex and racial differences in pharmacological response: Effect of route of administration and drug delivery system on pharmacokinetics, *J. Womens Health (Larchmt.)*, 14:30–37 (2005).

82. S. D. Roy and G. L. Flynn. Transdermal delivery of narcotic analgesics: pH, anatomical, and subject influences on cutaneous permeability of fentanyl and sufentanil, *Pharm. Res.*, 7:842–847 (1990).

83. P. C. Liu, J. A. S. Nightingale, and T. Kuriharabergstrom. Variation of human skin permeation *in vitro*—Ionic vs. neutral compounds, *Int. J. Pharm.*, 90:171–176 (1993).

84. F. K. Akomeah, G. P. Martin, and M. B. Brown. Variability in human skin permeability *in vitro*: Comparing penetrants with different physicochemical properties, *J. Pharm. Sci.*, 96:824–834 (2007).

85. S. S. Lin, H. Ho, and Y. W. Chien. Development of a new nicotine transdermal delivery system—*In vitro* kinetics studies and clinical pharmacokinetic evaluations in two ethnic groups, *J. Control. Release*, 26:175–193 (1993).

86. Contracept, http://www.contracept.org/docs/orthoevra.pdf (accessed October 8, 2010).

87. J. Singh, M. Gross, M. O'Connell, B. Sage, and H. I. Maibach. Effect of iontophoresis in different ethnic groups' skin function, *Proc. Int. Symp. Control. Rel. Bioact. Mater.*, 21:365–366 (1994).

88. R. C. Wester and H. I. Maibach. Percutaneous absorption of drugs, *Clin. Pharmacokinet.*, 23:253–266 (1992).

89. E. Berardesca and H. Maibach. Racial differences in skin pathophysiology, *J. Am. Acad. Dermatol.*, 34:667–672 (1996).

90. B. Berner and V. A. John. Pharmacokinetic characterization of transdermal delivery systems, *Clin. Pharmacokinet.*, 26:121–134 (1994).

91. C. Herkenne, I. Alberti, A. Naik, Y. N. Kalia, F. X. Mathy, V. Preat, and R. H. Guy. *In vivo* methods for the assessment of topical drug bioavailability, *Pharm. Res.*, 25:87–103 (2008).

92. A. Jadoul, C. Hanchard, S. Thysman, and V. Preat. Quantification and localization of fentanyl and TRH delivered by iontophoresis in the skin, *Int. J. Pharm.*, 120:221–228 (1995).

93. G. Li, A. Badkar, S. Nema, C. S. Kolli, and A. K. Banga. *In vitro* transdermal delivery of therapeutic antibodies using maltose microneedles, *Int. J. Pharm.*, 368:109–115 (2009).

94. W. G. Reifenrath, G. S. Hawkins, and M. S. Kurtz. Percutaneous penetration and skin retention of topically applied compounds: An *in vitro–in vivo* study, *J. Pharm. Sci.*, 80:526–532 (1991).

95. B. N'Dri-Stempfer, W. C. Navidi, R. H. Guy, and A. L. Bunge. Optimizing metrics for the assessment of bioequivalence between topical drug products, *Pharm. Res.*, 25:1621–1630 (2008).

96. J. J. Escobar-Chavez, V. Merino-Sanjuan, M. Lopez-Cervantes, Z. Urban-Morlan, E. Pinon-Segundo, D. Quintanar-Guerrero, and A. Ganem-Quintanar. The tape-stripping technique as a method for drug quantification in skin, *J. Pharm. Pharmaceut. Sci.*, 11:104–130 (2008).

97. L. M. Russell and R. H. Guy. Measurement and prediction of the rate and extent of drug delivery into and through the skin, *Expert. Opin. Drug Deliv.*, 6:355–369 (2009).

98. L. K. Pershing, J. L. Corlett, and J. L. Nelson. Comparison of dermatopharmacokinetic vs. clinicial efficacy methods for bioequivalence assessment of miconazole nitrate vaginal cream, 2% in humans, *Pharm. Res.*, 19:270–277 (2002).

99. S. Nicoli, A. L. Bunge, M. B. Delgado-Charro, and R. H. Guy. Dermatopharmacokinetics: Factors influencing drug clearance from the stratum corneum, *Pharm. Res.*, 26:865–871 (2009).

100. I. Alberti, Y. N. Kalia, A. Naik, and R. H. Guy. Assessment and prediction of the cutaneous bioavailability of topical terbinafine, *in vivo*, in man, *Pharm. Res.*, 18:1472–1475 (2001).

101. J. Lademann, U. Jacobi, C. Surber, H. J. Weigmann, and J. W. Fluhr. The tape stripping procedure—Evaluation of some critical parameters, *Eur. J. Pharm. Biopharm.*, 72:317–323 (2009).

102. A. Teichmann, U. Jacobi, M. Ossadnik, H. Richter, S. Koch, W. Sterry, and J. Lademann. Differential stripping: Determination of the amount of topically applied substances penetrated into the hair follicles, *J. Invest. Dermatol.*, 125:264–269 (2005).

103. C. S. Chaurasia, M. Muller, E. D. Bashaw, E. Benfeldt, J. Bolinder, R. Bullock, P. M. Bungay, E. C. DeLange, H. Derendorf, W. F. Elmquist, M. Hammarlund-Udenaes, C. Joukhadar, D. L. Kellogg, Jr., C. E. Lunte, C. H. Nordstrom, H. Rollema, R. J. Sawchuk, B. W. Cheung, V. P. Shah, L. Stahle, U. Ungerstedt, D. F. Welty, and H. Yeo. AAPS-FDA workshop white paper: Microdialysis principles, application and regulatory perspectives, *Pharm. Res.*, 24:1014–1025 (2007).

104. M. Muller, R. Schmid, O. Wagner, B. Vonosten, H. Shayganfar, and H. G. Eichler. *In vivo* characterization of transdermal drug transport by microdialysis, *J. Control. Release*, 37:49–57 (1995).

105. G. Stagni, D. O'Donnell, Y. J. Liu, D. L. Kellogg, T. Morgan, and A. M. M. Shepherd. Intradermal microdialysis: Kinetics of iontophoretically delivered propranolol in forearm dermis. *J. Control. Release,* 63:331–339 (2000).

106. V. Vemulapalli, Y. Yang, P. M. Friden, and A. K. Banga. Synergistic effect of iontophoresis and soluble microneedles for transdermal delivery of methotrexate, *J. Pharm. Pharmacol.*, 60:27–33 (2008).

107. T. Seki, A. Wang, D. Yuan, Y. Saso, O. Hosoya, S. Chono, and K. Morimoto. Excised porcine skin experimental systems to validate quantitative microdialysis methods for determination of drugs in skin after topical application, *J. Control. Release*, 100:181–189 (2004).

108. F. Sjogren, C. Svensson, and C. Anderson. Technical prerequisites for *in vivo* microdialysis determination of interleukin-6 in human dermis, *Br. J. Dermatol.*, 146:375–382 (2002).

109. A. J. Rosenbloom, D. M. Sipe, and V. W. Weedn. Microdialysis of proteins: Performance of the CMA/20 probe, *J. Neurosci. Methods*, 148:147–153 (2005).

110. R. J. Schutte, S. A. Oshodi, and W. M. Reichert. *In vitro* characterization of microdialysis sampling of macromolecules, *Anal. Chem.*, 76:6058–6063 (2004).

111. G. F. Clough. Microdialysis of large molecules, *AAPS J.*, 7:E686–E692 (2005).

112. F. X. Mathy, A. R. Denet, B. Vroman, P. Clarys, A. Barel, R. K. Verbeeck, and V. Preat. *In vivo* tolerance assessment of skin after insertion of subcutaneous and cutaneous microdialysis probes in the rat, *Skin Pharmacol. Appl. Skin Physiol.*, 16:18–27 (2003).

113. F. Netzlaff, K. H. Kostka, C. M. Lehr, and U. F. Schaefer. TEWL measurements as a routine method for evaluating the integrity of epidermis sheets in static Franz type diffusion cells *in vitro*. Limitations shown by transport data testing, *Eur. J. Pharm. Biopharm.*, 63:44–50 (2006).

114. K. De Paepe, E. Houben, R. Adam, F. Wiesemann, and V. Rogiers. Validation of the VapoMeter, a closed unventilated chamber system to assess transepidermal water loss vs. the open chamber Tewameter, *Skin Res. Technol.*, 11:61–69 (2005).

115. G. Li, A. Badkar, H. Kalluri, and A. K. Banga. Microchannels created by sugar and metal microneedles: Characterization by microscopy, macromolecular flux and other techniques, *J. Pharm. Sci.*, 99:1931–1941 (2009).

116. C. S. Kolli and A. K. Banga. Characterization of solid maltose microneedles and their use for transdermal delivery, *Pharm. Res.*, 25:104–113 (2008).

117. J. Levin and H. Maibach. The correlation between transepidermal water loss and percutaneous absorption: An overview, *J. Control. Release*, 103:291–299 (2005).

118. C. Curdy, A. Naik, Y. N. Kalia, I. Alberti, and R. H. Guy. Non-invasive assessment of the effect of formulation excipients on stratum corneum barrier function *in vivo*, *Int. J. Pharm.*, 271:251–256 (2004).

119. C. Curdy, Y. N. Kalia, and R. H. Guy. Non-invasive assessment of the effects of iontophoresis on human skin *in vivo*, *J. Pharm. Pharmacol.*, 53:769–777 (2001).

120. R. Alvarez-Roman, A. Naik, Y. N. Kalia, H. Fessi, and R. H. Guy. Visualization of skin penetration using confocal laser scanning microscopy, *Eur. J. Pharm. Biopharm.*, 58:301–316 (2004).

121. Y. Wu, Y. Qiu, S. Zhang, G. Qin, and Y. Gao. Microneedle-based drug delivery: Studies on delivery parameters and biocompatibility, *Biomed. Microdevices*, 10:601–610 (2008).

122. A. K. Banga. Microporation applications for enhancing drug delivery, *Expert. Opin. Drug Deliv.*, 6:343–354 (2009).

123. M. E. Meuwissen, J. Janssen, C. Cullander, H. E. Junginger, and J. A. Bouwstra. A cross-section device to improve visualization of fluorescent probe penetration into the skin by confocal laser scanning microscopy, *Pharm. Res.*, 15:352–356 (1998).

124. P. J. Caspers, G. W. Lucassen, E. A. Carter, H. A. Bruining, and G. J. Puppels. *In vivo* confocal Raman microspectroscopy of the skin: Noninvasive determination of molecular concentration profiles, *J. Invest. Dermatol.*, 116:434–442 (2001).

125. J. M. Waller and H. I. Maibach. Age and skin structure and function, a quantitative approach (I): Blood flow, pH, thickness, and ultrasound echogenicity, *Skin Res. Technol.*, 11:221–235 (2005).

126. J. M. Lagarde, J. George, R. Soulcie, and D. Black. Automatic measurement of dermal thickness from B-scan ultrasound images using active contours, *Skin Res. Technol.*, 11:79–90 (2005).

127. J. Welzel, C. Reinhardt, E. Lankenau, C. Winter, and H. H. Wolff. Changes in function and morphology of normal human skin: Evaluation using optical coherence tomography, *Br. J. Dermatol.*, 150:220–225 (2004).

128. S. A. Coulman, J. C. Birchall, A. Alex, M. Pearton, B. Hofer, C. O'Mahony, W. Drexler, and B. Povazay. *In vivo*, in situ imaging of microneedle insertion into the skin of human volunteers using optical coherence tomography, *Pharm Res.*, [Epub] (2010).

129. J. P. Medendorp, K. S. Paudel, R. A. Lodder, and A. L. Stinchcomb. Near infrared spectrometry for the quantification of human dermal absorption of econazole nitrate and estradiol, *Pharm. Res.*, 24:186–193 (2007).

130. R. K. Sivamani, B. Stoeber, G. C. Wu, H. Zhai, D. Liepmann, and H. Maibach. Clinical microneedle injection of methyl nicotinate: Stratum corneum penetration, *Skin Res. Technol.*, 11:152–156 (2005).

131. E. J. Droog and F. Sjoberg. Nonspecific vasodilatation during transdermal iontophoresis—The effect of voltage over the skin, *Microvasc. Res.*, 65:172–178 (2003).

132. E. W. Smith, J. M. Haigh, and R. B. Walker. Analysis of chromameter results obtained from corticosteroid-induced skin blanching. I: Manipulation of data, *Pharm. Res.*, 15:280–285 (1998).

133. X. Chen, T. W. Prow, M. L. Crichton, D. W. Jenkins, M. S. Roberts, I. H. Frazer, G. J. Fernando, and M. A. Kendall. Dry-coated microprojection array patches for targeted delivery of immunotherapeutics to the skin, *J. Control. Release*, 139:212–220 (2009).

3 Microporation-Mediated Transdermal Drug Delivery

3.1 INTRODUCTION

In recent years, there has been increasing interest in microporation technologies that create micron-sized microchannels in the skin (1). These "minimally invasive" technologies involve a temporary physical disruption of the skin barrier to create superficial pores that typically breach the stratum corneum and the remaining epidermis. These pores in the superficial layers of skin are temporary, because these skin layers are continuously replaced by the natural process of desquamation. The use of microneedles (typically <1 mm) to create these microchannels is discussed first in this chapter followed by a discussion of other technologies such as thermal and laser ablation to create microchannels. Delivery is expected to favor hydrophilic drug molecules, as the microchannels created by microporation are hydrophilic. These microchannels are several microns in dimension; therefore, there are no size limits on the molecules that can be delivered through them. This is a good finding, as many of the hydrophilic drugs that will gain from transdermal delivery are macromolecules, including peptides, proteins, oligonucleotides, vaccines, and DNA vaccines. Factors affecting delivery via microporated skin for a given drug include drug concentration, microchannel depth, and microchannel density, though additional factors may be involved which are specific to the microporation technology being used. Applications of microporation to delivery of vaccines are discussed in Chapter 9 and to delivery of biopharmaceuticals are discussed in Chapter 10. Various tools used to characterize the microchannels created by microneedles, such as confocal microscopy and transepidermal water loss (TEWL), were discussed in Chapter 2. It should be noted that microneedles have other applications in medicine which are beyond the scope of discussion in this chapter.

3.2 SKIN MICROPORATION BY MICRONEEDLES

3.2.1 POTENTIAL ADVANTAGES OF A MICRONEEDLE PATCH

Microneedles or other microporation approaches provide a minimally invasive, painless way of creating microchannels in skin which can then allow drug transport into the skin and systemic circulation (2–7). Microneedles were first described in a 1976 patent (8), but Henry et al. were the first to report the use of microneedles for transdermal drug delivery (9). Microneedles can be compared to the proboscis of mosquitoes, as they have similar dimensions and can penetrate skin in a similar manner due to the sharpness of their tips. The hydrophilic microchannels created by microneedles in the skin are filled with interstitial fluid and can allow the transport of hydrophilic molecules and macromolecules, which would otherwise not enter the skin. Using a hairless guinea pig model, Banks et al. (10) reported that microneedle-mediated delivery was increased 10-fold and lag time was reduced 10-fold for the hydrophilic HCl salt of naltrexone compared to its base form, and similar results were found using human skin. When naltrexol, an active metabolite of naltrexone, was used, it was seen that the ionized charged form of the drug had higher delivery due to the increased solubility in the formulation and not directly due to charge and ionization considerations.

As we know, passive patches on the market are generally for moderately lipophilic small drug molecules that can partition into (and out of) the skin. In contrast, microporation allows the delivery of hydrophilic molecules, thereby expanding the scope of transdermal delivery to enable delivery of biopharmaceuticals that tend to be hydrophilic in nature. Furthermore, hollow microneedles can be used to achieve injection at precise depths in the dermis; in a way, a microneedle patch takes the best of the advantages of parenteral and transdermal delivery. A rapid onset is therefore also possible unlike the passive diffusion patches on the market. There is no or reduced risk of needlestick injuries, and self-administration at home should be feasible, resulting in better compliance and reduced health-care costs. This can also be very useful for mass immunization in developing countries using vaccine-coated microneedles. Significant progress has been made to show that antigens can be coated onto microneedles in a stable dry formulation, thereby eliminating the cold chain required to refrigerate liquid vaccine formulations. A lyophilized vial cannot be used to accomplish this, because it will still need reconstitution before administration, adding to the need for trained personnel and facilities and thereby adding to the cost of vaccination programs. The cost of the patch is also expected to be minimal, ranging from a few cents per patch (11) to a few dollars for lifestyle applications (e.g., for smoking cessation) but may be more for cosmetic applications or for specialized therapy (12). A microneedle array patch is expected to be simple in design without the need for having any power supply or advanced microelectronics, unlike many other drug delivery systems.

3.2.2 Fabrication of Microneedles

This chapter has its focus on the applications of microneedles to drug delivery, and a detailed discussion of the extensive literature published on microneedle fabrication is therefore outside the scope of this chapter. However, some comments on microneedle fabrication will be made in this section. Microneedles may be fabricated in-plane or out-of-plane. The length can be controlled more accurately for in-plane needles, but out-of-plane needles offer the advantages of fabrication in two-dimensional arrays by wafer-level processing. Material selection is very important for fabrication of microneedles. In the past, silicon microneedles were mostly used. Several investigators have described fabrication of silicon microneedles (13–18), including those with a buried microchannel that can then be integrated to a microfluid chip that has a connection tube and capability to be attached to a syringe (19). Other means of making hollow silicon microneedles have also been described (20). Fabrication of silicon microneedles needs clean-room processing and expensive microelectromechanical systems (MEMS)-based (21–23) microfabrication techniques. Furthermore, silicon is brittle and may break off in the skin, or silicon grains may be left behind in the skin though the constant regeneration of epidermal layers will remove any such residue over time (20). More recently, several alternative materials are beginning to emerge and gain preference. McAllister et al. have described techniques to fabricate microneedles from various materials such as silicon, metal, polymer, and glass (24). A two-photon polymerization process has also been reported to fabricate three-dimensional micro-structured medical devices, including microneedles and tissue engineering scaffolds. Organic–inorganic hybrid materials were used, and femtosecond laser pulses from a titanium:sapphire laser were applied to break chemical bonds on a photo initiator. The radicalized starter molecules generated react with the monomers in the hybrid material to create polymolecules (25). More recently, product development efforts have centered on metal, polymer, or sugar microneedles. Metal microneedles may be made by laser patterning of desired microneedle design onto a metal surface followed by raising the microneedles out of plane (4). Polymeric microneedles can be fabricated from biocompatible plastics, biodegradable polymers, or water-soluble polymers. Polymeric microneedles can be made using fabrication techniques such as injection molding, but more typically, micromolding techniques are used. Fabrication of out-of-plane hollow polymeric microneedles with side openings and an integrated drug reservoir have been recently described (26). Hollow microneedles are typically more difficult to fabricate, though they offer usage advantages by allowing an active fluid flow through them into the skin (27). Also, hollow microneedles may be structurally weaker than solid microneedles, especially when silicon is used (4).

Maltose microneedles can be fabricated by a micromolding process. Powdered maltose is heated to 140°C, and the temperature is held for an hour to convert to a maltose–candy matrix that is then allowed to cool to room temperature before casting in a mold at 95°C, when it quickly forms microneedles (28). We characterized the microchannels created by sugar and metal microneedles by microscopy, macromolecular flux, and other techniques (29). Thread-forming polymers, dextrin, chondroitin sulfate, and albumin have been used to form microneedles that were then tested to deliver erythropoietin to mice. C_{max} values in the range of 96 to 138 mIU/ml were obtained, and bioavailabilities were in the 60% to 82% range (30). Dextrin microneedles have also been investigated for delivery of insulin (31). Dissolving microneedles made of carboxymethylcellulose (CMC) or amylopectin have been made by casting a viscous solution in a mold during centrifugation. For CMC, solutions were concentrated by evaporation under vacuum or by heating to produce a highly viscous solution that still had enough fluidity to fill the mold. Centrifugation avoided void formation during drying by continuously compressing the mold contents. It was shown that model drugs could be encapsulated in these dissolving microneedles during fabrication (32). Microneedles made of biodegradable polymers such as polylactic-co-glycolic acid (PLGA) have also been fabricated by micromolding using a polydimethylsiloxane (PDMS) mold and have been shown to enhance permeation of calcein and bovine serum albumin (BSA) across porated human cadaver skin. PDMS micromolds have been used to prepare micromold microdevices, as it is optically transparent and has a low surface energy and good thermal stability (33,34). In a subsequent study, calcein and BSA were encapsulated, separately, in PLGA microneedles, and the release kinetics were studied. After fabrication at high temperatures required for melted PLGA, about 90% of the BSA was found to be in the native state (11). PDMS micromolding technique has also been used to fabricate polymeric hollow microneedles (35). The economics of the fabrication method being used and the ability to scale up the method to mass produce inexpensive biocompatible and mechanically strong microneedles will determine the choice of the desired method of microneedle fabrication.

3.2.3   Mechanisms of Delivery and Type of Microneedles

Drug can be delivered into skin via microchannels created by microneedles by placing a patch on the microporated skin and relying on diffusion for delivery of the drug vertically through the microchannels. This is also the predominant mechanism by which drugs are delivered into skin using other microporation technologies, such as thermal, radiofrequency, or laser ablation. Vertical diffusion through the microchannel will be accompanied by some lateral diffusion into the surrounding tissue as well. However, in the case of microneedles, other mechanisms of delivery are also possible. For example, the drug may be directly coated onto microneedles or can be directly incorporated into microneedles that dissolve in the skin upon insertion. Alternatively, hollow microneedles can be used to directly infuse a drug formulation into the skin. Depending on the desired mechanism of delivery, different types of microneedles will be used, and these are discussed in more detail in this section.

3.2.3.1   Solid Microneedles

Solid microneedles can be made from various materials and with different geometries (Figure 3.1). They may also be coated or may dissolve in the skin, and these are discussed in more detail later. A microneedle device currently on the market is the DermaRoller®, which is used for cosmetic applications. In one configuration, it has several circular arrays with a total of 192 stainless steel needles arranged on a cylinder. DermaRoller models with needle lengths of 150, 500, or 1500 μm have been tested *in vitro* for creation of microchannels and permeation of model hydrophilic drugs using human skin to show their potential for drug delivery applications (36). The model with 1500-μm needles may not be appropriate for such studies, as it may reach too deep into the skin. We used the model (CIT8) with 500-μm needles and found that the needles had a conical geometry and created about 16 microchannels/cm^2 of skin. Using fluorescent microparticles and confocal microscopy, we have shown that the depth of penetration into skin is around 150 μm, and the average surface diameter of the pores is around 82 μm (29). The DermaRoller device avoids the bed of nails effect (see Section 3.2.4.1) as only one row penetrates the skin at one time. Zhou et al. investigated another type of microneedle rollers in 250-, 500-, and 1000-μm needle sizes and reported a pore diameter of around 70 μm with the 500-μm needles (37). In another configuration, the DermaRoller Company produced the DermaStamp®, which has a central microneedle with five radially arranged

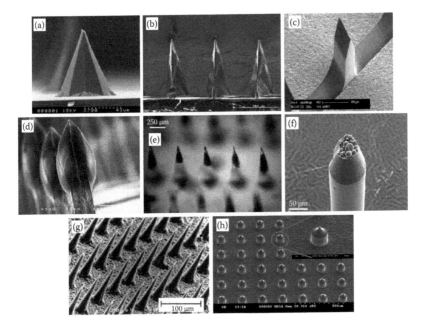

FIGURE 3.1 A single, short silicon microneedle. (Reprinted from *Int. J. Pharm.*, 389:122–129, L. Wei-Ze et al., Super-Short Solid Silicon Microneedles for Transdermal Drug Delivery Applications [65]. Copyright 2010, with permission from Elsevier.) (b) Maltose microneedles. (c) Polycarbonate microneedles. (Reprinted from *Eur. J. Pharm. Biopharm.*, 69:1040–1045, J. H. Oh et al., Influence of the Delivery Systems Using a Microneedle Array on the Permeation of a Hydrophilic Molecule, Calcein [69]. Copyright 2008, with permission from Elsevier.) (d) Coated microneedles. (Reprinted from M. Cormier et al., *J. Control. Release*, 97:503–511, Transdermal Delivery of Desmopressin Using a Coated Microneedle Array Patch System [129]. Copyright 2004, with permission from Elsevier.) (e) Tapered-cone PLGA microneedles and (f) biodegradable microneedles with encapsulated microparticles. (With kind permission from Springer Netherlands: *Pharm. Res.*, Polymer Microneedles for Controlled-Release Drug Delivery, 23:1008–1019, J. H. Park, M. G. Allen, and M. R. Prausnitz, 2006 [11].) (g) An array of conical-shaped silicon microneedles. (S. Henry et al.: Microfabricated Microneedles: A Novel Approach to Transdermal Drug Delivery, *J. Pharm. Sci.*, 1998, 87:922–925 [9]. Copyright Wiley-VCH Verlag GmbH & Co. Reproduced with permission.) (h) A single-crystal silicon microneedle array. (Reprinted from *Nanomedicine*, 1:184–190, Y. Xie, B. Xu, and Y. Gao, Controlled Transdermal Delivery of Model Drug Compounds by MEMS Microneedle Array [15]. Copyright 2005, with permission from Elsevier.) **(See color insert.)**

stainless steel microneedles of 300 μm length. This arrangement of microneedles on the DermaStamp® has been indicated to be suitable for holding one drop of a liquid formulation (such as vaccine formulations) after dipping in the formulation. CLSM has been utilized to visualize microchannels created by this device in human subjects, and by using a dye (0.2% sodium fluorescein), the depth of penetration was reported to be 150 μm (38).

3.2.3.1.1   Coated Solid Microneedles

Drug-coated microneedles offer the advantage of the drug being directly placed in the skin upon insertion, and delivery will not be dependent on diffusion from a patch placed on microporated skin. However, the amount of drug that can be coated on a microneedle array is limited, and it must be ensured that the drug is released from the microneedles and into the skin. A dip-coating process was developed to coat microneedles and was demonstrated to coat calcein, vitamin B, BSA, model proteins, plasmid DNA, modified vaccinia virus, and even microparticles onto microneedles. Coatings could be applied just to the needle shafts and dissolved quickly in skin (39,40). As microneedles do not penetrate the skin to their full height, the entire coating may not dissolve in the skin, and this needs to be taken into consideration. The concentration, viscosity, and surface tension of the coating solution are important to ensure proper wetting of the microneedle surface. The coating must be uniform, and the drug or antigen coated on the microneedles must be stable over time. Multiple coatings with appropriate drying time between coatings can be used to load more drug onto the microneedles (41). Vaccine coatings can also be applied by the dip-coating process (42). A gas-jet technique was also reported for applying a small amount of drug coating onto very short microneedles (43). Coating of parathyroid hormone (PTH) 1-34 on titanium microneedles has been described in the literature (44,45). A liquid drug formulation was developed to create a film with controlled thickness on a rotating drum that, in turn, was used to apply coating to microneedles in a continuous process. The viscosity of the coating solution has to be high to avoid dripping of the solution from the coated microneedles before drying can be accomplished. A surfactant was added to the coating solution to decrease the contact angle to facilitate coating on the titanium surface. Each microprojection was coated with 60 ng of solids having 30 ng of PTH 1-34 and thereby resulting in each array of approximately 1300 microprojections having a dose of 40 μg of PTH 1-34. When using coated microneedles, stability of the coated formulation over the duration of their shelf life must be ensured. Once inserted into skin, release of the coating should be consistent and rapid. The formulation developed to enable the drum-coating process that was used had a high concentration of PTH (15.5%), and potential problems of gelation and aggregation must be avoided. Aggregation and other problems related to formulation development and delivery of proteins were discussed by the author in another text (46).

3.2.3.1.2   Dissolving Microneedles

These microneedles will dissolve in the skin once inserted, creating microchannels before they dissolve away. We characterized water-soluble microneedles made from maltose (47), investigated the formation and closure of microchannels created by these microneedles (48), and demonstrated their use for delivery of low molecular weight heparin (LMWH) (49), methotrexate (50), nicardipine HCl (47), and IgG (29,51). These maltose microneedles are about 550 μm in length with a tip radius of 4 μm and create microchannels in the skin which are 60 μm wide and have an average depth of 160 μm when inserted manually. In a typical configuration, maltose microneedles were stacked in three layers, each layer having 27 microneedles. The array thus created 81 microchannels, and the pore density was 125/cm^2 of the skin (48). Others have used dissolving microneedles with drug incorporated in them, and these are discussed in the next section.

3.2.3.1.3 *Dissolving Microneedles with Drug Incorporated*

As discussed for drug-coated microneedles, these microneedles offer the same advantage that the drug will be directly placed in the skin once the microneedles are inserted, and delivery will not be dependent on diffusion from a patch placed on microporated skin. Drug will release from the microneedles as they dissolve in the skin. However, the amount of drug that can be incorporated into a microneedle array is still limited. A patch of 100 dissolving microneedles (each 500 μm long) made of thread-forming biopolymers, such as chondroitin sulfate and dextran, was fabricated in molds containing 100 inverted cone-shaped wells. Recombinant human growth hormone (rhGH) and desmopressin (DDAVP) were formulated in these needles in such a way that the drug was localized at the acral end of the microneedles, and they were then applied to rats manually. The total weight of the acral ends of the microneedles was 0.2 mg, and in the case of rhGH, about 0.1 mg could be formulated into this mass (52). Each microneedle may typically weigh around 10 μg. Assuming 10% loading, an array of 1000 microneedles will allow incorporation of just 1 mg of the drug. However, this dose will be enough for vaccine applications and even for some very potent therapeutic proteins such as interferons and erythropoietin (11). Generally, drug is incorporated into dissolving microneedles. However, it has been reported that even hollow silicon microneedles could be made to hold drug by sealing them with a thin 170-nm gold membrane that breaks when the microneedles are inserted into skin (53).

3.2.3.2 Hollow Microneedles

Hollow microneedles (Figure 3.2) have a lumen or internal bore that is utilized to pass drug dissolved in a formulation through the microneedle and into the skin. A pressure gradient or another driving force can be used to create this flow. Infusion rates have to be low (typically, 10 to 100 μL/min) to avoid back pressure created by the dense skin tissue. Infusion flow rates can be improved by partially retracting the microneedles after insertion into skin (54). Another possible approach could be use of the enzyme hyaluronidase as discussed in Section 10.5.1. The resistance to flow may decrease with an increase in microneedle density per unit area, though the bed of nails effect (see Section 3.2.4.1) should be avoided. Hollow microneedles that have openings at the tips may get clogged as they are inserted into the skin. Therefore, side openings with off-centered holes have been designed (20,26). Roxhed et al. described a microneedle patch having hollow

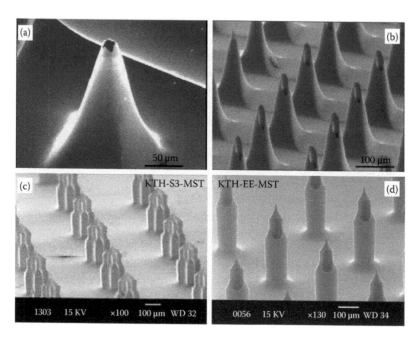

FIGURE 3.2 Different types of hollow microneedles: (a) Volcano-shaped design and (b) "Hypodermic" microneedle design. (Reprinted from *Sensors and Actuators A*, 114:267–275, E. V. Mukerjee et al., Microneedle Array for Transdermal Biological Fluid Extraction and *In Situ* Analysis [122]. Copyright 2004, with permission from Elsevier.) (c) Gold membrane-sealed side-open hollow microneedles. (With kind permission from Springer Science+Business Media: *Biomed. Microdevices*, Membrane-Sealed Hollow Microneedles and Related Administration Schemes for Transdermal Drug Delivery, 10:271–279, 2008, N. Roxhed, P. Griss, and G. Stemme [53]. Copyright 2008.) (d) Silicon-hollow microneedles with side-bore. (With kind permission from Springer Science+Business Media: *Pharm. Res.*, Novel Microneedle Patches for Active Insulin Delivery Are Efficient in Maintaining Glycaemic Control: An Initial Comparison with Subcutaneous Administration, 24:1381–1388, 2007, L. Nordquist, N. Roxhed, P. Griss, and G. Stemme [55]. Copyright 2007.)

silicon microneedles (400 μm) with an integrated liquid reservoir. The reservoir is covered by a composite that heats up and expands into the liquid reservoir when a small current is passed through lithographically defined heaters on its printed circuit board (Figure 3.3) (20,55). Micropyramidal hollow silicon microneedles have been designed by Nanopass technology in conjunction with Silex Microsystems (4,13). Hollow microneedles made of glass have been used to demonstrate delivery of model compounds into skin, and it was reported that the resistance to fluid flow offered by the dense dermal tissue can be offset to some degree by partially retracting the microneedles after insertion (56,57). A simple microneedle array patch (AdminPen™) is available in lengths of 600 to 1500 μm and can be connected to a regular syringe. This device does not have true hollow microneedles but is configured in such a way that it can possibly serve as an alternative to hollow microneedles for some studies (58).

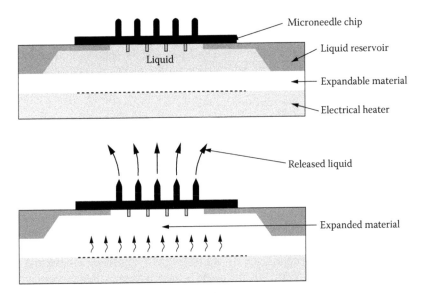

FIGURE 3.3 Hollow microneedle patch with a composite that heats up when a voltage is applied, thereby expanding into an integrated reservoir. (With kind permission from Springer Science+Business Media: *Pharm. Res.*, Novel Microneedle Patches for Active Insulin Delivery Are Efficient in Maintaining Glycaemic Control: An Initial Comparison with Subcutaneous Administration, 24:1381–1388, 2007, L. Nordquist, N. Roxhed, P. Griss, and G. Stemme [55]. Copyright 2007.)

3.2.4 FACTORS INFLUENCING MICRONEEDLE INSERTION

3.2.4.1 Insertion Kinetics

Metal microneedles are usually strong, but when other materials are used, it must be ensured that they have sufficient mechanical strength to penetrate the skin without fracture. Due to the elasticity of skin, some deformation of skin takes place before microneedles penetrate the skin barrier. Therefore, the insertion depth is usually less than the length of the microneedles (59). This is further affected by the insertion method (e.g., whether the insertion was manual or was by using an applicator or a more advanced electronic insertion device). The distance between the individual microneedles in the array is also important. If the needle-to-needle spacing is too short, the pressure exerted on each microneedle gets distributed over a large number of microneedles, thus resulting in insufficient pressure and failure of skin penetration. This results in the "bed-of-nails effect" (60), where a person can actually lie down on a bed of nails as the body weight is distributed over several nails and the weight distribution over any individual nail is not sufficient for it to penetrate the body. In the first published report on the use of microneedles, a 50-fold drop of skin resistance was reported, which was similar to the drop in resistance caused by a 30-gauge hypodermic needle (9). It was reported that an insertion force of about 0.1 to 3 N is required, which is low enough to allow manual insertion of microneedles (61). In another study, a microneedle array patch having microneedles with a length of 150 μm and a density of 484/cm² was inserted using an applicator that provided an insertion force of about 2 N. These microneedles had an octagonal micropyramid shape and a sharp tip, with a width less than 1 μm, and the applicator had a retainer ring to reduce the effect of skin deformation on the depth of penetration (59). A radius of curvature of less than 10 μm is typically desired for adequate sharpness to penetrate the skin.

3.2.4.2 Microneedle Length and Density

Microneedle length, and more precisely the insertion depth of the microneedle, will determine where the drug is initially delivered in the skin (Figure 3.4). Microneedles with a length ranging from 50 to 900 μm have been reported in the literature. The marketed device for influenza vaccine (see Section 10.5.2.1) actually has a 1.5-mm needle. Insertion of short (e.g., <200 μm) microneedles may typically require the use of an applicator or even an insertion device. On the other hand, longer microneedles (e.g., >400 μm) can usually be inserted manually if desired. Verbaan et al. assembled microneedle arrays from commercially available 30G hypodermic needles and demonstrated that lengths of 550 to 900 μm could be inserted manually into dermatomed human skin, as monitored by TEWL and tryptan blue staining. Skin was stained on the stratum corneum side after piercing, and subsequently, blue dots appeared on the dermal side of the skin, suggesting that the dye permeated though the microchannels. However, 300-μm long microneedles could not be inserted manually into skin (62). In subsequent studies, the authors reported that microneedles of length ≤300 μm are able to pierce dermatomed human skin when inserted using an electrically driven impact insertion applicator. This suggests that in addition to length, the speed of insertion is very important to determine the piercing properties of microneedles (60,63). It was suggested that vibratory actuation can reduce the force required for microneedle insertion by as much as 70% (64). It was also reported that short (70 to 80 μm) silicon microneedles could be inserted into skin manually by finger pressing and swaying against the backing layer of the microneedle array (65). Additionally, humidity of the environment, hydration level of skin, and site-to-site and subject-to-subject variability may also affect skin penetration efficiency. In a study designed to

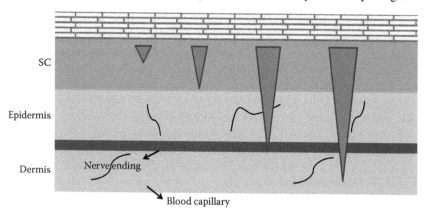

FIGURE 3.4 Placement of microneedle at different depths in the skin.

evaluate the effect of needle length and density on transdermal delivery, Yan et al. (66) reported that microneedle arrays with a length of 600 μm or more and a density of less than 2000 needles/cm² of skin were more effective, as indicated by electrical resistance of skin and flux of acyclovir across skin. Solid silicon microneedles with a length ranging from 100 to 1100 μm and a density ranging from 400 to 11,900 needles/cm² were used in this study. The study was done using human epidermal membrane and needles attached to a plastic syringe plunger were manually inserted. One concern with use of human epidermal membrane is that it may not have the elastic recoil that would be expected in full-thickness or dermatomed skin. This elastic recoil may lead to some degree of pore closure when microneedles are withdrawn from the skin. Another study also used the reverse end of a plastic syringe as an applicator and suggested that the bed of nails effect can be avoided by inserting the array in a rolling motion, to distribute the force over one row of microneedles at a time rather than inserting the array with a downward vertical pressure (67).

3.2.4.3   Microneedle Geometry

The effect of microneedle geometry on skin permeability was investigated by Davidson et al. using insulin-coated microneedles. They reported that the effective permeability was most affected by the depth of penetration of the microneedle array. Microneedle spacing also had an effect, but delivery was largely not affected by microneedle diameter and coating depth (68). Oh et al. also reported that delivery from a calcein gel was higher with 500-μm polycarbonate microneedles as compared to 200-μm microneedles. Permeation also increased as the density of microneedles was increased from 45 to 154/cm² (69). An optimization algorithm was reported in the literature to predict skin permeability following treatment with microneedles based on microneedle geometry. The center-to-center spacing between two microneedles (pitch) and the fractional area of the skin disrupted by the microneedles needs to be taken into consideration. The highest value of optimization is achieved for 17 microneedles per row in case of solid microneedles, or 5 microneedles per row in case of hollow microneedles. An increase in the number of microneedles does not necessarily lead to increased skin permeability. Aspect ratio that relates the pitch to the microneedle radius is important and should be less than 2 for optimum penetration. The material used for fabrication of the microneedles does not affect this algorithm (70).

3.2.4.4   Depth of Insertion and Pain

As stratum corneum has no nerves, microneedle insertion in skin is painless (9). Nerve fibers responsible for sensory perception and pain are present along with capillaries and lymphatic vessels in the papillary dermis (4). In reality, most microneedles will reach viable epidermis and even superficial dermis but are still painless at this depth of penetration because their small size reduces the chance of encountering a nerve or of stimulating a nerve. Microneedles of 430 μm length have been reported to be painless (27). It should be remembered that the entire length of the microneedle will generally not penetrate the skin, and it is the depth of penetration rather than the microneedle length which may be more important. The depth of penetration can be affected by several factors, including insertion (manual versus applicator or insertion device) method. In a study with 18 human subjects, TEWL, redness, and blood flow increased as microneedle (solid and hollow) length increased from 200 to 550 μm, but all sizes were painless and irritation lasted for less than 2 hours (71). Sivamani et al. used 200-μm long hollow microneedles and reported that the volunteers reported no pain from these microneedles (72). In a clinical study by Haq et al. using 180- and 280-μm long microneedles, subjects reported "pressing" and "heavy" sensation when microneedles were applied. In contrast, application of hypodermic needles was described as "sharp" and "stabbing" sensations (67). In another clinical study, metal microneedles of 620 μm length (<1 μm radius of curvature at the tip) were used, and pain was scored on a 0 to 100 visual analog pain scale. Microneedles resulted in a 6 ± 5 reading, while a positive control with a 5-mm hypodermic needle resulted in a score of 24 ± 16 (73). However, longer microneedles can be painful, though not as painful as hypodermic needles. Methods to assess damage to skin by needles have been described in the literature (74). In a study where microprojection arrays were applied to hairless guinea pig, it was reported that the skin tolerability reactions ranged from no detectable erythema to mild reactions that resolved within 24 hours. There were no significant reactions such as edema, bleeding, or infection (75).

3.2.5 Pore Closure and Safety Following Microporation

Using a hairless rat model porated by maltose microneedles, we reported that the skin starts to recover its barrier property within 3 to 4 hours of poration, though the pores take up to 15 hours to close completely. When the pores are occluded by a non-breathable film or by any solution, pore closure is delayed up to 72 hours, and pores close somewhere between 72 and 120 hours (see Figure 3.5). The creation and closure of pores was shown by us in this study by several techniques such as histology, confocal microscopy, calcein imaging, TEWL measurements, and methylene blue staining (48). It is believed that the increased TEWL immediately after microporation is the trigger for barrier recovery and pore closure, and occlusion delays pore closure as it interferes with this increased water loss from porated skin (76,77). Flux

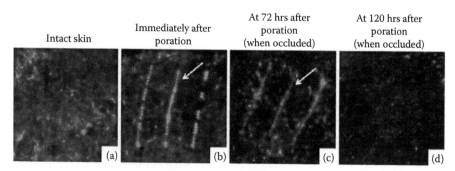

FIGURE 3.5 Calcein imaging studies to show pore closure in hairless rat skin. Fluorescent images showing (a) absence of microchannels before poration; presence of microchannels (b) immediately after poration; (c) at 72 hours after poration; and (d) at 120 hours after poration, under occluded conditions. (With kind permission from Springer Science+Business Media: *Pharm. Res.*, Formation and Closure of Microchannels in Skin Following Microporation, [Epub], H. Kalluri and A. K. Banga [48]. Copyright 2010.) **(See color insert.)**

studies have also been used as an indirect measure of pore closure. Lin et al. demonstrated continuous sustained delivery of an oligonucleotide for 24 hours across hairless guinea pig skin and suggested that the pore pathways are open during this period (78). Using a hairless guinea pig model and staining technique, Banks et al. also reported that altrexol could be delivered across microporated skin for 48 hours and that skin reseals rapidly following removal of occlusion. Based on TEWL data, the authors suggested that the healing of the skin is almost complete by 72 hours. Stainless steel microneedles of 750 μm length were used in this study (79). Clinical studies reported similar results. Haq et al. (67) used methylene blue staining and TEWL analysis to suggest that microchannels start to reseal and repair within 8 to 24 hours of creation (67). Using impedance spectroscopy, Gupta reported that occlusion delayed recovery of skin barrier following microneedle treatment in human subjects. Microneedle geometry has also been shown to be an important consideration for pore closure, with recovery being delayed by longer or more microneedles, as well as by larger cross-sectional area (80). All these findings look very promising for the future of microneedle technology, as they suggest that the pores stay open for a considerable time when a patch is applied on porated skin, thereby allowing constant infusion of drug into the bloodstream via microchannels. However, once the patch is removed, pores start to close due to removal of the occlusive conditions, thereby minimizing any safety risks associated with open pores on the skin being exposed to the environment. More studies are needed for a better mechanistic understanding of pore closure. Interestingly, occlusion has been reported to have similar effects on the recovery of the skin barrier following sonophoresis treatment (see Section 6.6).

One reason why a better understanding of pore closure is needed is due to the concern that microporated skin may be a portal for the entry of microbes in the interval between removal of the patch and complete closure of pores. However, these concerns of microbial infection have never been demonstrated. DermaRoller has sold more than 200,000 units without any reports of serious adverse effects (12). Similarly, companies like Altea (Atlanta, Georgia) and TransPharma (Lod, Israel) reported that their microporation devices have been well tolerated in clinical studies (see Section 10.5). It was suggested that microbial penetration followed by treatment with microneedle arrays will be much less as compared to that following treatment by hypodermic needles (81). Another study reported that skin treated with microneedles was not infected when incubated with *Staphylococcus aureus*; short (70 to 80 μm) silicon microneedles were used in this study. In contrast, rats treated with a macroneedle puncture or by abrasion were infected (65).

3.2.6 DELIVERY OF PARTICULATES THROUGH HAIR FOLLICLES AND THROUGH MICROPORATED SKIN

As mentioned earlier, the microchannels created in skin by microneedles are several microns in dimension; therefore, there are no size limits on the molecules that can be delivered through them. In fact, even small particulates can be delivered via these microchannels. There is precedent for delivery of particulates via skin. Commercially available nanoparticles with a diameter around 5 nm (Quantum Dots, Invitrogen, Carlsbad, California) have been reported to penetrate stratum corneum and localize in the epidermis and dermis within about 8 hours. This finding may also have toxicological implications, because nanoparticles are commonly present in many cosmetics (82–84). Solid lipid nanoparticles, which are typically larger, around 75 to 300 nm, have been reported to enhance drug transport into skin, but this may be due to indirect effects rather than their penetration into intact skin (85,86). Implication of hair follicles was suggested for immunization studies using a solution formulation or nanoparticles in the 180- to 200-nm size range (87,88). Larger particulates have also been known to enter hair follicles because hair follicles have diameters ranging from 30 to 80 μm. A detailed discussion of the literature relating to delivery of particulates via hair follicles is somewhat beyond the scope of this book. However, this literature can help to understand particle uptake by microchannels and hopefully stimulate more research to better understand it, as only limited research has been done in this area. Structural aspects of hair follicles were discussed in Section 1.2.1.1. Uptake of particles by hair follicles can be studied by several techniques, including a cyanoacrylate surface biopsy where a cyanoacrylate drop is allowed to polymerize on skin under a glass slide and then the hair shaft and cast of follicular infundibula is ripped off with the slide (89). A differential stripping process can be used where tape stripping is first performed before the cyanoacrylate stripping (90). Particles may preferentially accumulate near hair follicle openings as a function of time and may penetrate as deep as 1000 to 2300 μm (91,92). It seems that an optimum size for penetration into hair follicles is around 5 μm. Larger particles (>10 μm) do not penetrate into the follicular orifices (92,93). Similarly, uptake of particles into microchannels has been investigated. Delivery of 100-nm nanoparticles into the receptor chamber over 48 hours across human epidermal membrane treated with microneedles and mounted on Franz diffusion cells has been reported. No delivery was seen across intact skin (94). Because microchannels typically just reach the epidermis or the superficial layers of the dermis, the depth of penetration of particles across microchannels is expected to be much less than that achieved via penetration through the hair follicles. We have shown using confocal microscopy that 2-μm sized particulates (FluoSpheres, Invitrogen, Carlsbad, California) reached to a depth of around 110 μm in hairless rat skin treated with 500-μm long maltose microneedles (1).

3.3   SKIN MICROPORATION BY OTHER APPROACHES

There has been a surge of interest in recent years in using microneedles for skin microporation. As is evident from the preceding discussion, there is a wealth of literature published on fabrication and applications of microneedles to transdermal drug delivery. However, many companies are developing other microporation technologies, and there are several products in preclinical and clinical development stages. Some such technologies include thermal, radio-frequency, and laser microporation and the use of a powder jet injector.

3.3.1   Thermal Ablation

An increase in temperature by a few degrees increases transdermal delivery, and the utilization of this fact to launch a new product on the market, as well as the safety issues associated with heat, are discussed in Section 10.3. The focus of this section is on the use of very high-temperature pulses to cause thermal ablation of skin. Heating skin with short, high-temperature (>100°C) pulses can cause decomposition and vaporization of the stratum corneum to create micron-sized holes in the skin by removal of tissue (95). Altea Therapeutics (Atlanta, Georgia) is developing a thermal ablation technology to create microchannels in skin which are typically 50 to 200 μm in width and 30 to 50 μm in depth. An array of electrically resistive filaments is applied on the skin surface, and a short controlled pulse of electric current is passed through the filaments using a reusable handheld applicator. The filaments transfer thermal energy to the skin and ablate localized areas to create microchannels within 2 to 5 milliseconds. A PassPort® patch can then be applied to the microporated skin site, and the drug enters skin through these microchannels by diffusion. The single-use disposable patch has a regular patch attached to a film of metallic filaments. The part with the filaments is first activated by the handheld applicator. Once micro-channels are created in the skin, a fold-over design aligns the patch with the newly formed micropores. This device developed by Altea Therapeutics also has features to record or lock out doses and to set reminders for dosing (96). Badkar et al. showed that interferon α-2b can be delivered through these micropores (97). Bramson et al. found that reporter gene expression was increased 100-fold following the application of adenovirus vector to thermally microporated skin, as compared to an intact skin control (98).

3.3.2 LASER MICROPORATION

Medical lasers have been used for cosmetic and reconstructive surgery and have attracted attention in recent years as a potential approach for skin microporation. Pantec Biosolutions AG (Ruggell, Liechtenstein) is developing a painless laser epidermal system (P.L.E.A.S.E) that can microporate skin using a handheld erbium:YAG laser device. In a typical configuration, the device is applied to the inner forearm, activated for just a few seconds, followed by application of a drug patch to the site allowing the drug to diffuse through the microchannels created by laser microporation. One initial focus area is *in vitro* fertilization. A larger bench-top configuration of the device is also in development for use in clinics for dermatological conditions or immunization applications (99). This device from Pantec has received ISO certification as well as CE marking based on a rigorous audit to ensure safety, health, and environmental requirements. This CE marking will enable sale (after approval) to the European Economic Area (EEA). The erbium:YAG laser emits light at 2.94 μm, and this wavelength corresponds to a major water absorption peak; thus, water becomes the chromophore at this wavelength. The excitation of water molecules on skin leads to superheating and an explosive evaporation, which in turn results in the creation of micropores in the skin with minimal thermal effects. These micropores are about 150 to 200 μm in diameter, and their depth depends on the applied laser energy per unit area (fluence), with a typical depth being about 100 to 200 μm (100,101). A tolerability study on 12 volunteers indicated that there was minimal or no discomfort and no thermal damage to the tissue surrounding the pores. Slight to moderate erythema was observed which returned to baseline by day 5 of treatment. The TEWL returned to baseline levels by day 3 of treatment (101). Several other investigators used the erbium:YAG laser to facilitate delivery of small molecules, macromolecules, and vaccines (102–105). Visible radiation of Nd:YAG laser has also been used to increase the permeation of 5-Fluorouracil into skin with potential use to improve the efficacy of topical chemotherapy with this drug (106). Another company, Norwood Abbey (Victoria, Australia), received approval for laser-assisted delivery of a lidocaine formulation (see Section 10.5).

3.3.3 Radio-Frequency Ablation

Radio-frequency (RF) ablation is a medical technology that was adapted for skin microporation. Applying electric current in the radio-frequency range (100 to 500 kHz), this technology relies on cell ablation to use the RF scalpel in electrosurgery, e.g., to remove small tumors. This technology, consisting of a closely spaced array of microelectrodes, is being developed by TransPharma Medical (Lod, Israel) for skin microporation. When placed on the skin, application of high-frequency current causes ionic vibrations within the skin cells, which in turn leads to localized heating and cell ablation. The ViaDerm™ device is used to pretreat the skin using a handheld reusable electronic control unit in conjunction with a disposable microelectrode array containing hundreds of microelectrodes. Once micropores have been created in the skin, drug delivery profiles can be controlled by the type of patch utilized. A patch based on a dry formulation, which is then dissolved by the interstitial fluid from the microchannels, will typically produce a peak-drug profile similar to a subcutaneous injection (107). On the other hand, placing a drug reservoir, in a hydrogel or similar matrix form, on the microporated skin will only allow a slow drug diffusion through the micropores. This slow diffusion will simulate an intravenous infusion similar to that achieved by currently marketed conventional patches, except that hydrophilic molecules can be delivered through microporated skin. The ViaDerm™ device generates 144 microchannels over a 1.4-cm² area, with each microchannel being about 30 to 50 µm wide and 50 µm deep. These microchannels have been reported to allow delivery of gene therapy vectors and 100-nm diameter nanoparticles into the skin (108). In a study reported by Levin et al. (109), a printed patch with a thin layer of hGH in dry form was applied to skin following radio-frequency ablation. A high bioavailability was observed which was attributed to the dissolution of water-soluble hGH on the printed patch by the interstitial fluid from the microchannels, to create a high localized concentration on the surface of skin. Diffusion across this high concentration gradient resulted in a delivery profile similar to subcutaneous injection. The bioactivity of the hGH delivered into serum was demonstrated by elevated levels of systemic insulin-like growth factor-1 (IGF-I) after *in vivo* delivery of hGH to hypophysectomized rats (Figure 3.6). The ViaDerm device has also been investigated for delivery of fluorescein isothiocyanate (FITC)-dextrans (10, 40, or 70 kDa), diclofenac, granisetron, human growth hormone, insulin, and teriparatide (hPTH1-34). Granisetron delivery was studied in human volunteers, and variability was similar to that measured for oral delivery (107,110).

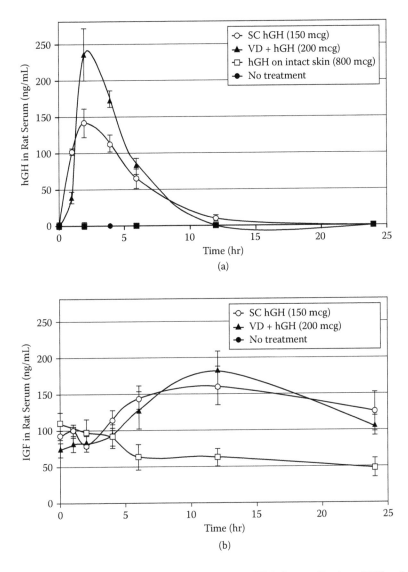

FIGURE 3.6 Serum levels (ng/ml) of (a) hGH and (b) IGF-I after application of 200 μg hGH on a 1.4 cm² ViaDerm-treated area, subcutaneous injection of 150 μg hGH, or no treatment (control). A negative control with 800 μg hGH applied to intact skin was also used. (With kind permission from Springer Science+Business Media: *Pharm. Res.*, Transdermal Delivery of Human Growth Hormone through RF-Microchannels, 22:550–555, 2005, G. Levin et al. [109]. Copyright 2005.)

3.3.4 OTHER MEANS OF SKIN MICROPORATION

The predominant technologies for skin microporation include microneedles, thermal microporation, radio-frequency microporation, and laser microporation. There are a few other technologies that have received limited attention and will be briefly mentioned first in this section. It has been suggested that high-intensity (around 20 W/sq cm) sonophoresis at 20 kHz frequency can actually create pores in the stratum corneum by a process termed *sonomacroporation*. This poration is induced by acoustic cavitation and can allow the transport of particles up to 25 μm into the skin (111). Micro-scissioning is another technique that uses 25 μm inert, sharp-sided, aluminum particles to create 50- to 200-μm-deep microchannels in skin by impacting skin with a flow of accelerated gas (112). The use of jet injectors may also be considered a type of microporation technology, which is discussed in Chapter 9 as they have been widely used and investigated for vaccine delivery applications.

3.4 COMBINATION APPROACHES

Iontophoresis (see Chapter 4) involves the application of low-level (typically, <0.5 mA/cm^2) electric current to drive charged (or neutral) drug molecules dissolved in an aqueous vehicle across the skin. Because microneedles create aqueous microchannels in the skin, a combination of these two techniques can drive hydrophilic drug molecules across the microchannels by an active force rather than by passive diffusion, and this combination has been investigated by several researchers. This can result in enhanced delivery of the drug across microporated skin. Also, this can allow for programmability of delivery, because the current can be modulated which, in turn, affects drug flux across microchannels. We have shown that methotrexate levels delivered to hairless rats were increased 14-fold by iontophoresis alone and 25-fold when a combination of iontophoresis and microneedles was used (50). We also showed that a combination of microneedles and iontophoresis was most effective in delivering a 13 kDa protein across skin (113,114). A combination of iontophoresis and microneedles has also been reported to increase the delivery of an oligonucleotide across hairless guinea pig skin (78). In another study, iontophoresis significantly enhanced the delivery of FITC-dextrans (3.8 to 200 kDa) across skin after pretreatment with microneedles. However, iontophoresis did not enhance the transport of deuterium oxide (D$_2$O) under similar conditions, suggesting that the increased FITC-dextran flux was not due to electroosmosis (115).

Microneedles have also been combined with other enhancement technologies and formulation approaches such as liposomes. Transdermal delivery of docetaxel loaded into elastic liposomes has been reported to be enhanced across microneedle-treated skin (116). Insulin loaded into nanovesicles has also been iontophoretically delivered across microchannels in pretreated skin (117). Using a combination of hollow silicon microneedles (100 μm long) and low-frequency (20 kHz) sonophoresis, *in vitro* transport of calcein and BSA across pig skin was demonstrated (118). Another study found that a combination of laser pretreatment followed by electroporation resulted in a higher skin permeation of methotrexate than either technique used alone (104). Badkar et al. showed that interferon α-2b could not be delivered *in vivo* across hairless rat skin by passive or even by iontophoretic delivery. However, it was delivered by thermal microporation and iontophoresis-enhanced delivery through the micropores, resulting in a twofold increase in the dose delivered (97).

3.5 APPLICATIONS OF MICROPORATION

Products that have been marketed or are in clinical development using one of the microporation technologies are discussed in Chapter 10. Delivery of insulin is discussed in Chapter 7. This section will discuss some of the other applications of microporation.

3.5.1 DELIVERY OF SMALL DRUG MOLECULES

We investigated delivery of nicardipine HCl across microneedle-treated hairless rat skin both *in vitro* and *in vivo*. Nicardipine HCl is a calcium channel blocker used in the treatment of hypertension which undergoes extensive hepatic first-pass metabolism and will gain from transdermal drug delivery. *In vitro* flux was increased from 1.72 for passive delivery to 7.05 μg/cm^2/hour after skin was pretreated with maltose microneedles. In the *in vivo* study, plasma levels following microneedle pretreatment reached a C_{max} of 56.45 ng/ml. In contrast, passive delivery did not result in detectable levels until 8 hours and the delivery was low (47). We also investigated delivery of methotrexate in a similar manner. Methotrexate is a folic acid antagonist used in the treatment of psoriasis and rheumatoid arthritis. Its systemic administration is associated with many side effects, so topical administration is desired, but methotrexate is hydrophilic (log P: 1.85) and ionized at physiological pH; it therefore has no passive permeation. We were able to deliver methotrexate into skin by microneedles and iontophoresis (50). Another drug investigated for delivery by microneedles is naltrexone (10,73). In a clinical study proof of concept study, naltrexone HCl could be delivered across microneedle-treated skin to achieve steady-state plasma levels within 2 hours. For this study, 50 stainless steel microneedles, each 620 μm long with a radius of curvature <1 μm at the tip, were arranged in a 5 × 10 array, and the patches were assembled in a laminar flow hood and then sterilized by ethylene oxide gas (73).

3.5.2 PHOTODYNAMIC THERAPY

In photodynamic therapy, a combination of a photosensitizing drug and the specific wavelength that excites the drug being used is utilized to generate cytotoxic singlet oxygen that can be used to kill cancer cells. For skin lesions, preformed photosensitizers such as porphyrins are not used, as they do not penetrate the skin due to their high molecular weight (MW). Precursors like 5-amino-levulinic acid (ALA; MW 167.6 Da) have been used instead. However, ALA does not penetrate deeper due to its hydrophilicity; therefore, it may be ineffective against deeper lesions. Delivery of ALA into skin has been reported to be facilitated by using chemical enhancers (119) or erbium:YAG laser (102). The use of microneedles to deliver a preformed photosensitizer has also been reported. Silicon microneedles (270 μm) were used in this study, and *in vitro* studies were conducted across excised mice and porcine skin, followed by *in vivo* studies in nude mice. Because microchannels are aqueous, some water solubility for the preformed photosensitizer is needed, and in this study, meso-tetra porphine tetra tosylate (TMP) was used (120).

3.5.3 DELIVERY OF LIPOPHILIC DRUGS

Most of the enhancement technologies such as iontophoresis or microporation are designed to enable delivery of hydrophilic drug molecules that do not normally cross the skin. However, it is not clear if lipophilic drugs can be delivered via coated microneedles or can be infused into skin in a formulation via hollow microneedles. Any such studies need to consider that lipophilic drugs will typically have some passive permeation, so a question needs to be asked as to what is being gained by formulating them for delivery via microporation technologies. However, it may be possible to change the depth of delivery, depot formation, or add programmability potential by using innovative transdermal delivery systems. Because many conventional drug molecules are lipophilic, there may be unique situations where these enhancement technologies can have a niche for a unique delivery profile of a particular lipophilic drug molecule. If simple enhancement of delivery is desired, a control for passive delivery should be run using an appropriate vehicle like propylene glycol or a mixture of cosolvents typically used for such formulations rather than formulating the drug in the same vehicle as the one being used for the enhancement technology. Formulating the drug into a particulate formulation (see Section 3.2.6) may provide another means to enable delivery of lipophilic drugs, especially if the intended site of action is at the base of the hair follicles.

3.5.4 DELIVERY OF MACROMOLECULES

Delivery of insulin by microporation is discussed in Chapter 7. We demonstrated the delivery of several proteins and other macromolecules into microchannels by microneedles or thermal microporation. Other drugs investigated for microneedle mediated delivery include LMWH (49), IgG (29,51), and daniplestim, a 13-kDa model protein (113,114) (see also Chapter 8). Low molecular weight heparins (LMWHs) are widely used as they are safer and have a greater antithrombotic activity as compared to heparin. However, they have poor oral bioavailability, and noninvasive administration routes are desired. Transdermal delivery would be effective, but they cannot be delivered passively as they are hydrophilic, charged macromolecules. We investigated delivery of a LMWH (MW 3 kDa) by iontophoresis, phonophoresis, and microneedles (49). Delivery of an antisense oligonucleotide across hairless guinea pig skin was also reported using a stainless steel microprojection array having 430 μm long microneedles and arranged at a density of 240 microneedles/cm^2 (78). Delivery of parathyroid hormone 1-34 by coated metal microneedles is discussed in Section 10.5.

3.5.5 DIAGNOSTIC AND OTHER APPLICATIONS

Sampling of interstitial fluid for glucose monitoring is being developed as an alternative to finger-stick blood sampling. The use of reverse iontophoresis technique to

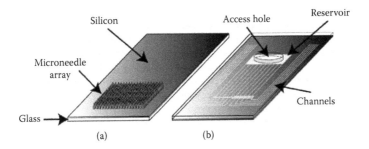

FIGURE 3.7 (a) Front and (b) backside of a microneedle array with hollow microneedles, integrated fluidic microchannels, and collecting reservoir for sampling of interstitial fluid. (Reproduced with permission from E. V. Mukerjee et al., *Sensors and Actuators A*, 114:267–275, 2004 [122].)

accomplish this is discussed in Section 4.8. Such sampling has also been investigated by microporation technologies. Use of thermal microporation to sample interstitial fluid for blood glucose monitoring in human diabetic subjects was reported (121). Mukerjee et al. (122) investigated use of a microneedle array having hollow silicon microneedles with integrated fluidic microchannels (Figure 3.7). Each microneedle of the 20 × 20 array consisted of 200- to 350-µm long needles with an off-centered borehole of 10 to 15 µm diameter, capable of drawing in fluid by capillary action. A chip having the microneedle array was held in place for 15 to 20 minutes by a spring clip to a human earlobe when fluid was seen to enter the back side channels of the microdevice and collect in the reservoir. Glucose levels were measured in this fluid, and a control was run on sweat to ensure that the collected fluid was interstitial fluid and not sweat. As mentioned earlier, microneedles also have other applications in medicine which are beyond the scope of discussion in this chapter. However, some of these applications are briefly mentioned here and include delivery of plasmid DNA for gene expression (123) and cutaneous gene delivery (2,124). Hollow microneedles (125) and solid-coated microneedles (126) have been used for intrascleral drug delivery to the eye. Microneedles have also been used as sensing electrodes (127) and as electrodes for electroporation (34). Use of dissolvable microneedles to deliver luciferase reporter plasmid and its expression in the cells of the ear, back, and footpad skin of mice has been described in the literature (128).

REFERENCES

1. A. K. Banga. Microporation applications for enhancing drug delivery, *Expert. Opin. Drug Deliv.*, 6:343–354 (2009).
2. D. V. McAllister, M. G. Allen, and M. R. Prausnitz. Microfabricated microneedles for gene and drug delivery, *Annu. Rev. Biomed. Eng.*, 2:289–313 (2000).
3. M. R. Prausnitz. Overcoming skin's barrier: The search for effective and user-friendly drug delivery, *Diabetes Technol. Ther.*, 3:233–236 (2001).
4. S. Coulman, C. Allender, and J. Birchall. Microneedles and other physical methods for overcoming the stratum corneum barrier for cutaneous gene therapy, *Crit. Rev. Ther. Drug Carrier Syst.*, 23:205–258 (2006).

5. R. K. Sivamani, D. Liepmann, and H. I. Maibach. Microneedles and transdermal applications, *Expert. Opin. Drug Deliv.*, 4:19–25 (2007).
6. J. Vandervoort and A. Ludwig. Microneedles for transdermal drug delivery: A minireview, *Front Biosci.*, 13:1711–1715 (2008).
7. H. Kalluri and A. Banga. Microneedles and transdermal drug delivery, *J. Drug Del. Sci. Tech.*, 19:303–310 (2009).
8. M. S. Gerstel and V. A. Place. Drug delivery device. [3,964,482]. 1976. Alza Corporation.
9. S. Henry, D. V. McAllister, M. G. Allen, and M. R. Prausnitz. Microfabricated microneedles: A novel approach to transdermal drug delivery, *J. Pharm. Sci.*, 87:922–925 (1998).
10. S. L. Banks, R. R. Pinninti, H. S. Gill, P. A. Crooks, M. R. Prausnitz, and A. L. Stinchcomb. Flux across microneedle-treated skin is increased by increasing charge of naltrexone and naltrexol *in vitro*, *Pharm. Res.*, 25:1677–1685 (2008).
11. J. H. Park, M. G. Allen, and M. R. Prausnitz. Polymer microneedles for controlled-release drug delivery, *Pharm. Res.*, 23:1008–1019 (2006).
12. M. R. Prausnitz, J. C. Birchall, and Co-Organizers. First International Conference on Microneedles, May 23–25, Atlanta, GA (2010).
13. H. J. G. E. Gardeniers, R. Luttge, E. J. W. Berenschot, M. J. de Boer, S. Y. Yeshurun, M. Hefetz, R. Oever, and A. Berg. Silicon micromachined hollow microneedles for transdermal liquid transport, *J. Microelectromechanical Syst.*, 12:855–862 (2003).
14. M. A. Teo, C. Shearwood, K. C. Ng, J. Lu, and S. Moochhala. *In vitro* and *in vivo* characterization of MEMS microneedles, *Biomed. Microdevices*, 7:47–52 (2005).
15. Y. Xie, B. Xu, and Y. Gao. Controlled transdermal delivery of model drug compounds by MEMS microneedle array, *Nanomedicine*, 1:184–190 (2005).
16. A. Rodriguez, D. Molinero, E. Valera, T. Trifonov, L. F. Marsal, J. Pallares, and R. Alcubilla. Fabrication of silicon oxide microneedles from macroporous silicon, *Sensors and Actuators B*, 109:135–140 (2005).
17. S. Rajaraman and H. T. Henderson. A unique fabrication approach for microneedles using coherent porous silicon technology, *Sensors and Actuators B*, 105:443–448 (2005).
18. N. Wilke, A. Mulcahy, S. Ye, and A. Morrissey. Process optimization and characterization of silicon microneedles fabricated by wet etch technology, *Microelectronics J.*, 36:650–656 (2005).
19. S. Paik, S. Byun, J. Lim, Y. Park, A. Lee, S. Chung, J. Chang, K. Chun, and D. Cho. In-plane single-crystal-silicon microneedles for minimally invasive microfluid systems, *Sensors and Actuators A*, 114:276–284 (2004).
20. N. Roxhed, B. Samel, L. Nordquist, P. Griss, and G. Stemme. Painless drug delivery through microneedle-based transdermal patches featuring active infusion, *IEEE Trans. Biomed. Eng.*, 55:1063–1071 (2008).
21. R. Shawgo, A. Grayson, Y. Li, and M. Cima. BioMEMS for drug delivery, *Curr. Opin. Solid State Mat. Sci.*, 6:329–334 (2002).
22. S. L. Tao and T. A. Desai. Microfabricated drug delivery systems: From particles to pores, *Adv. Drug Deliv. Rev.*, 55:315–328 (2003).
23. J. Z. Hilt and N. A. Peppas. Microfabricated drug delivery devices, *Int. J. Pharm.*, 306:15–23 (2005).
24. D. V. McAllister, P. M. Wang, S. P. Davis, J. H. Park, P. J. Canatella, M. G. Allen, and M. R. Prausnitz. Microfabricated needles for transdermal delivery of macromolecules and nanoparticles: Fabrication methods and transport studies, *Proc. Natl. Acad. Sci. U. S. A*, 100:13755–13760 (2003).

25. A. Doraiswamy, C. Jin, R. J. Narayan, P. Mageswaran, P. Mente, R. Modi, R. Auyeung, D. B. Chrisey, A. Ovsianikov, and B. Chichkov. Two photon induced polymerization of organic-inorganic hybrid biomaterials for microstructured medical devices, *Acta Biomater.*, 2:267–275 (2006).
26. D. Bodhale, A. Nisar, and N. Afzulpurkar. Structural and microfluidic analysis of hollow side-open polymeric microneedles for transdermal drug delivery applications, *Microfluid Nanofluid*, 8:373–392 (2010).
27. M. R. Prausnitz. Microneedles for transdermal drug delivery, *Adv. Drug Deliv. Rev.*, 56:581–587 (2004).
28. T. Miyano, Y. Tobinaga, T. Kanno, Y. Matsuzaki, H. Takeda, M. Wakui, and K. Hanada. Sugar micro needles as transdermic drug delivery system, *Biomed. Microdevices*, 7:185–188 (2005).
29. G. Li, A. Badkar, H. Kalluri, and A. K. Banga. Microchannels created by sugar and metal microneedles: Characterization by microscopy, macromolecular flux and other techniques, *J. Pharm. Sci.*, 99:1931–1941 (2009).
30. Y. Ito, J. Yoshimitsu, K. Shiroyama, N. Sugioka, and K. Takada. Self-dissolving microneedles for the percutaneous absorption of EPO in mice, *J. Drug Target.*, 14:255–261 (2006).
31. Y. Ito, E. Hagiwara, A. Saeki, N. Sugioka, and K. Takada. Feasibility of microneedles for percutaneous absorption of insulin, *Eur. J. Pharm. Sci.*, 29:82–88 (2006).
32. J. W. Lee, J. H. Park, and M. R. Prausnitz. Dissolving microneedles for transdermal drug delivery, *Biomaterials*, 29:2113–2124 (2008).
33. J. H. Park, M. G. Allen, and M. R. Prausnitz. Biodegradable polymer microneedles: Fabrication, mechanics and transdermal drug delivery, *J. Control. Release*, 104:51–66 (2005).
34. S. O. Choi, Y. C. Kim, J. H. Park, J. Hutcheson, H. S. Gill, Y. K. Yoon, M. R. Prausnitz, and M. G. Allen. An electrically active microneedle array for electroporation, *Biomed. Microdevices*, 12:263–273 (2010).
35. S. Kuo and Y. Chou. Novel polymer microneedle arrays and PDMS micromolding technique, *Tamkang J. Sci. Engg.*, 7:95–98 (2004).
36. M. M. Badran, J. Kuntsche, and A. Fahr. Skin penetration enhancement by a microneedle device (Dermaroller) *in vitro*: Dependency on needle size and applied formulation, *Eur. J. Pharm. Sci.*, 36:511–523 (2009).
37. C. P. Zhou, Y. L. Liu, H. L. Wang, P. X. Zhang, and J. L. Zhang. Transdermal delivery of insulin using microneedle rollers *in vivo*, *Int. J. Pharm.*, 392:127–133 (2010).
38. S. Bal, A. C. Kruithof, H. Liebl, M. Tomerius, J. Bouwstra, J. Lademann, and M. Meinke. *In vivo* visualization of microneedle conduits in human skin using laser scanning microscopy, *Laser Phys. Lett.*, 7:242–246 (2010).
39. H. S. Gill and M. R. Prausnitz. Coated microneedles for transdermal delivery, *J. Control. Release*, 117:227–237 (2007).
40. H. S. Gill and M. R. Prausnitz. Coating formulations for microneedles, *Pharm. Res.*, 24:1369–1380 (2007).
41. G. Widera, J. Johnson, L. Kim, L. Libiran, K. Nyam, P. E. Daddona, and M. Cormier. Effect of delivery parameters on immunization to ovalbumin following intracutaneous administration by a coated microneedle array patch system, *Vaccine*, 24:1653–1664 (2006).
42. Q. Zhu, V. G. Zarnitsyn, L. Ye, Z. Wen, Y. Gao, L. Pan, I. Skountzou, H. S. Gill, M. R. Prausnitz, C. Yang, and R. W. Compans. Immunization by vaccine-coated microneedle arrays protects against lethal influenza virus challenge, *Proc. Natl. Acad. Sci. USA*, 106:7968–7973 (2009).

43. X. Chen, T. W. Prow, M. L. Crichton, D. W. Jenkins, M. S. Roberts, I. H. Frazer, G. J. Fernando, and M. A. Kendall. Dry-coated microprojection array patches for targeted delivery of immunotherapeutics to the skin, *J. Control. Release*, 139:212–220 (2009).

44. M. Ameri, P. E. Daddona, and Y. F. Maa. Demonstrated solid-state stability of parathyroid hormone PTH(1-34) coated on a novel transdermal microprojection delivery system, *Pharm. Res.*, 26:2454–2463 (2009).

45. M. Ameri, S. C. Fan, and Y. F. Maa. Parathyroid hormone PTH(1-34) formulation that enables uniform coating on a novel transdermal microprojection delivery system, *Pharm. Res.*, 27:303–313 (2010).

46. A. K. Banga. *Therapeutic peptides and proteins: Formulation, processing, and delivery systems*, Taylor and Francis, London, 2006.

47. C. S. Kolli and A. K. Banga. Characterization of solid maltose microneedles and their use for transdermal delivery, *Pharm. Res.*, 25:104–113 (2008).

48. H. Kalluri and A. K. Banga. Formation and closure of microchannels in skin following microporation, *Pharm. Res.* [Epub] (2010).

49. S. S. Lanke, C. S. Kolli, J. G. Strom, and A. K. Banga. Enhanced transdermal delivery of low molecular weight heparin by barrier perturbation, *Int. J. Pharm.*, 365:26–33 (2009).

50. V. Vemulapalli, Y. Yang, P. M. Friden, and A. K. Banga. Synergistic effect of iontophoresis and soluble microneedles for transdermal delivery of methotrexate, *J. Pharm. Pharmacol.*, 60:27–33 (2008).

51. G. Li, A. Badkar, S. Nema, C. S. Kolli, and A. K. Banga. *In vitro* transdermal delivery of therapeutic antibodies using maltose microneedles, *Int. J. Pharm.*, 368:109–115 (2009).

52. K. Fukushima, A. Ise, H. Morita, R. Hasegawa, Y. Ito, N. Sugioka, and K. Takada. Two-layered dissolving microneedles for percutaneous delivery of peptide/protein drugs in rats, *Pharm. Res.* [Epub] (2010).

53. N. Roxhed, P. Griss, and G. Stemme. Membrane-sealed hollow microneedles and related administration schemes for transdermal drug delivery, *Biomed. Microdevices*, 10:271–279 (2008).

54. W. Martanto, J. S. Moore, T. Couse, and M. R. Prausnitz. Mechanism of fluid infusion during microneedle insertion and retraction, *J. Control. Release*, 112:357–361 (2006).

55. L. Nordquist, N. Roxhed, P. Griss, and G. Stemme. Novel microneedle patches for active insulin delivery are efficient in maintaining glycaemic control: An initial comparison with subcutaneous administration, *Pharm. Res.*, 24:1381–1388 (2007).

56. W. Martanto, J. S. Moore, O. Kashlan, R. Kamath, P. M. Wang, J. M. O'Neal, and M. R. Prausnitz. Microinfusion using hollow microneedles, *Pharm. Res.*, 23:104–113 (2006).

57. P. M. Wang, M. Cornwell, J. Hill, and M. R. Prausnitz. Precise microinjection into skin using hollow microneedles, *J. Invest. Dermatol.*, 126:1080–1087 (2006).

58. V. Yuzhakov. The AdminPen microneedle device for painless and convenient drug delivery, *Drug Del. Technol.*, 10:32–36 (2010).

59. Y. Wu, Y. Qiu, S. Zhang, G. Qin, and Y. Gao. Microneedle-based drug delivery: Studies on delivery parameters and biocompatibility, *Biomed. Microdevices*, 10:601–610 (2008).

60. F. J. Verbaan, S. M. Bal, D. J. van den Berg, J. A. Dijksman, M. van Hecke, H. Verpoorten, B. A. van den, R. Luttge, and J. A. Bouwstra. Improved piercing of microneedle arrays in dermatomed human skin by an impact insertion method, *J. Control. Release*, 128:80–88 (2008).

61. S. P. Davis, B. J. Landis, Z. H. Adams, M. G. Allen, and M. R. Prausnitz. Insertion of microneedles into skin: Measurement and prediction of insertion force and needle fracture force, *J. Biomech.*, 37:1155–1163 (2004).

62. F. J. Verbaan, S. M. Bal, D. J. van den Berg, W. H. Groenink, H. Verpoorten, R. Luttge, and J. A. Bouwstra. Assembled microneedle arrays enhance the transport of compounds varying over a large range of molecular weight across human dermatomed skin, *J. Control. Release*, 117:238–245 (2007).

63. Z. Ding, F. J. Verbaan, M. Bivas-Benita, L. Bungener, A. Huckriede, D. J. van den Berg, G. Kersten, and J. A. Bouwstra. Microneedle arrays for the transcutaneous immunization of diphtheria and influenza in BALB/c mice, *J. Control. Release*, 136:71–78 (2009).

64. M. Yang and J. D. Zahn. Microneedle insertion force reduction using vibratory actuation, *Biomed. Microdevices*, 6:177–182 (2004).

65. L. Wei-Ze, H. Mei-Rong, Z. Jian-Ping, Z. Yong-Qiang, H. Bao-Hua, L. Ting, and Z. Yong. Super-short solid silicon microneedles for transdermal drug delivery applications, *Int. J. Pharm.*, 389:122–129 (2010).

66. G. Yan, K. S. Warner, J. Zhang, S. Sharma, and B. K. Gale. Evaluation needle length and density of microneedle arrays in the pretreatment of skin for transdermal drug delivery, *Int. J. Pharm.*, 391:7–12 (2010).

67. M. I. Haq, E. Smith, D. N. John, M. Kalavala, C. Edwards, A. Anstey, A. Morrissey, and J. C. Birchall. Clinical administration of microneedles: Skin puncture, pain and sensation, *Biomed. Microdevices*, 11:35–47 (2009).

68. A. Davidson, B. Al-Qallfaf, and D. B. Das. Transdermal drug delivery by coated microneedles: Geometry effects on effective skin thickness and drug permeability, *Chem. Engg. Res. Design*, 86:1196–1206 (2008).

69. J. H. Oh, H. H. Park, K. Y. Do, M. Han, D. H. Hyun, C. G. Kim, C. H. Kim, S. S. Lee, S. J. Hwang, S. C. Shin, and C. W. Cho. Influence of the delivery systems using a microneedle array on the permeation of a hydrophilic molecule, calcein, *Eur. J. Pharm. Biopharm.*, 69:1040–1045 (2008).

70. B. Al Qallaf and D. B. Das. Optimizing microneedle arrays to increase skin permeability for transdermal drug delivery, *Ann. N. Y. Acad. Sci.*, 1161:83–94 (2009).

71. S. M. Bal, J. Caussin, S. Pavel, and J. A. Bouwstra. *In vivo* assessment of safety of microneedle arrays in human skin, *Eur. J. Pharm. Sci.*, 35:193–202 (2008).

72. R. K. Sivamani, B. Stoeber, G. C. Wu, H. Zhai, D. Liepmann, and H. Maibach. Clinical microneedle injection of methyl nicotinate: Stratum corneum penetration, *Skin Res. Technol.*, 11:152–156 (2005).

73. D. P. Wermeling, S. L. Banks, D. A. Hudson, H. S. Gill, J. Gupta, M. R. Prausnitz, and A. L. Stinchcomb. Microneedles permit transdermal delivery of a skin-impermeant medication to humans, *Proc. Natl. Acad. Sci. USA*, 105:2058–2063 (2008).

74. X. M. Wu, H. Todo, and K. Sugibayashi. Effects of pretreatment of needle puncture and sandpaper abrasion on the *in vitro* skin permeation of fluorescein isothiocyanate (FITC)-dextran, *Int. J. Pharm.*, 316:102–108 (2006).

75. J. A. Matriano, M. Cormier, J. Johnson, W. A. Young, M. Buttery, K. Nyam, and P. E. Daddona. Macroflux microprojection array patch technology: A new and efficient approach for intracutaneous immunization, *Pharm. Res.*, 19:63–70 (2002).

76. H. Zhai and H. I. Maibach. Occlusion vs. skin barrier function, *Skin Res. Technol.*, 8:1–6 (2002).

77. L. Kennish and B. Reidenberg. A review of the effect of occlusive dressings on lamellar bodies in the stratum corneum and relevance to transdermal absorption, *Dermatol. Online. J*, 11:7 (2005).

78. W. Lin, M. Cormier, A. Samiee, A. Griffin, B. Johnson, C. L. Teng, G. E. Hardee, and P. E. Daddona. Transdermal delivery of antisense oligonucleotides with microprojection patch (Macroflux) technology, *Pharm. Res.*, 18:1789–1793 (2001).

79. S. L. Banks, R. R. Pinninti, H. S. Gill, K. S. Paudel, P. A. Crooks, N. K. Brogden, M. R. Prausnitz, and A. L. Stinchcomb. Transdermal delivery of naltrexol and skin permeability lifetime after microneedle treatment in hairless guinea pigs, *J. Pharm. Sci.*, 99:3072–3080 (2010).

80. J. Gupta. Microneedles for transdermal drug delivery in human subjects, Ph.D. Thesis, Georgia Institute of Technology, (2009).

81. R. F. Donnelly, T. R. Singh, M. M. Tunney, D. I. Morrow, P. A. McCarron, C. O'Mahony, and A. D. Woolfson. Microneedle arrays allow lower microbial penetration than hypodermic needles *in vitro*, *Pharm. Res.*, [Epub] (2009).

82. J. P. Ryman-Rasmussen, J. E. Riviere, and N. A. Monteiro-Riviere. Penetration of intact skin by quantum dots with diverse physicochemical properties, *Toxicol. Sci.*, 91:159–165 (2006).

83. L. J. Mortensen, G. Oberdorster, A. P. Pentland, and L. A. Delouise. *In vivo* skin penetration of quantum dot nanoparticles in the murine model: The effect of UVR, *Nano. Lett.*, 8:2779–2787 (2008).

84. S. H. Jeong, J. H. Kim, S. M. Yi, J. P. Lee, J. H. Kim, K. H. Sohn, K. L. Park, M. K. Kim, and S. W. Son. Assessment of penetration of quantum dots through *in vitro* and *in vivo* human skin using the human skin equivalent model and the tape stripping method, *Biochem. Biophys. Res. Commun.*, 394:612–615 (2010).

85. V. Jenning, A. Gysler, M. Schafer-Korting, and S. H. Gohla. Vitamin A loaded solid lipid nanoparticles for topical use: Occlusive properties and drug targeting to the upper skin, *Eur. J. Pharm. Biopharm.*, 49:211–218 (2000).

86. H. Chen, X. Chang, D. Du, W. Liu, J. Liu, T. Weng, Y. Yang, H. Xu, and X. Yang. Podophyllotoxin-loaded solid lipid nanoparticles for epidermal targeting, *J. Control. Release*, 110:296–306 (2006).

87. H. Fan, Q. Lin, G. R. Morrissey, and P. A. Khavari. Immunization via hair follicles by topical application of naked DNA to normal skin, *Nat. Biotechnol.*, 17:870–872 (1999).

88. Z. Cui and R. J. Mumper. Chitosan-based nanoparticles for topical genetic immunization, *J. Control. Release*, 75:409–419 (2001).

89. N. Otberg, H. Richter, H. Schaefer, U. Blume-Peytavi, W. Sterry, and J. Lademann. Variations of hair follicle size and distribution in different body sites, *J. Invest. Dermatol.*, 122:14–19 (2004).

90. A. Teichmann, U. Jacobi, M. Ossadnik, H. Richter, S. Koch, W. Sterry, and J. Lademann. Differential stripping: Determination of the amount of topically applied substances penetrated into the hair follicles, *J. Invest. Dermatol.*, 125:264–269 (2005).

91. V. M. Meidan, M. C. Bonner, and B. B. Michniak. Transfollicular drug delivery—Is it a reality? *Int. J. Pharm.*, 306:1–14 (2005).

92. R. Toll, U. Jacobi, H. Richter, J. Lademann, H. Schaefer, and U. Blume-Peytavi. Penetration profile of microspheres in follicular targeting of terminal hair follicles, *J. Invest. Dermatol.*, 123:168–176 (2004).

93. S. Mordon, C. Sumian, and J. M. Devoisselle. Site-specific methylene blue delivery to pilosebaceous structures using highly porous nylon microspheres: An experimental evaluation, *Lasers Surg. Med.*, 33:119–125 (2003).

94. S. A. Coulman, A. Anstey, C. Gateley, A. Morrissey, P. McLoughlin, C. Allender, and J. C. Birchall. Microneedle mediated delivery of nanoparticles into human skin, *Int. J. Pharm.*, 366:190–200 (2009).

95. J. H. Park, J. W. Lee, Y. C. Kim, and M. R. Prausnitz. The effect of heat on skin permeability, *Int. J. Pharm.*, 359:94–103 (2008).

96. C. H. Dubin. Transdermal delivery—Making a comeback! *Drug Del. Technol.*, 10:24–28 (2010).

97. A. V. Badkar, A. M. Smith, J. A. Eppstein, and A. K. Banga. Transdermal delivery of interferon alpha-2B using microporation and iontophoresis in hairless rats, *Pharm. Res.*, 24:1389–1395 (2007).
98. J. Bramson, K. Dayball, C. Evelegh, Y. H. Wan, D. Page, and A. Smith. Enabling topical immunization via microporation: A novel method for pain-free and needle-free delivery of adenovirus-based vaccines, *Gene Ther.*, 10:251–260 (2003).
99. Pantec Biosolutions, http://www.pantec-biosolutions.com (accessed July 21, 2010).
100. Y. G. Bachhav, S. Summer, A. Heinrich, T. Bragagna, C. Bohler, and Y. N. Kalia. P.L.E.A.S.E: A promising tool for intraepidermal drug delivery, Presented at the Annual Meeting of the Controlled Release Society, July 2008, New York, (2008).
101. Y. N. Kalia, Y. G. Bachhav, T. Bragagna, and C. Bohler. Intraepidermal delivery: P.L.E.A.S.E., a new laser microporation technology, *Drug Del. Technol.*, 8:26–31 (2008).
102. J. Y. Fang, W. R. Lee, S. C. Shen, Y. P. Fang, and C. H. Hu. Enhancement of topical 5-aminolaevulinic acid delivery by erbium:YAG laser and microdermabrasion: A comparison with iontophoresis and electroporation, *Br. J. Dermatol.*, 151:132–140 (2004).
103. J. Y. Fang, W. R. Lee, S. C. Shen, H. Y. Wang, C. L. Fang, and C. H. Hu. Transdermal delivery of macromolecules by erbium:YAG laser, *J. Control. Release*, 100:75–85 (2004).
104. W. R. Lee, S. C. Shen, C. L. Fang, R. Z. Zhuo, and J. Y. Fang. Topical delivery of methotrexate via skin pretreated with physical enhancement techniques: Low-fluence erbium:YAG laser and electroporation, *Lasers Surg. Med.*, 40:468–476 (2008).
105. W. R. Lee, T. L. Pan, P. W. Wang, R. Z. Zhuo, C. M. Huang, and J. Y. Fang. Erbium:YAG laser enhances transdermal peptide delivery and skin vaccination, *J. Control. Release*, 128:200–208 (2008).
106. C. Gomez, A. Costela, I. Garcia-Moreno, F. Llanes, J. M. Teijon, and D. Blanco. Laser treatments on skin enhancing and controlling transdermal delivery of 5-fluorouracil, *Lasers Surg. Med.*, 40:6–12 (2008).
107. G. Levin. Advances in radio-frequency transdermal drug delivery. *Pharm. Technol.*, s12–s19 (2008).
108. J. Birchall, S. Coulman, A. Anstey, C. Gateley, H. Sweetland, A. Gershonowitz, L. Neville, and G. Levin. Cutaneous gene expression of plasmid DNA in excised human skin following delivery via microchannels created by radio frequency ablation, *Int. J. Pharm.*, 312:15–23 (2006).
109. G. Levin, A. Gershonowitz, H. Sacks, M. Stern, A. Sherman, S. Rudaev, I. Zivin, and M. Phillip. Transdermal delivery of human growth hormone through RF-microchannels, *Pharm. Res.*, 22:550–555 (2005).
110. A. C. Sintov, I. Krymberk, D. Daniel, T. Hannan, Z. Sohn, and G. Levin. Radiofrequency-driven skin microchanneling as a new way for electrically assisted transdermal delivery of hydrophilic drugs, *J. Control. Release*, 89:311–320 (2003).
111. L. J. Weimann and J. Wu. Transdermal delivery of poly-*l*-lysine by sonomacroporation, *Ultrasound Med. Biol.*, 28:1173–1180 (2002).
112. T. O. Herndon, S. Gonzalez, T. R. Gowrishankar, R. R. Anderson, and J. C. Weaver. Transdermal microconduits by microscission for drug delivery and sample acquisition, *BMC Med.*, 2:12 (2004).
113. S. Katikaneni, A. Badkar, S. Nema, and A. K. Banga. Molecular charge mediated transport of a 13 kD protein across microporated skin, *Int. J. Pharm.*, 378:93–100 (2009).
114. S. Katikaneni, G. Li, A. Badkar, and A. K. Banga. Transdermal delivery of a approximately 13 kDa protein—An *in vivo* comparison of physical enhancement methods, *J. Drug Target.*, 18:141–147 (2010).

115. X. M. Wu, H. Todo, and K. Sugibayashi. Enhancement of skin permeation of high molecular compounds by a combination of microneedle pretreatment and iontophoresis, *J. Control. Release*, 118:189–195 (2007).

116. Y. Qiu, Y. Gao, K. Hu, and F. Li. Enhancement of skin permeation of docetaxel: A novel approach combining microneedle and elastic liposomes, *J. Control. Release*, 129:144–150 (2008).

117. H. Chen, H. Zhu, J. Zheng, D. Mou, J. Wan, J. Zhang, T. Shi, Y. Zhao, H. Xu, and X. Yang. Iontophoresis-driven penetration of nanovesicles through microneedle-induced skin microchannels for enhancing transdermal delivery of insulin, *J. Control. Release*, 139:63–72 (2009).

118. B. Chen, J. Wei, and C. Iliescu. Sonophoretic enhanced microneedles array (SEMA)—Improving the efficiency of transdermal drug delivery, *Sensors and Actuators B: Chemical*, 145:54–60 (2010).

119. M. B. Pierre, E. Ricci Jr., A. C. Tedesco, and M. V. Bentley. Oleic acid as optimizer of the skin delivery of 5-aminolevulinic acid in photodynamic therapy, *Pharm. Res.*, 23:360–366 (2006).

120. R. F. Donnelly, D. I. Morrow, P. A. McCarron, W. A. David, A. Morrissey, P. Juzenas, A. Juzeniene, V. Iani, H. O. McCarthy, and J. Moan. Microneedle arrays permit enhanced intradermal delivery of a preformed photosensitizer, *Photochem. Photobiol.*, 85:195–204 (2009).

121. A. Smith, D. Yang, H. Delcher, J. Eppstein, D. Williams, and S. Wilkes. Fluorescein kinetics in interstitial fluid harvested from diabetic skin during fluorescein angiography: Implications for glucose monitoring, *Diabetes Technol. Ther.*, 1:21–27 (1999).

122. E. V. Mukerjee, S. D. Collins, R. R. Isseroff, and R. L. Smith. Microneedle array for transdermal biological fluid extraction and in situ analysis, *Sensors and Actuators A*, 114:267–275 (2004).

123. M. Pearton, C. Allender, K. Brain, A. Anstey, C. Gateley, N. Wilke, A. Morrissey, and J. Birchall. Gene delivery to the epidermal cells of human skin explants using microfabricated microneedles and hydrogel formulations, *Pharm. Res.*, 25:407–416 (2008).

124. F. Chabri, K. Bouris, T. Jones, D. Barrow, A. Hann, C. Allender, K. Brain, and J. Birchall. Microfabricated silicon microneedles for nonviral cutaneous gene delivery, *Br. J. Dermatol.*, 150:869–877 (2004).

125. J. Jiang, J. S. Moore, H. F. Edelhauser, and M. R. Prausnitz. Intrascleral drug delivery to the eye using hollow microneedles, *Pharm. Res.*, 26:395–403 (2009).

126. J. Jiang, H. S. Gill, D. Ghate, B. E. McCarey, S. R. Patel, H. F. Edelhauser, and M. R. Prausnitz. Coated microneedles for drug delivery to the eye, *Invest. Ophthalmol. Vis. Sci.*, 48:4038–4043 (2007).

127. J. C. Harper, S. M. Brozik, J. H. Flemming, J. L. McClain, R. Polsky, D. Raj, G. A. Ten Eyck, D. R. Wheeler, and K. E. Achyuthan. Fabrication and testing of a microneedles sensor array for p-cresol detection with potential biofuel applications, *ACS Appl. Mater. Interfaces.*, 1:1591–1598 (2009).

128. E. Gonzalez-Gonzalez, T. J. Speaker, R. P. Hickerson, R. Spitler, M. A. Flores, D. Leake, C. H. Contag, and R. L. Kaspar. Silencing of reporter gene expression in skin using siRNAs and expression of plasmid DNA delivered by a soluble protrusion array device (PAD), *Mol. Ther.*, 18:1667–1674 (2010).

129. M. Cormier, B. Johnson, M. Ameri, K. Nyam, L. Libiran, D. D. Zhang, and P. Daddona. Transdermal delivery of desmopressin using a coated microneedle array patch system, *J. Control. Release*, 97:503–511 (2004).

4 Iontophoretic Intradermal and Transdermal Drug Delivery

4.1 INTRODUCTION

An introduction to iontophoresis was presented in Chapter 1. In this chapter, details on iontophoretic delivery such as pathways, mechanism, theory, factors affecting delivery, and some examples will be presented. Other examples for delivery of proteins are presented in Chapter 8, and some case studies are discussed in Chapter 7. Several iontophoretic devices are on the market, and these are discussed in Chapter 10, Section 10.6, which provides a detailed discussion of the commercial development of iontophoresis and other skin enhancement technologies, including some of the studies conducted in human subjects. It is important to understand some of the very basic terminology used in the field. The electronic current from a power source is delivered to a solution where it is converted to a flow of ions taking place through the solutions and skin. The terminals leading the current into and out of the solution are electrodes, the positive pole being the anode and the negative pole being the cathode. The electrode may be smaller than the drug reservoir that is in contact with the skin. The current density is defined as the current intensity per unit cross-sectional area. In general, the current density will vary from point to point in any system, and the value calculated from the amperes divided by the surface area is just an average value at the treatment surface.

4.1.1 HISTORICAL ORIGIN OF IONTOPHORESIS

When the Greek physician, Etius, prescribed shocks of torpedo, an electric fish, for treatment of gout, he set the stage for biomedical applications of electricity even before the discovery of electricity. A series of experiments were conducted over the next two centuries which were rather dramatic but would not be considered scientific or ethical today due to more awareness of humane usage of animals or human subjects. These were previously reviewed by the author (1). In the latter part of the 19th century, Morton was interested in the electrical transport of drugs through skin. He conducted an experiment on himself in which finely powdered graphite was driven into his arm under positive electrode, producing small black spots that persisted for several weeks. Perhaps the first therapeutic use of iontophoresis came in the form of good news for patients with sweaty palms in 1936, when Ichihashi noted that sweating could be reduced by ion transfer of certain applied solutions, by electrophoretic techniques. This was the origin of the application of iontophoresis for the treatment of hyperhidrosis, a condition characterized by excessive sweating, which can be socially and occupationally distressing. The use of iontophoresis for treatment of hyperhidrosis is discussed in Section 4.12. While investigators were busy trying to reduce sweating, Gibson and Cooke had an altogether different problem. They wanted to increase sweating in order to get enough sweat for diagnosis of cystic fibrosis, as it was known that cystic fibrosis patients have a high concentration of sodium and chloride in their sweat. They used iontophoretic application of pilocarpine to induce sweating. The procedure was found to be painless and required only 5 minutes; rapid sweating was induced and continued for 30 minutes (2,3). Following the discovery of the use of iontophoresis to induce sweating, additional studies were conducted, and iontophoresis of pilocarpine was approved by the U.S. Food and Drug Administration (FDA) for diagnosis of cystic fibrosis (Section 4.12).

4.2   ELECTRICAL PROPERTIES OF SKIN

Electrical properties of the epidermal stratum corneum were investigated relatively early by tape stripping when it was observed that removal of the stratum corneum dramatically reduced the observed resistance (4). This suggests that stratum corneum forms the high electric resistance layer and is a very important element for skin impedance. The high resistance of this layer, in turn, is due in part to its lower water content (about 20%) as compared to the normal physiological level (about 70%). To understand the concept of impedance in rather simplistic terms, we need to first understand the term *capacitance*. An arrangement of parallel plates separated by a very small distance and connected to a battery allows electrons to distribute over the lower plate. The electrons on the lower plate then induce a positive charge on the upper plate so that more electrons can now flow into the lower plate from the battery. This arrangement of parallel plates acts as a capacitor or condenser, and this property is called capacitance. Thus, the capacitance of a capacitor is its ability to store an electric charge, flowing into it in the form of current. Biological tissues, such as skin tissue, also have a capacitance because of their ability to store electrical charge and are thus electrical capacitors. When an electric circuit contains both capacitive and resistant elements, it is said to be reactive, in contrast to the one containing resistant elements only, which is said to be resistive. The equivalent circuit model for the stratum corneum employs a parallel arrangement of a resistor and a capacitor or a resistor in series with a parallel combination of a resistor and a capacitor (4). A reactive circuit is said to present impedance rather than resistance. The impedance represents the total electrical opposition of the circuit to the passage of a current through it. Measurements of electrical impedance of the skin are important to understand electrically assisted transdermal delivery of drugs and also for *in vivo* electrical measurements on the body (5,6). Human skin reportedly shows a high impedance to the alternating current of low frequency, but this decreases as frequency is increased (7). Skin resistance also decreases with the application of square voltage or current pulses in the range of 1 to 10 mA, which are frequently used for transcutaneous electrical stimulation of nerves. Due to variations in skin resistance, the current intensity of voltage-regulated stimuli cannot be easily controlled so that only current regulated stimuli are suggested for use because the current through the tissue is the most significant stimulus parameter (8). The skin and the electrode, together with any gel or contact material used, constitute an interface that has an electrode-skin impedance that should be considered when measurements are made (9). The loss of skin resistance with application of current may be due to a reorientation of molecules along the ion transport pathways, such as the possible realignment of lipid molecules in hair follicles and sweat glands (10). Also, application of current will lead to an increase in the local ion concentration which will result in reduced resistance. The impedance of excised nude mouse skin, at 0.2 Hz, decreased by a factor of about 5 when exposed to iontophoretic current (0.16 mA/cm^2 for 1 hour) during a period of hydration (8 hours) as compared to skin which underwent only hydration (11). The electrical resistance of excised human skin was measured at 0.2 Hz and was found to decrease by an order of magnitude when a current of 0.16 mA/cm^2 was applied for 1 hour. The resistance then recovered, but the plateau value was lower than the resistance of the skin before the current was applied (12). The current–voltage relationship in skin has been known to be nonlinear (13) and has been more thoroughly investigated for excised skin (14,15).

4.3 PATHWAYS OF IONTOPHORETIC DRUG DELIVERY

It has been known for a long time that sweat glands play a role for transport during iontophoresis. An early study showed that pore patterns developed on the skin following iontophoretic transfer of basic and acidic dyes and metallic ions (16). For example, thorough rubbing and washing of the skin following the iontophoretic delivery of methylene blue revealed a remarkable pattern of channels traversed by the dye. The blue dots observed on the skin were found to be the sites of the pores of the skin which are the orifices of the coils of sweat glands, suggesting that the dye enters the skin via these pores. The pore patterns persisted for several weeks in many cases. Similarly, fluorescein dye has been shown to penetrate excised human skin upon applying a current density of 0.16 mA/cm^2 and appeared on the dermal surface as spots at pore sites (12). A comprehensive review of the pathways of iontophoretic current flow through mammalian skin has been published (17). The macropores and other conductive pathways in skin have been studied by scanning electrochemical microscopy during iontophoresis, and by theoretical considerations. The current maximum was found near the exit of a hair follicle, and the determined micropore radius was 9.2 to 14.1 μm. The macropore was considered to be a long cylindrical tube that was closed at one end (18,19). During iontophoresis, the greatest concentration of ionized species is expected to move into some regions of the skin where the skin is damaged, or along the sweat glands and hair follicles, as the diffusional resistance of the skin to permeation is lowest in these regions. Thus, a pore pathway is generally assumed for iontophoretic delivery. Iontophoresis of pilocarpine is used to induce sweating in the diagnosis of cystic fibrosis (Section 4.12), suggesting that some drug probably travels down the eccrine duct. In a study with desglycinamide-arginine vasopressin (DGAVP) as a model peptide, its transport across human stratum corneum and snake skin was compared to assess the role of appendages such as hair follicles, sweat, and sebaceous glands which are present in human skin but absent in shed snake skin. Although the initial resistance of both human and snake skin were in the same order of magnitude (about 25 $k\Omega.cm^2$), the steady-state ion-tophoretic DGAVP flux across human stratum corneum was about 140 times larger than through shed snake skin. Also, the average lag time across human stratum corneum was 0.7 hour, while that across shed snake skin was 2.5 hours. Azone pretreatment of the skin led to a large increase in transport across snake skin but not human skin. This suggests that the intercellular lipid pathway contributes very little to the iontophoretic flux across human skin but is very important for snake skin (20). Using special electrodes, it was suggested that the dominant pathway for flow of electric current through skin is through the sweat ducts. This study used a very thin (0.15 mm diameter) wire that was thin enough to distinguish between most pores and a very thin (0.1 μm) metal film electrode. The film electrode was placed on the skin and was permanently marked by the pathways of current flow so that dots developed after some seconds at places with sweat duct units (21). The utilization of aqueous pathway for iontophoretic delivery has been reinforced by a study that observed the transport kinetics of an anion (salicylate), cation (phenylethylamine), polar neutral compound of low molecular weight (mannitol), and polar neutral compound of high molecular weight (inulin). Using both intact and stripped dermatomed excised human skin, iontophoresis enhanced the delivery of all compounds relative to passive transport, and the skin was shown to be both ion and size selective (22).

It should be noted that the pore pathway for delivery does not necessarily imply skin appendages only. Using a scanning electrochemical microscope, reddish-brown spots have been visualized in the skin for transdermal flow of iron. These spots may represent precipitation or complexation of iron within a localized pore-type region but were not associated with a skin appendage. However, there is no evidence for any transcellular transport during iontophoresis (17). Similarly, electron micrographs for the iontophoretic *in vivo* transport of mercuric chloride in pig skin revealed that the primary pathway is via an intercellular route. Even with follicular transport, it should be noted that the final pathway is still intercellular between hair follicles and epidermal cells (23). Also, the iontophoretic transport of pindolol and calcitonin through both guinea pig skin and human skin equivalent was identical. Because living skin equivalent does not contain any appendages, this suggests that an appendageal pathway is not necessary for iontophoretic transport to occur (24). Skin behind the ears of guinea pigs has no hair follicles or sebaceous glands and could therefore be used in studies of shunt pathways, though the small size of the site would be a limitation (25). Iontophoretic transport of calcein through human stratum corneum was also shown to occur through intercellular route using scanning confocal fluorescence microscopy. Some involvement of even transcellular path was shown, but no appendageal transport was observed (26). It was shown that in the presence of an applied electric field, the stratum corneum lipid lamellae become more accessible to water and ions. This suggests that ion and water transport during iontophoresis is at least partly associated with stratum corneum lipid lamellae (27). Scanning electrochemical microscopy has been used for direct imaging of the ionic flux of $Fe (CN)_6^{4-}$ through pores of hairless mice skin. Activation of low resistance pores was shown to occur during iontophoresis, with spatial density of current-carrying pores increasing from 0 to 100 to 600 pores/cm² during the first 30 to 60 minutes of iontophoresis. The pore density reaches a quasi-steady-state value in proportion to the applied current density, with its contribution to total skin conductance increasing from 0% to 5% to 50% to 95% (28,29). Iontophoresis increases the concentration of the drug in the part of the skin that is the rate-limiting barrier during passive diffusion. Thus, for lipophilic drugs such as fentanyl, it will increase its concentration in viable skin, while for hydrophilic drugs such as thyrotropin-releasing hormone, it will increase its concentration in stratum corneum (30). This is because hydrophilic drugs have difficulty partitioning into stratum corneum, and lipophilic drugs have difficulty partitioning out of stratum corneum into viable skin.

4.4 THEORETICAL BASIS OF IONTOPHORESIS

Some of the theoretical considerations developed for analysis of transdermal delivery under iontophoretic transport are discussed in this section. However, it may be noted that the techniques for electrical enhancement of percutaneous absorption, in realistic terms, constitute a rather complex area with a large number of operating variables, and the results depend on the drug candidate being studied. Many of the theoretical equations for iontophoresis have been derived based on experimentation with simple ions such as sodium transport studies. These equations may not be applicable to many drugs, especially to drugs that bind to the skin or for macromolecules. The situation may be further complicated for peptide drugs that have the potential to undergo enzymatic degradation during transport through the skin and also undergo other losses such as by adsorption and self-aggregation. The basic equation also needs a modification for electroosmotic flow which will be discussed in Section 4.7.

In general, the flux of an ionic species, i, is given as:

$$\text{Flux} = \text{concentration} \times \text{mobility} \times \text{driving force}$$

or

$$J_i = C_i \times m \times \text{driving force} \tag{4.1}$$

The driving force on the species i is its chemical potential gradient. Thus,

$$J_i = C_i\, m\, (-du_i/dx) \tag{4.2}$$

The thermodynamic expression for the electrochemical potential, u_i, is given as:

$$u_i = u_{i(o)} + RT \ln C_i + z_i\, F\, E \tag{4.3}$$

where $u_{i(o)}$ is the standard chemical potential, and z_i is the valence of the species i, F is Faraday's constant, and E is the electrostatic potential. Assuming $u_{i\,(o)}$ is constant and substituting Equation 4.3 into du_i/dx in Equation 4.2, we get:

$$J_i = -C_i\, m\, (RT\, 1/C_i\, dC_i/dx + z_i\, F\, dE/dx) \tag{4.4}$$

The mobility, m, is related to diffusivity, D, as:

$$m = D_i/RT \qquad (4.5)$$

Substituting Equation 4.5 for m in Equation 4.4, we get:

$$J_i = -D_i \, dC_i/dx - z_i \, m \, F \, C_i \, dE/dx \qquad (4.6)$$

This is a fundamental relationship, called the Nernst–Planck equation, and is widely used to describe the membrane transport of ions (31). Several more rigorous theoretical models have been developed for iontophoretic delivery, which were reviewed and discussed for those interested in a more detailed treatment (32–37). A better appreciation of the meaning of this equation may be achieved by considering the case of a nonelectrolyte, in which the charge, $z = 0$. In this case, the Nernst–Planck equation is reduced to:

$$J = -D \, dC/dx \qquad (4.7)$$

which is Fick's first law of diffusion. On the other hand, for an ion with a uniform concentration throughout the system ($dC_i/dx = 0$), the Nernst–Planck equation becomes

$$J = -z_i \, m \, F \, C_i \, dE/dx \qquad (4.8)$$

which is the equation for electrophoresis. The Nernst–Planck equation may be thus interpreted as implying that when concentration gradient and an electric field both exist, the ionic flux is a linear sum of the fluxes that would arise from each effect alone. Though the permeability of ionized drugs through skin is low, it cannot be assumed to be negligible (38); thus, the potential contribution of passive flux needs to be considered. An expression was derived (35,39) for macroscopic membranes which takes into account the lag time. This expression was derived for effective lag time and enhancement ratio under the application of a uniform electric field to a uncharged homogeneous membrane. The enhancement ratio (E.R.), which is the ratio of the steady-state flux with applied voltage divided by steady-state flux by passive diffusion alone, was given as:

$$\text{E.R.} = J(v)/J(o) = v/1 - e^{-v}, \text{ where } v = zFE/RT \qquad (4.9)$$

E.R. can also be expressed as a ratio of permeability coefficients by iontophoretic transport (P_{AE}) over passive diffusion (P_o):

$$\text{E.R.} = P_{AE}/P_o \qquad (4.10)$$

Again, the assumptions made in the derivation are not likely to be valid for a complex membrane like the skin that is known to be a charged heterogeneous structure with pH_{iso} of 3 to 4. This will lead to some deviation from the theory. However, the model may be a useful tool for analyzing the details of iontophoresis experiments. The electric current (I) is related to the algebraic sum of the positive and negative ionic fluxes as follows:

$$I = F\left(\Sigma_i J_{i+} - \Sigma_k J_{k-}\right) \tag{4.11}$$

where J_{i+} and J_{k-} are the flux of the ionic species I^+ and k^-, respectively. By substituting Equation 4.6 for J_i and integrating, we can derive the following expression for membrane potential:

$$\Sigma = I\, dx/F^2(u + v) + RT/F\, d(U - V)/(U + V) \tag{4.12}$$

where $U = \Sigma_i m_{i+}\, C_{i+}$ and $V = \Sigma_k m_{k-}\, C_{k-}$, respectively. m_{i+} and m_{k-} are the mobilities for ionic species I^+ and k^-. From Equation 4.12, we can see that membrane potential is dependent on the concentration and mobility of all ions in the membrane. The first term of Equation 4.12 is an IR drop across the membrane, and the second term is a diffusion potential. Therefore, Equation 4.12 indicates that even when the current is turned off, the diffusion potential remains and acts as a driving force. Thus, after an iontophoresis treatment, the potential gradient in the Nernst–Planck equation (Equation 4.6) may not be zero but will rather depend on the concentration and mobility of all the ions in the membrane. It has been shown that the flux of an ion would be expected to remain elevated for a period of time after the current is removed, because dC/dx at the membrane–receptor interface resulting from an iontophoresis treatment is significantly greater than that by passive diffusion alone.

4.5 FACTORS AFFECTING IONTOPHORETIC DELIVERY

There is a complex multitude of factors operating during iontophoresis (40–43). To understand the delivery profiles and be able to commercially use this technology, it is important to realize that various formulation and electrochemical and biological factors are involved in the process (44). Statistical techniques such as factorial design and response surface methodology can be used to minimize the number of experiments required to optimize the transdermal iontophoretic delivery of a drug under different operational conditions (45–48).

4.5.1 ELECTRIC CURRENT

If the transport pathways across the skin are current dependent, then an increase in current density is expected to increase the amount of drug that will be delivered (17,49,50). According to Faraday's laws of electrolysis, the transport of one molar concentration of a univalent ion requires the passage of 96,485 Coulombs of electricity, if the ion has a transport number of unity. Hence, the maximum rate of transport,

$$J_{max} = (MW)\, I/96{,}485 \tag{4.13}$$

where (MW) is the molecular weight of the ion, and I is the current.

The more general case for a species I is:

$$J_i = t_i\, I_T\, (MW)/z\, F \tag{4.14}$$

where F is Faraday's constant, z is the number of charges per drug molecule (valency), and t_i is the transport number. The transport number (or transference number parameter), t_i, of an ionic species is the fraction of total applied current carried by that ionic species and may be calculated as:

$$t_i = I_i/I_T \qquad (4.15)$$

This suggests that the transference number parameter, t_i, can be assigned a fixed value for a drug ion under a set of experimental conditions even though this value could be considerably less than unity because of competition from Na^+ or Cl^- ions in the body. At low electric fields, iontophoretic enhancement is in reasonable agreement with the predictions of the constant field model for electrodiffusion. However, at higher power levels, the drug transference number in the membrane needs to be taken into account (39). If t_i is known, then theoretical flux can be calculated because all other parameters are known. Thus, a good prediction for each drug candidate can be made as to whether it may be a good candidate for iontophoretic delivery or not. Alternatively, the iontophoretic skin permeation (mg/hr) can be predicted as follows:

$$\text{Iontophoretic Skin Permeation (mg/hr)} =$$

$$\frac{MW * Current * Current\ efficiency * 3600}{Molecular\ charge * Faraday's\ constant} \qquad (4.16)$$

The steady-state plasma levels can then be calculated by dividing the steady-state iontophoretic skin permeation by plasma clearance. As expected from Equation 4.14, a linear dependence of the flux (J) on the total current density (I_T) applied at steady state is expected. Several published reports support this expected result. The release of neuropeptide, angiotensin, by microiontophoresis as well as its skin permeation by iontophoresis was found to be proportional to the current density applied (51,52). The drug delivery rate for various inorganic ions through various types of excised skin has been shown to have a linear relationship with current density (53). Similarly, the flux of acetate ion across excised hairless mouse skin increased as the current density was increased (54). Flux of arginine vasopressin across excised hairless rat skin was found to be a linear function of current density and duration of application (55). Thus, we can see that iontophoretic drug delivery is proportional to the applied current. Because the current is easily controlled by electronics, this technology provides a convenient means to control the rate of delivery of drugs (56). However, the current density and current intensity cannot be indefinitely increased as it will irritate or damage the skin, and also produce unpleasant electrical sensation. Also, once a limiting transport number is reached, further increase in current density may not increase drug flux (57). Additionally, different factors affecting iontophoresis may influence each other. For example, we used a response surface statistical design and observed an interaction between drug concentration and current density for delivery of tacrine hydrochloride across skin. Delivery was found to increase with increasing drug concentration at high current density but not at low current density (Figure 4.1). Tacrine is a lipophilic cation (log P 3.3) that associates with the skin to

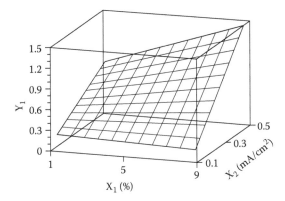

FIGURE 4.1 Response surface plot eliciting the effect of X_1 (drug concentration) and X_2 (current density) on Y_1 (cumulative tacrine delivered in 24 hours). (With kind permission from Springer Netherlands: *Pharm. Res.*, Response Surface Methodology to Investigate the Iontophoretic Delivery of Tacrine Hydrochloride, 21:2293–2299, 2004, R. S. Upasani and A. K. Banga [48]. Copyright 2004.)

reduce electroosmotic flow and a complex interrelationship of various factors affecting delivery was observed, which would have been hard to predict from the classical one factor at a time experimental approach (48). The maximum tolerable current increases with electrode area in a nonlinear fashion. Although a relationship has been described in the literature (58,59), the situation is understood to be much more complex. In general, 0.5 mA/cm² is often stated to be the maximum current density that should be used on humans without specifying the surface area. However, a current of 0.3 mA/cm² or less is perhaps more practical to avoid skin irritation and burns, especially if the current will be applied for prolonged periods.

4.5.2 MODALITY OF CURRENT

In direct current mode, direction of current flow continues unchanged in one direction, while in the case of alternating current, it changes periodically. Direct current (DC) may also be interrupted in a periodic manner, to produce what is called a DC pulse or periodic current. Iontophoresis is usually carried out by a continuous direct current, though pulsed DC has also been used. The wave form, current intensity, duty cycle (or on–off ratio) and frequency of periodic current may be altered. The duty cycle and the frequency of this current can be adjusted for optimum results. Also, the current can be delivered under different waveforms such as sinusoidal, square, triangular, trapezoidal, and so forth. A pulse waveform supposedly allows the skin to depolarize and return to its initial state before the onset of the next pulse. This is because stratum corneum acts as a capacitance, and this polarization may reduce the magnitude of a current applied as a constant current. It has also been suggested that pulse current will be less irritating to skin, so that patients can tolerate higher levels of current if pulse DC at high frequency is used. Also, it is hoped that higher drug fluxes could be achieved with pulsed current as compared to equivalent DC. Using time-dependent Nernst–Planck equations for electrodiffusion through homogeneous and structureless membranes, it has been suggested that average flux for constant current DC versus pulse current is about the same for high frequencies of pulse current (60). It has been suggested that pulse DC can result in lower skin resistance and higher drug delivery if the steady-state current during the "on" phase of the pulse is very small and the frequency is low enough to allow depolarization of the skin during the "off" phase (61). In an *in vitro* study on the iontophoretic transport of morphine hydrochloride across hairless mouse skin, the use of pulsed current resulted in lower drug flux as compared to direct current. However, pulsed current was considered to be less damaging to the skin based on a comparison of the passive flux of water across skin before and after iontophoresis for DC versus pulse current (62). Direct current has been reported to induce a higher transdermal flux than pulsed current across hairless rat skin for both fentanyl (63) and sufentanil (64). For delivery of thyrotropin-releasing hormone across excised rabbit pinna skin, pulsed iontophoretic flux has been reported to be higher than that obtained with continuous current. However, the frequency of the pulsed current had no significant effect on the flux (65). The molecular weight may possibly be involved in the difference in efficiency between current profiles, with larger molecules being more efficiently delivered by pulsed current, though this has not been established. Differences in blood pressure reductions for DC, monophasic rectangular pulsed and monophasic trapezoidal constant current iontophoretic delivery of captopril were not found to be significant (66). Similarly, Na^+ flux across excised full-thickness nude mouse skin by using pulsed current was the same as obtained by an equivalent continuous current and enhanced skin depolarization at high frequency actually decreased transport (67). *In vitro* iontophoretic transport of lysine and glutamic acid across hairless mouse skin was investigated as a function of several current profiles. Pulse DC at comparable charge delivered less than uninterrupted DC, but iontophoretic transport responded to and was well regulated by the use of different current profiles. The fluxes were directly proportional to the unipolar square-wave (positive) duty cycle. When bipolar AC (positive and negative) was used, fluxes were linear only after a threshold of 50% positive duty cycle. The frequency (2.5 to 2500 Hz) did not affect the delivery of these small molecules (68). Thus, it is not established that pulse DC leads to better flux enhancement, contrary to what is sometimes believed, and if it is better, the mechanism is more complex than just a lowering of skin impedance. Several recent studies from the University of Utah, have investigated alternating current (AC) for iontophoresis. These studies suggest that AC iontophoresis provides less variability in delivery as compared to constant current DC iontophoresis (69–72).

4.5.3 PHYSICOCHEMICAL PROPERTIES OF THE DRUG

The charge, size, structure, and lipophilicity of the drug will all influence its potential to be an iontophoresis candidate. Ideal candidates for iontophoresis should be water-soluble, potent drugs that exist in their salt form with high charge density (42,73,74). Conductivity experiments can be done to speculate on which drugs may be the best candidates for iontophoresis and to select the optimum pH for maximum delivery. The commonly used local anesthetic hydrochloride salts show excellent conductivity, with the best values around pH 5 (73). The conductivity of a drug can also be used to estimate the competitive transport between the drug and other extraneous ions during iontophoretic transport. In a study on the anodal transport of a model cation across excised human skin, the ratio of the specific conductance of the cation in deionized-distilled water to that of the solution applied in the donor compartment was used. It was observed that a linear relationship between iontophoretic flux and specific conductance was observed, but only after a certain threshold conductivity value (75,76). More recently, it was reported that the effective mobility calculated from capillary zone electrophoresis experiments at pH 6.5 can be used to predict electromigration component of transdermal iontophoretic flux of peptides (77).

Structure-transport relationships are hard to predict due to a multitude of factors that are involved in iontophoretic delivery (59). The molecular size of the solute is a major factor determining the feasibility of iontophoretic delivery and the amount transported. The efficiency of delivery of carboxylate ions showed the following rank order: acetate > hexanoate > dodecanoate, suggesting that smaller and more hydrophilic ions are transported faster than larger ions (54,78). Using a series of positive, negative, or uncharged solutes, the data for their iontophoretic delivery across excised human skin were best described by a linear relationship between the logarithm of the iontophoretic permeability coefficient and the molal volume, as predicted by the "free volume" model. However, predictions based on such models may not agree with the literature reports that have delivered much larger molecules (e.g., calcitonin) than what would be predicted based on the model (79). In one study, the mobility of seven medications suitable for iontophoretic administration was investigated at pH 5, 7, and 9. Three positive basic drugs, acyclovir, lidocaine hydrochloride, and minoxidil, were used. The other four drugs, methylprednisolone sodium succinate, dexamethasone sodium phosphate, adenine arabinoside monophosphate, and metronidazole, had a negative charge. Acyclovir ionized by protonation of the imidazole nitrogen's at positions 3 and 7 on the guanine rings. At its first pK_a of 2.27, it was 50:50 dication:monocation, and at its second pK_a of 9.25, it was 50:50 monocation:neutral species. Its mobility was observed to be best at pH 7, suggesting that mobility is best when maximum amount of monocation and minimum amount of free base is available (80). The salt form of the drug can also play an important role to control delivery efficiency if a protonated drug is being used and is the primary current-carrying species in the formulation. At neutral pH, cationic drugs often exist as a mixture of protonated and unprotonated species. As the protonated drug migrates toward the skin under electric field, an imbalance is created with more protonated drug in the boundary layer. This, in turn, lowers the pH in the boundary layer which results in a relatively higher proton transport. The pH may even drop below the pK of the acid. The problem will not occur with positively charged drugs that are not protonated (e.g., quaternary ammonium salts). This problem can be avoided by using a weak acid salt of the drug, such as acetate rather than hydrochloride. For instance, the succinate salts of verapamil, gallopamil, and nalbuphine all had a higher flux and current efficiency than the corresponding hydrochloride salts (58). Nonionized molecules will also be typically delivered by iontophoresis due to electroosmosis. However, if the molecule has a good passive permeability, then there may be no advantage to use iontophoresis. Using a series of n-alkanols, it was shown that iontophoresis hindered the lipoidal transport pathways. The iontophoretic enhancement values decreased linearly with increasing alkyl chain length, with the transport being even less than passive diffusion at alkyl chain lengths of greater than six (81). The upper size limit for iontophoretic delivery is not known. However, *in vitro* electrotransport of macromolecules up to 14 kDa size has been reported (see Chapter 8). Several studies have investigated iontophoretic delivery of insulin (Section 7.4), which could have existed with a molecular weight as high as 36,000, because it normally exists as a hexamer and the MW of the monomer is about 6000. The molecular volume may be more important than the molecular weight, because a compact folded molecule such as a globular protein may be able to pass through pores more readily than an open extended fibrous one, such as an unfolded globular or a fibrous protein. Thus, the tertiary and quaternary structure of a protein will play a role in the overall feasibility and efficiency of delivery.

4.5.4 FORMULATION FACTORS

The drug concentration, pH, and ionic strength of the formulation will all affect the iontophoretic delivery of the drug. An increase in the drug concentration in the formulation will typically result in higher iontophoretic delivery, as seen with acetate ions (54), or metoprolol (49). Increasing concentration will increase iontophoretic delivery up to some point, but at still higher concentrations, the flux may become independent of concentration. This could be because the boundary layer gets saturated with the drug while the bulk donor solution is still not saturated. Therefore, the concentration within the skin membrane does not increase by adding more drug in the donor compartment or formulation (57,58). In order to avoid pH changes, a buffer system is usually used. The iontophoretic delivery of drug will be reduced by these buffer ions as they will compete with the drug for carrying the current. As buffer ions are usually small, mobile, and highly charged, they will usually be more efficient in carrying the current. HEPES (4-(2-hydroxyethyl)piperazine-1-ethanesulfonic acid) buffer has been used in iontophoresis research as it has a high buffering capacity at pH 7.4 as it is zwitterionic at this pH and thus has reduced charge-carrying ability. Addition of sodium chloride provides one primary cation (Na^+) and one primary anionic (Cl^-) charge carrier. The pH of the buffer can be raised to 7.2 to 7.4 with tetrabutyl-ammonium hydroxide. Tetrabutylammonium hydroxide is preferably used for pH adjustment as it avoids the addition of small positive ions that would otherwise carry a large fraction of the charge (11,82). Chloride also participates in the electrochemical reaction to form silver chloride. The buffer should be optimized for the system for which it is being used (e.g., use of 25 mM HEPES along with about 75 mM of sodium chloride and 0.02 % of sodium azide has been reported) (83). An ethanolamine/ethanolamine HCl buffer has also been used for human studies with iontophoresis and provides the advantage of not having small mobile competitive ions and not having to add additional chloride ions (84). Buffers used as receptor fluids for *in vitro* studies should be deaerated prior to use by sonicating at 37°C or slightly higher temperature to prevent bubble formation on the tissue. Alternatively, bubbles can be removed by filtration or another tested method. An increase in ionic strength will decrease delivery as the extraneous ions will compete with the drug for the current. Because many drugs such as peptides are very potent, they are used in low concentrations so that a small amount of additives can have a rather large negative influence on delivery efficiency. Furthermore, peptides are macromolecules and have low mobility to start with. The co-ions from buffering agents are usually more mobile than the drug and will thus reduce the fraction of current carried by the drug ion. This, in turn, will reduce the iontophoretic delivery. Extraneous co-ions can also be introduced at the electrodes, such as the generation of hydronium and hydroxide ions when platinum electrodes are used. These ions are even more harmful to transport efficiency, their mobility being three to five times greater than small inorganic ions such as sodium, potassium, and chloride (85). The ionic strength of the buffer should thus be a compromise to achieve just adequate buffer capacity to avoid pH drifts but not be too high to minimize the competition for current. An increase in solution viscosity may also decrease the iontophoretic flux by hindering the mobility of the drug (49,50). The efficiency of drug delivery will be determined by the concentration of extraneous ions and the mobility of the drug ion in the skin relative to the mobility of these other ions (53). For a discussion of optimizing the amount of halide ion in the formulation to maintain the electrochemistry with silver electrodes, the reader is referred to Section 4.6. Another factor that will have a significant influence on delivery is the pH of the formulation. The pH can determine whether or not the drug is charged or it can affect the ratio of the charged and uncharged species (86). In the case of polypeptides, the type of charge is also controlled by the formulation pH relative to the isoelectric point of the polypeptide. Iontophoretic delivery of a drug may be hindered by the presence of high concentrations (>15% v/v) of cosolvents (such as propylene glycol) in the formulation. This could be due to a decrease in the conductivity of the drug solution as well as a decrease in electroosmotic flow (87).

4.5.5 BIOLOGICAL FACTORS

For small ions, iontophoretic delivery may not be affected by the type of skin being used. For delivery of lithium through human, pig, and rabbit skin, the iontophoretic flux was nearly identical even though the passive flux differed by more than an order of magnitude (53). However, most drug molecules have complex structures that may interact with the skin in various ways. Thus, their iontophoretic delivery profile will have to be evaluated on a case-by-case basis. The factors involved in the selection of skin are discussed in Section 2.2. The dermal blood supply determines the systemic and underlying tissue solute absorption during iontophoresis. However, the blood supply does not appear to affect the epidermal penetration fluxes during iontophoretic delivery. This was suggested by the observation that the solute concentration in the upper layers of skin following iontophoresis was comparable in anesthetized rats and sacrificed rats. Because the latter had no blood supply, it is presumed that the blood supply did not affect penetration through the epidermis (88).

4.6 PH CONTROL IN IONTOPHORESIS RESEARCH

4.6.1 SELECTION OF ELECTRODE MATERIAL

Electrochemistry at the electrode–solution interface where electronic current is converted to ionic current is controlled by the electrode material used. Appropriate choice of electrodes is a factor that is critical to successful iontophoretic delivery of a drug. The electrode material in an iontophoretic device is very important as it determines the type of electrochemical reaction taking place at the electrodes. Unless special electrodes or other suitable mechanisms are used, iontophoresis is accompanied by electrolysis of water. Electrodes such as platinum, stainless steel, or carbon graphite do not participate in the electrochemistry. The inert electrochemistry thus forces the water in the reservoir to become fuel for the electrochemistry (89). As oxidation occurs at the anode and reduction at the cathode, the following reactions take place:

$$H_2O \rightarrow 2H^+ + 2\,O_2 + 2\,e^- \text{ (at anode)}$$
$$2H_2O + 2e^- \rightarrow H_2 + 2OH^- \text{ (at cathode)}$$

As hydrogen and hydroxyl ions are produced, this leads to a pH drop in the solution containing the anode and a rise in the pH at the solution containing the cathode. These ions generated by hydrolysis of water are very mobile ions and may also slowly alter the conductivity of the skin. Changes in pH may also lead to irritation and burns. Furthermore, the drug may degrade at the electrode. Significant degradation of propranolol HCl was observed when a high current was passed through platinized electrodes and the solutions in both receptor and donor compartments were discolored at the end of the *in vitro* experiment (90). The possibility of introducing metallic ions into the skin must also be carefully considered because some metals such as chromium and nickel are well-known dermal allergens. Analysis of the fluid surrounding an anode made of brass or medical grade steel by atomic absorption spectrophotometer revealed the presence of small amounts of copper, nickel, and chromium, which are known dermal allergens. The presence of these metals was found with current density as low as 400 μA/cm^2 during 2 to 4 minutes (91).

4.6.2 SILVER/SILVER CHLORIDE ELECTRODES

A better choice of electrode material with respect to pH changes is silver for anode and chloridized silver for cathode. Silver/silver chloride or reversible electrodes prevent pH drifts as they are consumed by the active electrochemistry, thus avoiding the use of water in the electrochemistry. However, if the current density at the silver electrode surface exceeds a limiting value for the formation of silver chloride, pH shifts might occur. At anode, the silver oxidizes under the influence of an applied potential, and when chloride ion is present, it reacts to form insoluble silver chloride which precipitates on the anode surface, and an electron is released to the electrical circuit. Thus, the use of silver anode prevents electrolysis of water, and the use of chloride salt eliminates silver ion migration (74). Simultaneously, the silver chloride cathode is reduced, using an electron from the circuit, to silver metal that precipitates at the electrode surface, and the resulting chloride ion is free to migrate into the body:

$$Ag + Cl^- \rightarrow AgCl + e^- \text{ (at anode)}$$
$$AgCl + e^- \rightarrow Ag + Cl^- \text{ (at cathode)}$$

Thus, the use of silver for the anode and chloridized silver for the cathode is desired. The amount of chloride added to the buffer can be calculated to drive the electrochemistry, as any excess will provide avoidable competition to the drug for transport. However, enough ions must be present to ensure that the desired electrochemistry is actually occurring. The ideal case would be where the drug exists as a hydrohalide salt (e.g., lidocaine hydrochloride) in which case the drug will provide the ion for electrochemistry, resulting in high current efficiency (89). However, because the dose is typically low for peptides (and most transdermal delivery candidates), the fraction of current carried by the peptide is low. The counterion of the peptide thus may not be enough to drive the electrochemistry, so that some chloride (e.g., as sodium chloride) may have to be added to the formulation. This addition will adversely affect delivery efficiency because the salt has its own counterion (Na^+) that will compete for the current. Nevertheless, a hydrohalide salt will still be beneficial. Also, the silver salts of chloride or bromide are highly insoluble so that they will not migrate. However, it has been noted that most peptide drugs are available as acetate salts. Because silver acetate is much more soluble than other silver salts, there is a possibility of silver migration and skin discoloration (56).

The reaction at anode in the absence of chloride ion may be written as:

$$Ag \rightarrow Ag^+ + e^-$$

In the absence of sodium chloride, the donor solution (containing anode) may become cloudy as the silver anode gradually dissolves in solution. If silver ions are created, they can migrate to the skin and cause it to discolor.

4.6.3 PREPARATION AND CARE OF SILVER/SILVER CHLORIDE ELECTRODES

Silver/silver chloride matrix electrodes are commercially available, and they can also be prepared from silver wire, if desired. It is important to use good-quality, double-distilled water for preparation of all solutions. For the formation of a silver chloride layer, two solutions are commonly used, KCl or HCl, electrolysis being carried out at moderate current densities. These solutions should be bromide-free as 0.01 mole% of bromide can cause the silver/silver chloride electrodes to behave erratically. Ag/AgCl electrodes can be prepared by electrolysis in a 0.1 M HCl solution at 0.4 mA/cm² (92). Another group lightly sanded their Ag wires before placing them in 1 M HCl for 10 minutes at 50°C (82). After rinsing, these were plated with AgCl by applying a current of 0.2 mA using 0.5 M KCl solution for 12 hours. Electrodes were then also plated with platinum black. Using a similar technique, Ruddy and Hadzija (93) electroplated their silver wire in 0.5 M KCl upon application of 1 mA current. Prior to electroplating, they lightly sanded the silver wire with steel wool, immersed it in 1 M HCl for 30 minutes, rinsed with purified water and methanol and then dried. Following electroplating, the finished electrodes were dried and fitted with nonconductive polyethylene sheath so as to allow only the tips to be exposed. Alternatively, silver wire also chloridized by immersing in 133 mM NaCl solution (using Pt-cathode) for about 3 hours at an applied current of 0.5 mA (94). Ag/AgCl electrodes were also prepared by dipping silver wire into molten AgCl. After cooling, the noncoated Ag wire was protected from contact with the electrolyte solution by using shrink-to-fit, salt-resistant, insulating tubing (95). Electrodes must be thoroughly cleaned to prevent damage to the Ag/AgCl layer by washing under flowing water immediately after use, rinsing with distilled water, and air drying before storage. Continuous use of Ag/AgCl electrodes may result in a black deposit of silver chloride on top of the gray Ag/AgCl layer and can be removed by a five-to-one dilution of dilute ammonium hydroxide with distilled water. If a layer of bright silver appears, the electrode can be rechlorinated, but a homogeneous sintered Ag/AgCl layer will not exist in this case (96). Due to the porous nature of some electrodes, it may be necessary to soak the electrodes for at least 24 hours following an experiment to avoid problems of residual radioactivity or residual drug leaching into the solution in the next experiment (97). Finally, small currents should be used to minimize the depletion of AgCl present on the Ag/AgCl electrodes (67) and to minimize any potential damage to skin.

4.6.4 ALTERNATIVE METHODS FOR pH CONTROL

As noted above, the addition of sodium chloride to maintain the electrochemistry for a reversible electrode will reduce drug flux, and special device design considerations need to be considered. Other alternative methods to control pH drifts are also feasible. Alternate methods may also be desired if silver/silver chloride electrodes cause drug precipitation, as has been suggested in the case of some peptides (98). Also, drug can degrade under the electrode, as discussed. Drug may also bind to the electrode (e.g., insulin-mimetic peroxovanadium compound has been reported to bind to Ag/AgCl electrodes) (99). An alternative to the silver/silver chlorides may be the use of intercalation electrodes. These electrodes can adsorb or desorb alkali metal ions such as sodium or potassium into their structure as the electrodes are oxidized or reduced. Ion-exchange membranes and resins can be used for pH control. An anion-exchange membrane can be used to separate the buffer from the drug solution and will prevent the buffer cations from entering the drug solution (58,100). Ion-exchange resins have also been used for pH control under iontophoresis electrodes (101,102). The following is an example of the pH buffering action of a polymeric ion-exchange resin that does not electromigrate under an applied electric field. Two weakly acidic cation-exchange resins, Amberlite IRP-64 and IRP-88, are mixed in a 50:50 ratio. Both resins are copolymers of methacrylic acid and divinyl benzene, with IRP-64 being the free acid form, while IRP-88 is a potassium salt. For a drug with a metal ion (M^+) (e.g., Na^+ from dexamethasone sodium phosphate), the pH buffering reaction at the cathode in presence of IRP-64 (P-COOH) is:

$$P\text{-}COOH + M^+ \quad \rightarrow \quad P\text{-}COOM + H^+$$
$$H^+ + OH^- \quad \rightarrow \quad H_2O$$

Thus, the OH^- ions generated at the cathode by electrolysis are removed and the pH-control mechanism can be used to deliver negatively charged drugs. The reaction at the anode can be used for the dispersive or inactive electrode. The presence of IRP-88 (P-COOK) results in generation of K^+, which is much less mobile than the H^+ that would have been generated in absence of the resin. This should help to minimize the possibility of an electrochemical burn under the anode electrode:

$$P\text{-}COOK + H^+ \quad \rightarrow \quad P\text{-}COOH + K^+$$

Electrodes clinically being used for topical delivery are discussed separately in Chapter 10. To prevent degradation or adsorption of drugs under or on the electrodes, various researchers have also used salt bridges to apply current (92,99,103,104). These salt bridges typically contain 3% w/v agar in 1 M NaCl and allow electrodes to be isolated from the donor and receptor solutions.

4.7   ELECTROOSMOTIC FLOW

If a voltage difference is applied across a charged porous membrane, bulk fluid flow, or volume flow, called *electroosmosis*, occurs in the same direction as flow of counterions. This flow is not diffusion and involves a motion of the fluid without concentration gradients (105). This bulk fluid flow by electroosmosis is a significant factor in iontophoresis and was found to be of the order of microliters/hour/cm^2 of hairless mouse skin (106). Because skin is a permselective membrane with negative charge at physiological pH (82), the counterions are usually cations, and electroosmotic flow occurs from anode to cathode, thus enhancing the flux of positively charged (cationic) drugs. The stratum corneum is thus often referred to as a cation permselective membrane. The major cation transported through epidermis, Na$^+$ has a transport number of 0.6, which is about twice the transport number of Cl$^-$, so that cations are transported more readily across the epidermis. The transport number of ions in skin can be calculated from potentiometric measurements (107). Investigations with scanning electrochemical microscopy have shown that electroosmotic flow is localized to shunt pathways, and the entire shunt pathway structure is required for the flow to take place (e.g., electroosmotic flow will not take place if iontophoresis is applied to just the stratum corneum) (108,109).

Due to electroosmotic flow, it usually is also possible to deliver neutral drugs under anode as the bulk fluid flow from anode to cathode can carry the neutral drug (110,111). The term *iontohydrokinesis* was used in early literature to describe this water transport during iontophoresis (112). Electroosmotic flow can also hinder drug flux in a situation where a negatively charged drug (anion) or a neutral drug is being delivered under the cathode. In such cases, the flux may actually increase after the current is stopped. For example, the cathodal mannitol flux through hairless mouse skin was retarded relative to passive transport due to net volume flow in the opposite direction. The transport of mannitol increased significantly after the termination of current (113). If the skin reverses its charge such as at a pH below its isoelectric point, the direction of electroosmotic flow will also reverse. The pH value of skin surface is about 3 to 4, which is about the isoelectric point of keratin in the stratum corneum layer (114). As the solution pH is decreased toward 4, electroosmotic flow decreases in magnitude and eventually will reverse direction as the charge reverses in the "pores" of the skin somewhere between pH 3 and pH 4. Some positively charged peptides may actually associate with the skin to reduce or neutralize its negative charge. In these cases, the cation permselectivity of skin may be lost, resulting in a reversal of electroosmotic flow to the cathode-to-anode direction. This was seen for the iontophoretic delivery of nafarelin through hairless mouse and human skin (see Chapter 8). The same phenomenon was also observed for another luteinizing hormone-releasing hormone (LHRH) analogue, leuprolide (115). Using ^{14}C-labeled mannitol, the direction of electroosmotic flow in the anode-to-cathode direction was first confirmed, and this was dramatically reduced or reversed as nafarelin was added to the anode chamber. Iontophoresis over a 12-hour period resulted in delivery of nafarelin to the receptor phase, but a significant amount associated with the skin during this time and then desorbed passively over the next 12 hours (103). Positively charged poly-L-lysines have also been observed to reduce electroosmotic flow in a concentration dependent fashion. Some dependence on molecular weight was also observed. At a concentration of 10 mg/ml, the 2.7 and 8.2 kDa poly-L-lysines decreased electroosmotic flow by five- to sixfold, while the decrease was 30-fold when the 20-kDa molecule was used at the same concentration (116). A series of positively charged β-blockers were also investigated for their potential to reduce electroosmotic flow. The three most lipophilic compounds (propranolol, timolol, and metoprolol) significantly reduced electroosmotic flow while the two least hydrophobic (atenolol and nadolol) did not (117,118). Similarly, doxorubicin has also been recently reported to interact with the negative charges of the skin to reduce electroosmotic flow (119).

Electroosmotic flow is also referred to as *water transference number*, t_w, which is defined as the number of moles of water transported per equivalent of electricity passed. Though both excised human skin and excised hairless mouse skin give significant electroosmotic flow, water transference numbers in human skin are stated to be at least five times greater than in hairless mouse skin at the same salt concentration. However, this may not translate into higher flux for neutral species due to the failure of water transference numbers for human skin measured at high current density (1 mA/cm^2) to extrapolate to lower current densities (105). The contribution of electroosmotic flow to drug flux enhancement during iontophoresis depends on the sign of the membrane charge, concentration of charges in the membrane, pore radius, stokes radius of the drug species, and the ionic concentration in the membrane. For high molecular weight species such as proteins, the overall flux will be low even though the relative contribution of electroosmotic flow to flux enhancement will be large as stokes radius is large and diffusion is slow. For a protein with a small net negative charge, it is possible that delivery may be higher under wrong polarity (i.e., under anode rather than cathode because of the higher contribution of electroosmotic flow). For example, the flux enhancement was found to be greater for anodic delivery than for cathodic delivery for the high molecular weight anionic species, carboxy inulin and bovine serum albumin (97). For a pore radius of about 2.5 nm, the ratio of electroosmotic flow to ionic flow is greater than unity for species larger than about 1 nm (105). In another study, the flux enhancement due to electroosmosis for model permeants with a molecular weight range of 60 to 504 (urea, mannitol, sucrose, and raffinose) was measured across a model synthetic membrane. The flux enhancement ratio was found to depend on molecular weight, being about four times greater for raffinose than for urea. This was followed up by studying transport of urea and sucrose across human epidermal membrane. The flux enhancement ratio was observed to be three times greater for sucrose than for urea (120). Iontophoresis literature often suggests that as ions move during ionto-phoresis, their water of hydration accounts for some or all of the electroosmotic flow, but this concept cannot explain the reversal of direction of electroosmotic flow when the membrane charge is reversed, and it has been shown that its contribution to electroosmotic flow is negligible (105). Acetaminophen, a hydrophilic neutral molecule, has often been used as a marker to understand the contribution of elec-troosmosis to iontophoretic delivery (121).

4.7.1 Theoretical Treatment of Electroosmotic Flow

Based on the above discussion, we can see that the classical Nernst–Planck equa-tion must be modified to include an electroosmosis component. The origin of elec-troosmotic flow lies in the realm of nonequilibrium or irreversible thermodynamics. Intuitively, one may visualize electroosmotic flow as occurring because there are immobile charges in the membrane that require the flow of mobile counterions to maintain electroneutrality in the membrane. Although theoretical models have been developed to explain electroosmotic flow, it is unrealistic to expect that all of the complexities of a natural membrane like skin can be accurately calculated by the rel-atively simple models the theories must assume (35,105). The Nernst–Planck equa-tion assumes that the only driving force on an ionic species is the negative gradient of chemical potential for that species alone; it neglects the potential coupling of gradients of chemical potentials for other species j with the flux of species i (31). As discussed, an example of such coupling that is important in iontophoretic delivery is the *solvent drag*, where the gradient of the chemical potential for water not only activates a flow of water (electroosmosis), but also yields a flow of solute dissolved in the water. Thus, a modification to the Nernst–Planck equation (Equation 4.6) is necessary as follows:

$$J_i = -D_i \, dC_i/dx - z_i \, m \, F \, C_i \, dE/dx \pm C_i J_v \qquad (4.17)$$

where J_v is the velocity of convective flow (i.e., volume flow per unit time per unit area).

4.8 REVERSE IONTOPHORESIS

As the name suggests, reverse iontophoresis is the back iontophoretic extraction (by electroosmotic flow) of a molecule from the body rather than its forward iontopho-retic delivery into the body. This technique can have important applications in medi-cal diagnostics, as it can allow noninvasive sampling of biological fluids (122,123). Reverse iontophoresis can thus be used to perform clinical chemistry without blood sampling. In addition to the advantage of being a noninvasive technique, it will allow for samples that are filtered by skin and are thus free of particulates or larger macro-molecules (124). This approach has been used for iontophoretic extraction of glucose from subcutaneous tissue in an attempt to develop an alternative for the commonly used invasive and inconvenient finger-stick technique. In preliminary *in vitro* stud-ies using hairless mouse skin, the amount of glucose extracted at the electrodes by 2 hours of iontophoresis at a current density of 0.36 mA/cm^2 was found to be proportional to the concentration of glucose solution bathing the dermal side (125).

In a subsequent study in human subjects, iontophoresis at 0.25 mA/cm^2 for 60 minutes was applied to the ventral forearm surface to show that iontophoretic sampling of glucose is feasible. It was found that the stratum corneum contains some glucose as a product of lipid metabolism within the skin which does not relate to the systemic glucose concentration so that sampling over short periods of 15 minutes or less cannot be used to predict blood glucose levels. However, glucose extracted at the electrode within 1 hour was expected to be representative of the subcutaneous tissue concentration. Extraction at the cathode was most efficient, though measurable amounts were also seen at the anode. Variations among individuals were seen, suggesting that specific calibration may be required for each person (126). As discussed, samples extracted by reverse iontophoresis are filtered by skin and are thus free of particulates or even macromolecules. This is an advantage for the long-term use of sensors that are faced with these problems when used in implantable detectors (127).

Sonophoresis has also been utilized for extraction of interstitial fluid for glucose monitoring (see Chapter 6). A glucose monitoring device based on the principles of reverse iontophoresis, the GlucoWatch Biographer, was marketed but later withdrawn from the market and is discussed in Section 10.6. Detection of glucose can be done by using glucose oxidase in the saline gel. Several other clinical chemistry applications should also be feasible based on the concepts of reverse iontophoresis. Noninvasive sampling of lactic acid as a model endogenous compound through hairless rat skin by iontophoresis has also been demonstrated using chloride ion in the body as an internal standard. Concentration of lactic acid extracted at the electrode was found to reflect the concentration of lactic acid on the dermis side (128). Other molecules that have been extracted by reverse iontophoresis from the subdermal space include valproate (129), urea (130–132), and lithium (133). Several other molecules have also been extracted by reverse iontophoresis, and iontophoretic extraction is generally linearly correlated with subdermal concentration once any reservoir in skin is depleted (42). Factors affecting electroosmotic flow during reverse iontophoresis have been investigated. Because electroosmotic flow normally takes place from anode to cathode, extraction at the cathode will be most efficient and is preferred. This will be true as long as skin is negatively charged, which would be the case when physiological pH is used. If the pH is lowered, charge on the skin may be neutralized as discussed before so that extraction at anode can become feasible. Extraction was found to be enhanced if the ionic strength in the electrode chambers was reduced. Use of a pulsed current or periodic alternation of electrode polarity did not reduce the efficiency of the extraction process. Electroosmotic flow can be increased to increase extraction efficiency at either anode or cathode by the use of some excipients such as divalent ions or EDTA in the electrode formulations, respectively. The mechanism is believed to be modification of the net negative charge of the skin (134,135).

For diagnostic applications of reverse iontophoresis, a sensitive analytical detection method for the target material is required because the amount extracted is likely to be very low. Though the approach can sample several materials at the same time, the need to have separate detection methods for each will pose a challenge. For measurement of glucose, calibration procedures for day-to-day use will need to be refined. Also, the collection formulation will need to be optimized to improve the efficiency of extraction, and the "on-board" sensor will need to be perfected (124).

4.9 TREATMENT PROTOCOLS AND FORMULATIONS

Iontophoresis treatment protocols vary from clinician to clinician. Typically, patients are treated every other day, terminating after six to eight visits if significant relief is not obtained (136–138). Setting up the unit takes about 5 minutes, after which constant supervision is usually not required. Typical treatment duration is 10 to 20 minutes using a current of 4 mA or less applied via an electrode such that the current density is less than 0.5 mA/cm^2. Prior to the application of the electrode, skin is prepared by cleaning with isopropyl alcohol to remove any oily secretions or dead cells and also to disinfect. The dispersive or inactive electrode should be placed at about 15-cm distance from the active or drug electrode to prevent formation of edge currents from one electrode to the other. Any excessive hair may be clipped but not shaved. The electrodes are prepared according to manufacturer's package instructions and should be applied over intact skin, not broken skin. Iontophoresis should not be done over damaged or denuded skin or other recent scar tissue. Delivery electrodes and the return electrodes are applied in the appropriate anatomic locations. The iontophoretic device is connected such that the clinician's choice of electrodes, cathode or anode, is attached to the delivery electrode. The device is switched on and dosage or timer is set, depending upon the manufacturer of the iontophoretic device. Current density should be increased slowly, and the patient should be informed that he or she may experience a tingling sensation initially. The current amperage is increased to either a maximum of 4 mA or the maximum tolerated by the patient. Iontophoretic treatment continues until the dosage is delivered, after which a slight erythema (redness) under the electrode is normal, and we found that it typically disappears in a few hours (139). The use of iontophoresis is contraindicated for patients with known adverse reactions or sensitivity to application of current or drug and those wearing cardiac pacemakers or other electrically sensitive devices. It is also contraindicated for treatment across the right and left temporal regions and for the treatment of the orbital region. The devices should obviously not be used if the patient is hypersensitive to the drug being used.

The major clinical use of iontophoresis is for the localized delivery of the anti-inflammatory agent, dexamethasone-21-phosphate. The pharmaceutic preparation of dexamethasone utilized is the same as that for injection. Some clinicians utilize dexamethasone phosphate (4 mg/ml) in combination with lidocaine (Xylocaine[7], 10 mg/ml) according to the clinical protocols of Harris (136) and Bertolucci (140). The use of lidocaine in the iontophoretic solutions was proposed by Petelenz et al. (141) to maintain pH. The dexamethasone solution is utilized either as formulated from the vendor, or diluted with an equal volume of sterile 0.9% saline. Iontophoretic dosage varies but ranges from 40 to 80 mA*min at 4 mA or less. Both the dosage range and amperage vary depending upon the clinician, anatomic site being treated, or tolerance of the patient to iontophoretic treatment. Clinically, dexamethasone phosphate is delivered from cathode or anode, the latter electrode having been proposed to deliver dexamethasone by electroosmosis (141). An appropriate-sized electrode is selected. The dexamethasone phosphate solution is dispensed, usually 2.5 ml for medium electrodes of most manufacturers, and applied to the delivery electrode. Iontophoresis is then conducted according to the treatment protocols described earlier. Although formulations for iontophoresis are simple to make, some pharmacies will provide customized iontophoresis medications for patients to be used with prescription products used by physical therapists.

4.10   COMBINED USE OF IONTOPHORESIS AND OTHER ENHANCEMENT TECHNIQUES

Combined use of iontophoresis with other enhancement techniques such as electroporation, sonophoresis, and microneedles has been reviewed (142) and has been discussed in other chapters in this book. Combined use of chemical and electrical enhancement to facilitate transdermal drug delivery may offer advantages of synergism (143–147) and is briefly discussed here. A combination of several penetration enhancers with iontophoresis has been investigated for the transdermal delivery of nonivamide acetate through rat skin. Cetylpyridinium chloride and isopropyl myristate were the most effective, but any potential clinical use of the latter will be limited, as it induced severe changes in the histological structure of the skin (148). In another *in vitro* study on transdermal delivery of sotalol, enhancers were as effective as iontophoresis. A combination of enhancers and iontophoresis did not further increase permeation but actually decreased it slightly as compared to enhancers or iontophoresis alone (149). The effect of ethanol on *in vitro* iontophoretic transport of a dopamine agonist has also been investigated. It was observed that ethanol increased the solubility of the agonist in water as much as fivefold, but this only resulted in a modest increase in the amounts transported across skin into the receptor compartment. A significant amount of the agonist was found to be retained in the skin (150). Any potential interaction between chemical and electrical enhancement should be investigated by techniques such as impedance spectroscopy (94) or X-ray diffraction (151).

4.11   APPLICATIONS OF IONTOPHORESIS

Delivery of peptides and proteins by iontophoresis and other enhancement modalities is discussed separately in Chapter 8. Similarly, several case studies are discussed in Chapter 7 and include delivery by iontophoresis. Clinical applications of iontophoresis will be discussed later in Section 4.12 in this chapter, and commercial development of iontophoresis products is discussed in Chapter 10. In this section, the focus is on a few conventional drugs that have been investigated for systemic delivery for potential use in wearable patches. Several other nonpeptide drugs have been investigated for systemic delivery, and examples of isolated studies are discussed at several places in this book.

4.11.1   IONTOPHORETIC DELIVERY TO THE NAIL

Onychomycosis, a fungal infection of the nail plate, is a common but difficult to treat condition, as drugs applied topically to the nail do not reach the deep-seated infection. Most of the products on the market are given systemically and have significant adverse effects. Lipid content of nail is only about 0.5% in contrast to the 10% lipid content of stratum corneum. Therefore, penetration enhancers, many of which act on the lipid domain, are usually ineffective to increase topical delivery. Keratolytic and thiolytic agents have been used but still are of limited effectiveness. The nail structure was described as a keratinized hydrogel (keratin pI approximately 5), and the nail plate has iontophoretic permselectivity similar to human skin. Therefore, iontophoresis has been widely investigated in recent years for treatment of onychomycosis and to understand the mechanisms of transport involved in nail iontophoresis (42,152–154). Nails loaded with terbinafine by delivering it iontophoretically have been shown to release the drug slowly over 60 days (155). Another recent study also investigated iontophoretic delivery of terbinafine across porcine and human nails (156). As the soft tissue surrounding the nail and the nail bed is skin, some delivery will take place across skin as well, and we investigated the iontophoretic delivery of terbinafine into skin for this reason (157,158). Nail resistance and hydration as well as the ionic strength of the formulation have been shown to be important factors controlling delivery. Nail hydration has been shown to be independent of pH within the range 3 to 11 but was significantly increased above pH 11. Permeability coefficients of water transport across nail were similar in the pH 1 to 10 range but much higher at pH 13. Nail permselectivity was also affected under extremely acidic conditions such as pH 1 (159–161). Transungual or nail iontophoresis has been typically carried out by using nail holders mounted on Franz diffusion cells. An *ex vivo* model using healthy intact toes from human cadavers has also been used (155). Human studies have also been carried out. In a study on 38 otherwise healthy patients, iontophoretic delivery of terbinafine was found to be efficacious for treatment of nail onychomycosis (162). Another study in six subjects investigated increased ionic transport into nail by iontophoresis. Using voltage and transonychial water loss (TOWL) measurements, this study suggested that nail DC iontophoresis is a feasible and likely a safe technique (163).

4.11.2 NICOTINE

Nicotine patches have been commercially available for several years and have been shown to be effective as an aid to smoking cessation (164,165). Pharmacokinetics of disposition of nicotine in healthy volunteers following transdermal delivery has been characterized in many studies (166–171). Nicotine has relatively high skin permeability, so the drug delivery patch should control the delivery (172). For example, iontophoretic delivery to achieve higher blood levels is not required. However, iontophoretic delivery of nicotine may provide the advantage of achieving pulsatile drug delivery across skin. This may be desirable as the absorption of nicotine from lungs during smoking is in pulsatile pattern. It has been shown (173) that a 1-mg "dose" of nicotine can be delivered from a reasonably sized system within 30 minutes using iontophoresis. By delivering this cigarette-equivalent bolus of nicotine, the lag time of passive absorption (2 to 4 hours) can be minimized, and patients may not crave a nicotine "high," particularly shortly after waking.

4.11.3   SYNTHETIC NARCOTICS: SUFENTANIL

Fentanyl and sufentanil are synthetic narcotics of the 4-anilinophenylpiperidine class. A detailed discussion of transdermal delivery of fentanyl is presented in Section 7.4. Sufentanil has a potency about 10 times that of fentanyl and has similar skin permeability as fentanyl. In a study on the passive permeation of fentanyl and sufentanil through human cadaver skin, neither drug influenced the permeation of the other when they were administered concurrently (174). Both fentanyl (MW 336.5) and sufentanil (MW 387.5) are relatively lipophilic molecules, with octanol/water partition coefficients of 717 and 2842, respectively (175). Thus, it seems that delivery is actually hindered by the viable skin rather than stratum corneum because clearance from the stratum corneum replaces diffusion through the stratum corneum as the rate-limiting step. In an *in vitro* study with freshly excised abdominal hairless rat skin, the quantity of fentanyl detected in various depths of stratum corneum (as determined by tape stripping) following iontophoresis showed a similar distribution profile as passive diffusion. An increase in the duration of current application also did not affect the quantity of fentanyl in stratum corneum, but larger quantities were detected in the viable tissue after iontophoresis (30). Iontophoretic delivery of sufentanil across hairless rat skin has been investigated. After a 2-hour lag time, iontophoretic delivery was linear with time with a flux of 219.9 ± 34.8 for direct current treatment, but only 0.8 ± 0.5 ng/cm^2-hr for passive diffusion (64).

4.11.4   IONTOPHORETIC DELIVERY OF DOPAMINE AGONISTS

Levodopa has been traditionally used for Parkinson's disease, but now many newer dopamine agonists including rotigotinine and ropinirole are also used. Parkinson's disease affects around 1 million people in the United States, and the disease is characterized by tremors, which makes it difficult for patients to self-administer oral dosage forms. It has been suggested that iontophoretic delivery of apomorphine for Parkinson's will allow patients to control the amount needed based on their symptoms by changing the current to change input, thereby having an on-demand delivery system (176). *In vitro* studies were conducted for iontophoretic delivery of R-apomorphine, a potent dopamine agonist, and it was found that delivery was enhanced when skin was pretreated with a nonionic surfactant (177,178). It was suggested that pretreatment of skin by surfactant followed by iontophoresis results in swelling of corneocytes and the presence of water pools in the intercellular regions of stratum corneum. These changes in the ultrastructure of stratum corneum are responsible for facilitating iontophoretic transport of R-apomorphine (179). This surfactant was later tested in healthy volunteers with a follow-up iontophoretic current application of 0.25 mA/cm^2 of skin for 3 hours. Skin irritation was observed to be mild (180). This was followed by a study in patients with advanced Parkinson's disease with the same current settings, and clinical improvements were observed. The bioavailability increased from around 10% to 13% when surfactant pretreatment was used. In contrast, bioavailability following oral administration is only about 1.7% due to first-pass effect and degradation in the gastrointestinal (GI) tract. The drug is currently administered by parenteral route (181).

Iontophoretic delivery of the hydrochloride salt of rotigotine, another dopamine agonist, across human stratum corneum was investigated (182). Rotigotine (log P 4.03) was marketed as a passive transdermal patch but was later withdrawn due to crystallization of the drug within the patch formulation. It is expected that the patch will return to market after reformulation, or other dopamine agonists will be developed. Similarly, 5-OH-DPAT, a potent dopamine agonist, has potential use to treat Parkinson's disease, but it has a very low oral bioavailability. Its transdermal iontophoretic delivery was investigated (121,183,184), and it was found that delivery increased linearly with current density and concentration, thus providing a potential advantage of dose titration (185). Another dopamine agonist, ropinirole, was also investigated for iontophoretic delivery. It undergoes extensive first-pass metabolism upon oral administration and is therefore a good transdermal candidate. Dose titration is required for its use in Parkinson's disease, and iontophoresis could therefore offer a promising approach. The hydrochloride salt of ropinirole has been delivered iontophoretically to the hairless rat model, and it has been suggested that therapeutic doses can be delivered (186).

4.11.5   MISCELLANEOUS

Zidovudine, an anti-AIDS drug, is poorly absorbed upon oral administration and has side effects related to excessive plasma concentration immediately after administration. Thus, transdermal delivery would be useful if therapeutic levels can be delivered. This was found feasible by iontophoretic delivery based on extrapolation of *in vitro* data with hairless mouse skin (187). Iontophoretic delivery of iron was investigated to avoid GI side effects and the inconvenience and discomfort of parenteral administration. Using ferric pyrophosphate (745 Da), an iron compound undergoing clinical investigations, flux data were extrapolated to predict that a 20-cm^2 iontophoretic patch will deliver about 2.5 mg iron in 12 hours, which is close to the dose required for iron deficiency (188). Transdermal delivery of capsaicin and its synthetic derivatives has been extensively investigated by passive (189–191) and iontophoretic (148,192,193) means by one research group. Capsaicin, the pungent principle extracted from red pepper, has antinociceptive, hypotensive, and hypolipidemia activities, but its skin toxicity and burning pain sensation limits its use. Synthetic analogues of capsaicin, nonivamide, and sodium nonivamide acetate have been made available to reduce or eliminate this skin irritation (190). Iontophoresis increased the transdermal penetration flux of sodium nonivamide acetate, and various factors affecting iontophoretic transport have been investigated (192,193). Another drug, methylphenidate, was also investigated for its iontophoretic delivery by using the hydrochloride salt. Based on the delivery of protonated drug across excised human cadaver skin, it seems that iontophoresis can deliver therapeutic amounts of this drug (194,195). In a different approach, solid lipid nanoparticles having triamcinolone acetonide acetate and formulated in a carbopol gel were reported to enhance the penetration of drug across porcine ear skin when delivered iontophoretically, though the authors have not discussed the mechanisms of such transport (196).

4.11.6 ALTERNATE APPLICATIONS FOR IONTOPHORESIS

It should be realized that iontophoresis is useful for more than just skin (topical dermatological or transdermal) applications. Though only skin applications are discussed in this chapter, these alternate applications are briefly introduced here. Ocular iontophoresis has been widely investigated for delivery of drugs such as gentamicin, cefazolin, fluorescein, tobramycin, lidocaine, epinephrine, timolol maleate, idoxuridine, dexamethasone, and others for several potential clinical applications (197–205). Transcorneal iontophoresis has been used for drug delivery to the anterior segment of the eye and transscleral iontophoresis for delivery to the posterior eye segment. Topical application of drugs does not lead to delivery to the back of the eye in posterior tissues such as choroid and retina. Therefore, iontophoresis offers promise in diseases that affect the back of the eye, such as age-related macular degeneration, diabetic retinopathy, posterior uveitis, and retinitis due to glaucoma (206,207). Transnasal iontophoresis has also been investigated for enhancing drug delivery to the brain (208).

Iontophoresis of fluoride has been used for tooth desensitization (209,210). Also, microiontophoresis is widely used in neuropharmacological and basic studies with neurons (211–214). Microiontophoresis involves the controlled ejection of drugs from micropipettes (tip diameter about 1 μm), thus allowing the study of effects and interactions of drugs on a very restricted area of tissue (215). Microiontophoresis can thus be used to administer neuroactive compounds such as neurotransmitters to single neurons to observe their effect on firing parameters. This allows the observation of effects on single neurons without affecting the whole nervous system, such as when the drug is given via systemic administration. It can also be used to deposit dyes or tracers into the cytoplasm of a cell or into the intercellular space for subsequent histological examination. The development and improvements in microiontophoretic electrodes (216) are beyond the scope of this book. Iontophoresis can also find applications for modulation of drug delivery from medical devices that may be implanted or introduced into the body. The regional nature of several cardiac diseases can benefit from localized drug delivery from an implant, which may be modulated by iontophoresis (217). Iontophoretic cardiac implants have been successfully used in dogs for delivery of the antiarrhythmic agent sotalol. This can allow for timely and localized delivery to the heart, avoiding problems of drug toxicity and bioavailability limitations (218–220). Suitable membranes to modulate iontophoretic release from such implantable systems are under development (221). Iontophoresis also has potential applications in wound healing, as electric fields have been reported to enhance skin wound repair (222,223).

4.12 CLINICAL APPLICATIONS OF IONTOPHORESIS FOR INTRADERMAL DELIVERY

Iontophoresis technology has been successfully used in clinical medicine to achieve intradermal delivery of drugs for several decades. Topical or intradermal delivery by iontophoresis is defined as local dermatological delivery in this chapter, including delivery to the epidermis or deeper layers of the dermis. The devices and electrodes used in clinical settings for dermatological delivery of drugs by iontophoresis will be discussed in Chapter 10. Iontophoresis has found widespread clinical dermatological applications (224,225). Most of the clinical applications of iontophoresis currently used in physical therapy involve the use of lidocaine and dexamethasone (226). However, several other drugs have been investigated for their iontophoretic delivery in clinical studies, and these will be briefly discussed. Because iontophoretic delivery takes place via an appendageal pathway, the technique may be useful for treatment of a number of follicular diseases, such as acne or androgenetic alopecia, associated with sebaceous gland activity (227,228). It should be realized that many of the applications discussed in this chapter are potential applications or those on which investigations have been carried out and reported in the published or patent literature. As also discussed in Section 7.4, absorption of lidocaine by the cutaneous microvasculature may be reduced by coiontophoresis of the vasoconstrictor norepinephrine (229). Another approach recently reported to reduce dermal clearance and allow more drug to penetrate deeper, is to reduce the regional blood perfusion (i.e., temperature-induced vasoconstriction). Blood flow to the skin is reduced when the temperature drops below 31°C. Using what the authors termed as "ChilDrive" iontophoresis, it was shown that the bioavailability of diclofenac sodium and prednisolone sodium phosphate was improved in the synovial fluid of the knee-joint region of the hind limb in rats when iontophoresis was applied to skin maintained at a temperature of about 20 to 25°C. Drug levels were monitored by intraarticular microdialysis and were 6- to 12-fold higher than passive delivery and two- to fourfold higher than iontophoresis in the absence of chilling (230). Theoretically, deeper penetration can also be achieved if the rate of delivery of the drug is higher than its dermal clearance, but this will be hard to control and may be dependent on the physicochemical properties of the drug molecule as well as on the intricacies of the enhancement technique being employed, if any. Iontophoretic wearable patches for systemic delivery are in development and are discussed in Chapter 10.

4.12.1   Delivery of Dexamethasone

Dexamethasone sodium phosphate converts partially to dexamethasone in skin, but then gets completely converted to dexamethasone in the bloodstream (231). Dexamethasone sodium phosphate, at its first pK_a of 1.89, is 50:50 neutral species:monoanion. At its second pK_a of 6.4, it is 50:50 monoanion:dianion. It has its maximum mobility around pH 7 (80). Dexamethasone, being an anionic drug, has been shown to be delivered better under the cathode, though delivery under the anode may be feasible under certain conditions (141) due to electroosmosis that represents the water flow that accompanies iontophoresis. Electroosmotic water movement is from the anode to cathode, and thus water moves into the skin at the anode. This becomes important because dexamethasone is often delivered along with lidocaine, and the latter may be better delivered under anode. In these cases, selection of clinical treatment protocol depends on the desired therapeutic effect by possibly switching polarity of the electrode during treatment. Switching of polarity may allow administration of drugs with opposite charge from the mixture. A major application of iontophoresis, typically utilized in physical therapy clinics, is in the treatment of acute musculoskeletal inflammation. Iontophoresis or phonophoresis of dexamethasone, often in combination with lidocaine, is commonly performed in physical therapy clinics to treat local inflammatory musculoskeletal conditions such as bursitis, tendinitis, arthritis, carpal tunnel syndrome, temporomandibular joint dysfunction, and so forth (232). Some early studies established that iontophoresis is an effective mode for delivering dexamethasone sodium phosphate to patients for treatment of inflamed tissues (136,140). Formulation and delivery considerations were discussed in the pharmaceutical literature (233–235). In a pilot study with two monkeys, it was established that dexamethasone sodium phosphate was transported from positive electrodes placed on several sites to the underlying tissue. After sacrificing the animal, a core of tissue was obtained from under each electrode, and detectable levels of dexamethasone were seen in all layers, as deep as tendinous structures and cartilaginous tissue (236). In a subsequent double-blind study on 53 patients with tendonitis at the shoulder joint, patients below 45 years responded to the iontophoretic administration of steroid. Patients over 45 years did not respond well to iontophoresis. However, this latter group also did not respond well to administration by local injection (140). In a study on 18 female subjects using the Phoresor device, a single treatment of dexamethasone iontophoresis following muscle soreness was found to slow the progression of muscle soreness (237). Iontophoretic delivery of dexamethasone has also found use in rheumatoid arthritis as an alternative to oral and injected delivery of corticosteroids because of its noninvasive, nontraumatic, painless, and site-specific delivery (238,239). Rheumatoid arthritis is a chronic progressive disease of the autoimmune system which is a major cause of work disability (240,241). Iontophoretic delivery of dexamethasone combined with a supervised exercise program was reported to be effective for treatment in a patient with rheumatoid arthritis (238). Iontophoresis of dexamethasone sodium phosphate may also offer an alternative to steroid injections for carpal tunnel syndrome, which has become an epidemic of the industrialized world. In one prospective, nonrandomized study, 23 hands of 18 patients with varying occupations were treated with wrist splinting with nonsteroidal anti-inflammatory medications and iontophoresis of dexamethasone using Phoresor II system and TransQ electrodes. A success rate comparable with splinting plus injection of dexamethasone into the carpal tunnel space was achieved. In a 6-month follow-up on patients who failed splints plus nonsteroidal anti-inflammatory medications alone, a success rate of 58% was found if iontophoresis was performed (137). Iontophoresis or injections of steroids are also used to reduce acute inflammation of the lateral epicondyle, the bony ridge on the outer portion of the elbow, for treatment of lateral epicondylitis (tennis elbow). Tennis elbow results from performing repetitive activities such as tennis, housework, cooking, physical labor, or computer operation, and iontophoresis can allow penetration of the drug to the lateral epicondyle (242). Iontophoresis of dexamethasone and lidocaine has also been found to be more efficient and effective in relieving pain and inflammation and to promote healing in patients with infrapatellar tendinitis as compared to established protocol of modalities and transverse friction massage (243). In contrast to the studies discussed thus far, some studies have suggested that iontophoretically administered dexamethasone was no more effective than saline placebo in alleviating the symptoms of patients with musculoskeletal dysfunction (244,245).

4.12.2 Delivery of Other Steroids

Enhanced penetration of corticosteroids such as hydrocortisone into skin may be useful for various topical or systemic diseases. For instance, topical applications can include treatment of inflammatory skin diseases (246,247). Chemical enhancers have been used to facilitate skin permeation of hydrocortisone (248). Iontophoretic delivery of hydrocortisone has also been investigated. The iontophoretic flux, even in the presence of surfactants, was low due to the relatively poor aqueous solubility of hydrocortisone and only limited capacity of surfactants to solubilize poorly soluble drugs (249). Salt forms such as hydrocortisone sodium succinate and hydrocortisone sodium phosphate will result in higher iontophoretic flux, but these salt forms tend to be unstable. The salt forms of hydrocortisone have ester linkages at C21 which may undergo hydrolysis during iontophoresis, especially when pH of donor and receptor solutions shifts, such as when platinum electrodes are used (250). Successful iontophoretic delivery of other steroids such as prednisolone through skin has also been reported, though a reservoir was observed to be formed in the skin. Iontophoresis of prednisolone through nail was also found to be feasible after soaking the nail in soapy water for 5 minutes to make it conductive (251). Iontophoretic delivery of steroids has also been investigated for treatment of Peyronie's disease (252). However, dexamethasone remains the drug of choice due to its much greater anti-inflammatory effect.

4.12.3 Treatment of Hyperhidrosis

Hyperhidrosis is a condition of excessive or abnormal sweating, far higher than what is required to maintain constant body temperature and not related to exercise or resulting from another underlying cause. It can occur in feet (plantar hyperhidrosis), armpits (axillary hyperhidrosis), or hands (palmar hyperhidrosis). Generally, palmar hyperhidrosis is most distressing, as hands are much more exposed in social and professional activities than other parts of the body. For hyperhidrosis without known cause (primary or essential hyperhidrosis), iontophoresis is often tried if antiperspirants do not lead to desired results (253). Use of iontophoresis for hyperhidrosis was documented as early as 1952 but studied more systematically in human subjects in the 1980s when it was reported to be effective for palmar hyperhidrosis (254,255). Today, tap water iontophoresis is considered to be the method of choice to initiate treatment for hyperhidrosis, as it has been shown to be very effective and has little to no side effects (256). The treatment consists of applying a direct current on skin of sufficient magnitude and duration apparently to obstruct sweat glands, though the exact mechanism is not known. Current is typically applied in 10- to 20-minute sessions, which need to be repeated two to three times per week, followed by a maintenance program of treatments at 1- to 4-week intervals, depending on the patient's response. Iontophoresis of just tap water has been reported to be effective for hyperhidrosis of palms, soles, and axillae (257–259). The addition of some chemicals to improve the effectiveness of tap water iontophoresis was also investigated (260,261). Injections of botulinum toxin type A (e.g., BOTOX, Allergan Inc., Irvine, California) have also been used to control hyperhidrosis.

Botulinum molecule is very large, but a case report was published where the molecule was delivered by iontophoresis (Figure 4.2) and was reported to be effective for severe palmar hyperhidrosis. Unlike tap water iontophoresis that needs multiple treatments, iontophoresis of the toxin can produce a sustained effect that lasts 3 months after only one treatment (262).

4.12.4 DIAGNOSIS OF CYSTIC FIBROSIS

Iontophoretic devices also have a use in the diagnosis of cystic fibrosis, and the historical origin of this use is discussed in Section 4.1. Iontophoretic devices are prescription devices approved for the diagnosis of cystic fibrosis by iontophoresis of pilocarpine. Iontophoresis of pilocarpine for about 15 minutes results in profuse sweating that continues for about 30 minutes following the treatment. Sweat is collected during this period and analyzed for chloride. The finding of a high sweat chloride value (above 60 mEq/L) on at least two occasions along with the presence of clinical features of cystic fibrosis is consistent with the diagnosis of the disease. In a typical protocol, electrodes are removed following iontophoresis of pilocarpine, the skin site is thoroughly cleaned, and a preweighed filter paper is placed on the skin area that was exposed to pilocarpine and sealed with a plastic wrap. Following the sweating period, the paper is removed, weighed immediately, and used for sweat analysis. No color change indicates that cystic fibrosis is not indicated, and a complete color change would result if cystic fibrosis is indicated. The system is based on a paper patch that collects sweat and shows color change when chloride levels exceed 40 to 50 mEq/l. In a study with 66 patients with cystic fibrosis, there were no false negatives (263). Another study examined the relation between the chloride concentration and the complexed chloride of the CF Indicator® using a digitizer and computer. It was found that the sweat chloride concentration can be calculated with reproducibility equal to that of the Gibson-Cooke sweat test (264). The use of this cystic fibrosis indicator was evaluated in a five-center study. This device can deliver pilocarpine iontophoretically and has a disposable chloride sensor patch that absorbs a specified volume of sweat. The device was found to be comparable to other tests and was suggested to be potentially useful for physicians offices, in clinics, and in similar settings (265). A different commercially available device can now be obtained from Wescor, Inc. (Logan, Utah). It consists of a Webster sweat inducer with pilogel[7] discs for iontophoresis of pilocarpine. By activating a start switch, an optimal quantity of pilocarpine is delivered for gland stimulation, equivalent to 5 minutes of iontophoresis at 1.5 mA. The pilogel[7] iontophoresis discs are hydrophilic hydrated gels containing pilocarpine (266). Sweat is collected on a Macroduct[7] collector and can be analyzed with a Sweat.Chek™ analyzer. The sweat flows between skin and the concave undersurface of the Macroduct collector, and a water-soluble blue dye on the collection surface allows visualization of the accumulated volume. The analyzer is based on conductivity measurements that have been suggested to be as reliable as analysis for chloride. It has been recommended that hospital departments that perform sweat tests and pediatricians should inform their patients about the small risk of minor burns during iontophoresis of pilocarpine. However, with the improvements in electrode design, chemical burns should be very infrequent

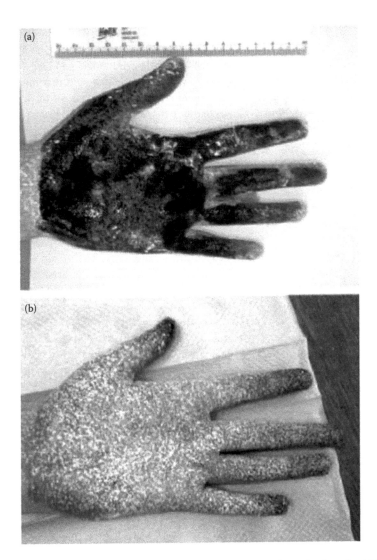

FIGURE 4.2 A starch–iodine test to show hyperhidrotic area in patient's palm (a) before and (b) 2 weeks post-BOTOX® iontophoresis. (Reproduced with permission from (G. M. Kavanagh, C. Oh, and K. Shams: BOTOX Delivery by Iontophoresis, *Br. J. Dermatol.*, 151:1093–1095, 2004 [262]. Copyright Wiley-VCH Verlag GmbH & Co. Reproduced with permission.) **(See color insert.)**

(267). Optimization of factors affecting iontophoretic delivery of pilocarpine has been attempted using response surface methodology (45).

4.12.5 IONTOPHORESIS OF HISTAMINE AND ANTIHISTAMINES

Skin wheals induced by intradermal injection of histamine cannot be completely inhibited by H_1 blockade. The prick technique is less traumatic and leads to the formation of a more specific H_1 dependent wheal. It was suggested that iontophoresis of histamine can be an even less traumatic and useful replacement for intradermal injections or prick tests to evaluate the activity of antihistamine drugs in skin by inhibiting the dermal reaction induced by histamine. Following iontophoresis (1.4 mA/cm^2 for 30 seconds) of histamine to humans, flares and wheals developed rapidly and remained for at least 3 hours. Thus, it seems that histamine iontophoresis can be a valuable tool to evaluate the activity of H_1 blocking agents (268). It has been reported that 4 hours after administration of the potent H_1 blocker cetirizine, there was a 100% inhibition of the wheal in 9 of 10 human volunteers. This suggests that histamine iontophoresis could be the best method for inducing a 100% H_1-dependent wheal (269). However, if an itch model with sustained itch half-life is desired, then the skin prick test was shown to produce a stronger and longer-lasting reproducible itch sensation as compared to iontophoretic application. During the brief iontophoresis period (10 seconds), histamine apparently passed the most superficial pruritoceptive C fibers too quickly to induce long-lasting itch sensations. When the prick test was used, some histamine deposited at the dermal–epidermal junction from where it was released throughout the time period when measurements were made (270). A pencil-shaped device for local iontophoretic delivery of the antihistamine, diphenhydramine hydrochloride was investigated in six subjects. The pencil-shaped aluminum cartridge served as the cathode and had an external diameter of 14 mm and was 92 mm long. The system also had an electronic display indicating when the system is in use. At its lower end, it had a plastic isolator with an agar gel–containing drug that served as the anode. The system was meant for local application of antihistamines to skin, such as for the short-term treatment of acute skin irritations caused by insect bites. Release of the drug from the system on short-term use (5 minutes) was demonstrated followed by a study in six subjects showing that a closed circuit is formed with an electric field in which drug transport can occur. The mean current flow was 86 μA, and the mean electrical resistance of the body was measured at 116 kΩ (271).

4.12.6 DELIVERY OF ANTIVIRAL AGENTS

Iontophoresis can potentially be used for iontophoretic delivery of antiviral agents. Transdermal delivery of acyclovir is discussed in Chapter 7 as a case study. The delivery of three antiviral agents, iododeoxyuridine, adenine arabinoside monophosphate, and phosphonoacetic acid, in mice was demonstrated several years ago (272). Another study showed that the antiviral drug, idoxuridine (5-Iodo-2′-deoxyuridine) can be delivered iontophoretically to mice. Because both anodal and cathodal iontophoresis were found to assist its delivery through skin, it was inferred that electroosmosis is an important factor in delivery (273). In a subsequent study on six subjects, iontophoresis of idoxuridine was conducted under the anode to 14 recurrent herpes labialis lesions. Immediate relief of discomfort and swelling and accelerated healing were reported, though the study was not blinded (274). Other studies showed that iontophoresis can be a promising technique to deliver antiviral agents for benefit in herpes treatment (275–278). Clinical research supports the iontophoretic application of antiviral agents in postherpetic neuralgia (279), and vinblastine has been delivered by iontophoresis for cutaneous Kaposi's sarcoma lesions (280).

4.12.7 Delivery of Antibiotic/Anti-Infective Agents

Iontophoresis can also be used to introduce a bactericidal agent into skin prior to taking the patient's blood, to reduce the effect of contaminants responsible for large number of false positives (281) or to kill microorganisms on a surface (282,283). Iontophoresis can also deliver antibiotics such as penicillin across the burn eschar into the underlying vascular tissues, with the drug levels being 200-fold higher than when no current is applied. Bactericidal concentrations can be achieved in areas considered to be major sites of the origin of infection (284). Similarly, iontophoresis of antibiotics has been used for treatment of burned ear, because systemic administration of antibiotics is not effective due to avascularity of the ear cartilage. In contrast, iontophoresis was able to deliver high doses of the antibiotics to the avascular infected ear cartilage (285,286). In another study, five rabbits with ear burns were treated with gentamicin iontophoresis in one ear and gentamicin-soaked gauze on the other. Analysis of the ear cartilage demonstrated a 20-fold increase in the levels of gentamicin in the iontophoresis-treated ear, in comparison to only low levels in the gauze-treated ears (287). In a retrospective study of patients admitted in one hospital over a 16-year period, it was found that the incidence of ear infection was virtually eliminated when managed by iontophoresis of antibiotics (288). Bacterial infections can occur easily in topical wounds, and iontophoresis could be useful to deliver antibiotics to prevent or treat such infections.

4.12.8 Other Applications of Iontophoresis

Tap water iontophoresis for treatment of hyperhidrosis was described earlier. Tap water iontophoresis using a hydrogalvanic bath is also used to achieve analgesia and hyperemia of the treated region (289). Several other isolated examples of potential uses of iontophoresis are discussed in this section. Iontophoresis has applications for the delivery of narcotic analgesics. Iontophoretic delivery of morphine for postoperative analgesia in patients who underwent total knee or hip replacement resulted in reduced need for the patient to take intravenous meperidine via the patient-controlled analgesic device. Also, levels of morphine observed in the plasma were high enough to provide early postoperative pain relief (290). A formulation of morphine citrate salts suitable for iontophoresis has been developed (291). Another narcotic analgesic, fentanyl (see Section 7.4), has been successfully delivered to humans in clinically significant doses following iontophoresis for 2 hours (292). Because hair follicles are a major pathway for conventional DC iontophoretic drug delivery, the method may be ideally suited for delivery of minoxidil to hair follicles to stimulate hair growth.

Minoxidil is a basic compound with a pK_a of 2.62 and most likely ionizes at the pyridine oxide. At pH 5, it exists mostly as the monocation (80). However, it has been said that since minoxidil has a low solubility in water with no net ionic charge, its cationic derivatives have been synthesized for iontophoretic delivery. Each of the cationic derivatives was synthesized by reacting minoxidil parent compound with an organic or an inorganic acid to form the cationic derivative (293). Similarly, iontophoresis may have potential use for treatment of acne, boils, and other skin disorders characterized by closed, blocked appendages in the epidermis of skin. Iontophoresis under the negative polarity apparently drives the ions generated by electrolysis into the skin to disrupt the blockage and to establish drainage from the appendages (294). Other isolated studies on clinical applications of iontophoresis include lithium for gouty arthritis (295), zinc for ischemic skin ulcers (296), acetic acid for calcium deposits (297), and traumatic myositis ossificans (298), cisplatin for basal and squamous cell carcinoma of skin (299), vinca alkaloids for chronic pain syndrome (300), and silver for chronic osteomyelitis (301). Iontophoresis has also found application in allergy testing. Patients allergic to grasses were tested iontophoretically with extracts of different grasses. The results demonstrated that the biologically active constituents of orchard, red top, sweet vernal, and June grasses can be transported readily into the skin of persons allergic to timothy grass. The skin reactions produced by iontophoresis, in general, are found to parallel the dermal responses observed in the usual skin tests (scratch and intradermal). Iontophoresis of chromium in guinea pigs has also been investigated in order to test the role of the technique in allergy testing for chromium (302). Iontophoretic application of potassium iodide to human knees has also been investigated. It was found that the iodide is taken up only when electric current is applied. About 10% of the iodide applied was noted to penetrate the skin, while X-ray fluorescence scan of the volunteer's thyroid gland showed that the average iodine content in the gland is increased by more than 30% (303). Iodine iontophoresis has also been used to reduce scar tissue in a patient (304).

REFERENCES

1. Y. W. Chien and A. K. Banga. Iontophoretic (transdermal) delivery of drugs: Overview of historical development, *J. Pharm. Sci.*, 78:353–354 (1989).
2. L. E. Gibson and R. E. Cooke. A test for the concentration of electrolytes in sweat in cystic fibrosis of the pancreas utilizing pilocarpine by iontophoresis, *Pediatrics*, 23:545–549 (1959).
3. L. E. Gibson. Iontophoretic sweat test for cystic fibrosis: Technical details, *Pediatrics*, 39:465 (1967).
4. T. Yamamoto and Y. Yamamoto. Electrical properties of the epidermal stratum corneum, *Med. Biol. Eng.*, 14:151–158 (1976).
5. L. Emtestam and S. Ollmar. Electrical impedance index in human skin: Measurements after occlusion, in 5 anatomical regions and in mild irritant contact dermatitis, *Contact Dermatitis*, 28:104–108 (1993).
6. F. Pliquett and U. Pliquett. Passive electrical properties of human stratum corneum *in vitro* depending on time after separation, *Biophys. Chem.*, 58:205–210 (1996).
7. R. Plutchik and H. R. Hirsch. Skin impedance and phase angle as a function of frequency and current, *Science*, 141:927–928 (1963).

8. A. V. Boxtel. Skin resistance during square-wave electrical pulses of 1 to 10 mA, *Med. Biol. Eng. Comput.*, 15:679–687 (1977).
9. J. H. Calderwood. Electrode-skin impedance from a dielectric viewpoint, *Physiol. Meas.*, 17:A131–A139 (1996).
10. Y. N. Kalia and R. H. Guy. The electrical characteristics of human skin *in vivo*, *Pharm. Res.*, 12:1605–1613 (1995).
11. R. R. Burnette and T. M. Bagniefski. Influence of constant current iontophoresis on the impedance and passive Na^+ permeability of excised nude mouse skin, *J. Pharm. Sci.*, 77:492–497 (1988).
12. R. R. Burnette and B. Ongpipattanakul. Characterization of the pore transport properties and tissue alteration of excised human skin during iontophoresis, *J. Pharm. Sci.*, 77:132–137 (1988).
13. Stephens, W. G. S. The current-voltage relationship in human skin. *Med. Electron. Biol. Engg.* 1, 389–399 (1963).
14. G. B. Kasting and L. A. Bowman. DC electrical properties of frozen, excised human skin, *Pharm. Res.*, 7:134–143 (1990).
15. G. B. Kasting and L. A. Bowman. Electrical analysis of fresh, excised human skin: A comparison with frozen skin, *Pharm. Res.*, 7:1141–1146 (1990).
16. H. A. Abramson and M. H. Gorin. Skin reactions. IX. The electrophoretic demonstration of the patent pores of the living human skin; its relation to the charge of the skin, *J. Phys. Chem.*, 44:1094–1102 (1940).
17. C. Cullander. What are the pathways of iontophoretic current flow through mammalian skin? *Adv. Drug Del. Rev.*, 9:119–135 (1992).
18. K. C. Melikov and I. A. Ershler. Localization of conductive pathways in human stratum corneum by scanning electrochemical microscopy, *Membr. Cell Biol.*, 10:459–466 (1996).
19. P. I. Kuzmin, A. S. Darmostuk, Y. A. Chizmadzhev, H. S. White, and R. O. Potts. A mechanism of skin appendage macropores electroactivation during iontophoresis, *Membr. Cell Biol.*, 10:699–706 (1996).
20. W. H. M. C. Hinsberg, J. C. Verhoef, L. J. Bax, H. E. Junginger, and H. E. Bodde. Role of appendages in skin resistance and iontophoretic peptide flux: Human versus snake skin, *Pharm. Res.*, 12:1506–1512 (1995).
21. S. Grimnes. Pathways of ionic flow through human skin *in vivo*, *Acta. Derm. Venereol. (Stockh.)*, 64:93–98 (1984).
22. P. Singh, M. Anliker, G. A. Smith, D. Zavortink, and H. I. Maibach. Transdermal iontophoresis and solute penetration across excised human skin, *J. Pharm. Sci.*, 84:1342–1346 (1995).
23. N. A. Monteiro-Riviere, A. O. Inman, and J. E. Riviere. Identification of the pathway of iontophoretic drug delivery: Light and ultrastructural studies using mercuric chloride in pigs, *Pharm. Res.*, 11(2):251–256 (1994).
24. D. F. Hager, F. A. Mancuso, J. P. Nazareno, J. W. Sharkey, and J. R. Siverly. Evaluation of a cultured skin equivalent as a model membrane for iontophoretic transport, *J. Control. Release*, 30:117–123 (1994).
25. A. C. Williams. *Transdermal and topical drug delivery*, Pharmaceutical Press, London, 2003.
26. M. R. Prausnitz, J. A. Gimm, R. H. Guy, R. Langer, J. C. Weaver, and C. Cullander. Imaging regions of transport across human stratum corneum during high-voltage and low-voltage exposures, *J. Pharm. Sci.*, 85:1363–1370 (1996).
27. L. A. R. M. Pechtold, W. Abraham, and R. O. Potts. The influence of an electric field on ion and water accessibility to stratum corneum lipid lamellae, *Pharm. Res.*, 13:1168–1173 (1996).

28. E. R. Scott, A. I. Laplaza, H. S. White, and J. B. Phipps. Transport of ionic species in skin: Contribution of pores to the overall skin conductance, *Pharm. Res.*, 10:1699–1709 (1993).
29. E. R. Scott, J. B. Phipps, and H. S. White. Direct imaging of molecular transport through skin, *J. Invest. Dermatol.*, 104:142–145 (1995).
30. A. Jadoul, C. Hanchard, S. Thysman, and V. Preat. Quantification and localization of fentanyl and TRH delivered by iontophoresis in the skin, *Int. J. Pharm.*, 120:221–228 (1995).
31. A. Finkelstein and A. Mauro. Physical principles and formalisms of electrical excitability. In E. R. Kandel (ed), *Handbook of physiology*, American Physiological Society, Bethesda, MD, 1977, pp. 161–213.
32. K. Kontturi and L. Murtomaki. Mechanistic model for transdermal transport including iontophoresis, *J. Control. Release*, 41:177–185 (1996).
33. G. B. Kasting and J. C. Keister. Application of electrodiffusion theory for a homogeneous membrane to iontophoretic transport through skin, *J. Control. Release*, 8:195–210 (1989).
34. J. Garrido, S. Mafe, and J. Pellicer. Generalization of a finite difference numerical method for the steady state and transient solutions of the Nernst–Planck flux equations, *J. Memb. Sci.*, 24:7–14 (1985).
35. J. C. Keister and G. B. Kasting. Ionic mass transport through a homogeneous membrane in the presence of a uniform electric field, *J. Memb. Sci.*, 29:155–167 (1986).
36. G. B. Kasting. Theoretical models for iontophoretic delivery, *Adv. Drug Del. Rev.*, 9:177–199 (1992).
37. A. K. Nugroho, O. Della-Pasqua, M. Danhof, and J. A. Bouwstra. Compartmental modeling of transdermal iontophoretic transport II: *In vivo* model derivation and application, *Pharm Res.*, 22:335–346 (2005).
38. J. Swarbrick, G. Lee, J. Brom, and P. Gensmantel. Drug permeation through human skin II: Permeability of ionizable compounds, *J. Pharm. Sci.*, 73:1352–1355 (1984).
39. G. B. Kasting, E. W. Merritt, and J. C. Keister. An *in vitro* method for studying the iontophoretic enhancement of drug transport through skin, *J. Memb. Sci.*, 35:137–159 (1988).
40. W. I. Higuchi, S. K. Li, A. H. Ghanem, H. Zhu, and Y. Song. Mechanistic aspects of iontophoresis in human epidermal membrane, *J. Control. Release*, 62:13–23 (1999).
41. B. Mudry, R. H. Guy, and M. Begona Delgado-Charro. Prediction of iontophoretic transport across the skin, *J. Control. Release*, 111:362–367 (2006).
42. A. Sieg and V. Wascotte. Diagnostic and therapeutic applications of iontophoresis, *J. Drug Target.*, 17:690–700 (2009).
43. P. Batheja, R. Thakur, and B. Michniak. Transdermal iontophoresis, *Expert. Opin. Drug Deliv.*, 3:127–138 (2006).
44. P. Singh and H. I. Maibach. Iontophoresis: An alternative to the use of carriers in cutaneous drug delivery, *Advan. Drug Delivery Rev.*, 18:379–394 (1996).
45. Y. Y. Huang, S. M. Wu, C. Y. Wang, and T. S. Jiang. A strategy to optimize the operation conditions in iontophoretic transdermal delivery of pilocarpine, *Drug Develop. Ind. Pharm.*, 21:1631–1648 (1995).
46. Y. Y. Huang, S. M. Wu, and C. Y. Wang. Response surface method: A novel strategy to optimize iontophoretic transdermal delivery of thyrotropin-releasing hormone, *Pharm. Res.*, 13:547–552 (1996).
47. A. C. Hirsch, R. S. Upasani, and A. K. Banga. Factorial design approach to evaluate interactions between electrically assisted enhancement and skin stripping for delivery of tacrine, *J. Control. Release*, 103:113–121 (2005).
48. R. S. Upasani and A. K. Banga. Response surface methodology to investigate the iontophoretic delivery of tacrine hydrochloride, *Pharm. Res.*, 21:2293–2299 (2004).
49. S. Thysman, V. Preat, and M. Roland. Factors affecting iontophoretic mobility of metoprolol, *J. Pharm. Sci.*, 81:670–675 (1992).

50. D. L. Chu, H. J. Chiou, and D. P. Wang. Characterization of transdermal delivery of nefopam hydrochloride under iontophoresis, *Drug Dev. Ind. Pharm.*, 20:2775–2785 (1994).

51. J. W. Harding and D. Felix. Quantification of angiotensin iontophoresis, *J. Neurosc. Methods*, 19:209–215 (1987).

52. M. Clemessy, G. Couarraze, B. Bevan, and F. Puisieux. Mechanisms involved in iontophoretic transport of angiotensin, *Pharm. Res.*, 12:998–1002 (1995).

53. J. B. Phipps, R. V. Padmanabhan, and G. A. Lattin. Iontophoretic delivery of model inorganic and drug ions, *J. Pharm. Sci.*, 78:365–369 (1989).

54. L. L. Miller and G. A. Smith. Iontophoretic transport of acetate and carboxylate ions through hairless mouse skin. A cation exchange membrane model, *Int. J. Pharm.*, 49:15–22 (1989).

55. P. Lelawongs, J. Liu, and Y. W. Chien. Transdermal iontophoretic delivery of arginine-vasopressin (II): Evaluation of electrical and operational factors, *Int. J. Pharm.*, 61:179–188 (1990).

56. B. H. Sage, C. R. Bock, J. D. Denuzzio, and R. A. Hoke. Technological and developmental issues of iontophoretic transport of peptide and protein drugs. In V. H. L. Lee, M. Hashida, and Y. Mizushima (eds), *Trends and future perspectives in peptide and protein drug delivery*, Harwood Academic, Chur, Switzerland, 1995, pp. 111–134.

57. Y. N. Kalia, A. Naik, J. Garrison, and R. H. Guy. Iontophoretic drug delivery, *Adv. Drug Deliv. Rev.*, 56:619–658 (2004).

58. J. E. Sanderson, S. D. Riel, and R. Dixon. Iontophoretic delivery of nonpeptide drugs: Formulation optimization for maximum skin permeability, *J. Pharm. Sci.*, 78:361–364 (1989).

59. N. H. Yoshida and M. S. Roberts. Structure-transport relationships in transdermal iontophoresis, *Adv. Drug Del. Rev.*, 9:239–264 (1992).

60. J. L. Harden and J. L. Viovy. Numerical studies of pulsed iontophoresis through model membranes, *J. Control. Release*, 38:129–139 (1996).

61. M. J. Pikal and S. Shah. Study of the mechanisms of flux enhancement through hairless mouse skin by pulsed DC iontophoresis, *Pharm. Res.*, 8:365–369 (1991).

62. M. Clemessy, G. Couarraze, B. Bevan, and F. Puisieux. Preservation of skin permeability during *in vitro* iontophoretic experiments, *Int. J. Pharm.*, 101:219–226 (1994).

63. S. Thysman, C. Tasset, and V. Preat. Transdermal iontophoresis of fentanyl—Delivery and mechanistic analysis, *Int. J. Pharm.*, 101:105–113 (1994).

64. V. Preat and S. Thysman. Transdermal iontophoretic delivery of Sufentanil, *Int. J. Pharm.*, 96:189–196 (1993).

65. Y. Y. Huang and S. M. Wu. Transdermal iontophoretic delivery of thyrotropin-releasing hormone across excised rabbit pinna skin, *Drug Develop. Ind. Pharm.*, 22:1075–1081 (1996).

66. C. A. Zakzewski, J. K. J. Li, D. W. Amory, J. C. Jensen, and E. Kalatzis-Manolakis. Design and implementation of a constant-current pulsed iontophoretic stimulation device, *Med. Biol. Eng. Comput.*, 34:484–488 (1996).

67. T. Bagniefski and R. R. Burnette. A comparison of pulsed and continuous current iontophoresis, *J. Control. Release*, 11:113–122 (1990).

68. J. Hirvonen, F. Hueber, and R. H. Guy. Current profile regulates iontophoretic delivery of amino acids across the skin, *J. Control. Release*, 37:239–249 (1995).

69. S. K. Li, W. I. Higuchi, H. Zhu, S. E. Kern, D. J. Miller, and M. S. Hastings. *In vitro* and *in vivo* comparisons of constant resistance AC iontophoresis and DC iontophoresis, *J. Control. Release*, 91:327–343 (2003).

70. G. Yan, S. K. Li, and W. I. Higuchi. Evaluation of constant current alternating current iontophoresis for transdermal drug delivery, *J. Control. Release*, 110:141–150 (2005).

71. S. Pacini, T. Punzi, M. Gulisano, F. Cecchi, S. Vannucchi, and M. Ruggiero. Transdermal delivery of heparin using pulsed current iontophoresis, *Pharm Res.*, 23:114–120 (2006).

72. G. Yan, Q. Xu, Y. G. Anissimov, J. Hao, W. I. Higuchi, and S. K. Li. Alternating current (AC) iontophoretic transport across human epidermal membrane: Effects of AC frequency and amplitude, *Pharm Res.*, 25:616–624 (2008).

73. L. P. Gangarosa, N. H. Park, B. C. Fong, D. F. Scott, and J. M. Hill. Conductivity of drugs used for iontophoresis, *J. Pharm. Sci.*, 67:1439–1443 (1978).

74. G. A. Lattin, R. V. Padmanabhan, and J. B. Phipps. Electronic control of iontophoretic drug delivery, *Ann. N. Y. Acad. Sci.*, 618:450–464 (1991).

75. N. H. Yoshida and M. S. Roberts. Role of conductivity in iontophoresis. 2. Anodal iontophoretic transport of phenylethylamine and sodium across excised human skin, *J. Pharm. Sci.*, 83:344–350 (1994).

76. N. H. Yoshida and M. S. Roberts. Prediction of cathodal iontophoretic transport of various anions across excised skin from different vehicles using conductivity measurements, *J. Pharm. Pharmacol.*, 47:883–890 (1995).

77. Y. Henchoz, N. Abla, J. L. Veuthey, and P. A. Carrupt. A fast screening strategy for characterizing peptide delivery by transdermal iontophoresis, *J. Control. Release*, 137:123–129 (2009).

78. L. L. Miller, G. A. Smith, A. Chang, and Q. Zhou. Electrochemically controlled release, *J. Control. Release*, 6:293–296 (1987).

79. N. H. Yoshida and M. S. Roberts. Solute molecular size and transdermal iontophoresis across excised human skin, *J. Control. Rel.*, 25:177–195 (1993).

80. S. S. Kamath and L. P. Gangarosa. Electrophoretic evaluation of the mobility of drugs suitable for iontophoresis, *Meth. Find. Exp. Clin. Pharmacol.*, 17:227–232 (1995).

81. S. D. Terzo, C. R. Behl, and R. A. Nash. Iontophoretic transport of a homologous series of ionized and nonionized model compounds: Influence of hydrophobicity and mechanistic interpretation, *Pharm. Res.*, 6:85–90 (1989).

82. R. R. Burnette and B. Ongpipattanakul. Characterization of the permselective properties of excised human skin during iontophoresis, *J. Pharm. Sci.*, 76:765–773 (1987).

83. A. K. Banga, S. Kulkarni, and R. Mitra. Selection of electrode material and polarity in the design of iontophoresis experiments, *Int. J. Pharm. Adv.*, 1:206–215 (1995).

84. A. Jadoul, J. Mesens, W. Caers, F. Beukelaar, R. Crabbe, and V. Preat. Transdermal permeation of alniditan by iontophoresis: *In vitro* optimization and human pharmacokinetic data, *Pharm. Res.*, 13:1348–1353 (1996).

85. J. B. Phipps and J. R. Gyory. Transdermal ion migration, *Adv. Drug Del. Rev.*, 9:137–176 (1992).

86. C. Cullander and R. H. Guy. (D) Routes of delivery: Case studies (6): Transdermal delivery of peptides and proteins, *Adv. Drug Del. Rev.*, 8:291–329 (1992).

87. A. Jadoul, V. Regnier, and V. Preat. Influence of ethanol and propylene glycol addition on the transdermal delivery by iontophoresis and electroporation, *Pharm. Res.*, 14:S-308–S-309 (1997).

88. S. E. Cross and M. S. Roberts. Importance of dermal blood supply and epidermis on the transdermal iontophoretic delivery of monovalent cations, *J. Pharm. Sci.*, 84:584–592 (1995).

89. B. H. Sage. Iontophoresis. In J. Swarbrick and J. C. Boylan (eds), *Encyclopedia of pharmaceutical technology*, Marcel Dekker, New York, 1993, pp. 217–247.

90. A. D'Emanuele and J. N. Staniforth. An electrically modulated drug delivery device. III. Factors affecting drug stability during electrophoresis, *Pharm. Res.*, 9:312–315 (1992).

91. L. E. Linblad and L. Ekenvall. Electrode material in iontophoresis, *Pharm. Res.*, 4:438 (1987).

92. L. L. Miller, C. J. Kolaskie, G. A. Smith, and J. Rivier. Transdermal iontophoresis of gonadotropin releasing hormone (LHRH) and two analogues, *J. Pharm. Sci.*, 79:490–493 (1990).

93. S. B. Ruddy and B. W. Hadzija. The role of stratum corneum in electrically facilitated transdermal drug delivery. 1. Influence of hydration, tape-stripping and delipidization on the DC electrical properties of skin, *J. Control. Release*, 37:225–238 (1995).

94. Y. N. Kalia and R. H. Guy. Interaction between penetration enhancers and iontophoresis: Effect on human skin impedance *in vivo*, *J. Control. Release*, 44:33–42 (1997).

95. P. G. Green, R. S. Hinz, C. Cullander, G. Yamane, and R. H. Guy. Iontophoretic delivery of amino acids and amino acid derivatives across the skin *in vitro*, *Pharm. Res.*, 8:1113–1120 (1991).

96. P. Singh and H. I. Maibach. Iontophoresis in drug delivery: Basic principles and applications, *Crit. Rev. Ther. Drug Carr. Syst.*, 11:161–213 (1994).

97. M. J. Pikal and S. Shah. Transport mechanisms in iontophoresis. III. An experimental study of the contributions of electroosmotic flow and permeability change in transport of low and high molecular weight solutes, *Pharm. Res.*, 7:222–229 (1990).

98. P. Lelawongs, J. Liu, O. Siddiqui, and Y. W. Chien. Transdermal iontophoretic delivery of arginine-vasopressin (I): Physicochemical considerations, *Int. J. Pharm.*, 56:13–22 (1989).

99. R. M. Brand, G. Duensing, and F. G. Hamel. Iontophoretic delivery of an insulin-mimetic peroxovanadium compound, *Int. J. Pharm.*, 146:115–122 (1997).

100. J. E. Sanderson and S. R. Deriel. Methods and apparatus for iontophoretic drug delivery. 828,794 [U.S. Pat. 4,722,726]. 1988. Florida, USA.

101. T. J. Petelenz, S. C. Jacobsen, R. L. Stephen, and J. Janata. Methods and apparatus for iontophoresis applications of medicaments at a controlled pH through ion exchange. 64,769 [U.S. Pat. 4,915,685]. 1990. Utah, USA.

102. M. T. V. Johnson and N. H. Lee. pH buffered electrodes for medical iontophoresis. Assignee: Empi. 402,880 [U.S. Pat. 4,973,303]. 1990. Minnesota, USA.

103. M. B. Delgadocharro and R. H. Guy. Iontophoretic delivery of nafarelin across the skin, *Int. J. Pharm.*, 117:165–172 (1995).

104. R. M. Brand and P. L. Iversen. Iontophoretic delivery of a telomeric oligonucleotide, *Pharm. Res.*, 13:851–854 (1996).

105. M. J. Pikal. The role of electroosmotic flow in transdermal iontophoresis, *Adv. Drug Del. Rev.*, 9:201–237 (1992).

106. M. J. Pikal and S. Shah. Transport mechanisms in iontophoresis. II. Electroosmotic flow and transference number measurements for hairless mouse skin, *Pharm. Res.*, 7:213–221 (1990).

107. J. D. Denuzzio and B. Berner. Electrochemical and iontophoretic studies of human skin, *J. Control. Release*, 11:105–112 (1990).

108. B. D. Bath, E. R. Scott, J. B. Phipps, and H. S. White. Scanning electrochemical microscopy of iontophoretic transport in hairless mouse skin. Analysis of the relative contributions of diffusion, migration, and electroosmosis to transport in hair follicles, *J. Pharm. Sci.*, 89:1537–1549 (2000).

109. O. D. Uitto and H. S. White. Electroosmotic pore transport in human skin, *Pharm. Res.*, 20:646–652 (2003).

110. M. J. Pikal. Transport mechanisms in iontophoresis. I. A theoretical model for the effect of electroosmotic flow on flux enhancement in transdermal iontophoresis, *Pharm. Res.*, 7:118–126 (1990).

111. E. Manabe, S. Numajiri, K. Sugibayashi, and Y. Morimoto. Analysis of skin permeation-enhancing mechanism of iontophoresis using hydrodynamic pore theory, *J. Control. Release*, 66:149–158 (2000).

112. L. P. Gangarosa, N. Park, C. A. Wiggins, and J. M. Hill. Increased penetration of non-electrolytes into mouse skin during iontophoretic water transport (iontohydrokinesis), *J. Pharmacol. Exp. Ther.*, 212:377–381 (1980).
113. A. Kim, P. G. Green, G. Rao, and R. H. Guy. Convective solvent flow across the skin during iontophoresis, *Pharm. Res.*, 10:1315–1320 (1993).
114. R. Y. Lin, Y. C. Ou, and W. Y. Chen. The role of electroosmotic flow on *in vitro* transdermal iontophoresis, *J. Control. Release*, 43:23–33 (1997).
115. A. J. Hoogstraate, V. Srinivasan, S. M. Sims, and W. I. Higuchi. Iontophoretic enhancement of peptides: Behaviour of leuprolide versus model permeants, *J. Control. Release*, 31:41–47 (1994).
116. J. Hirvonen and R. H. Guy. Attenuation of electroosmotic flow during transdermal iontophoresis. I. Effect of poly-L-lysines. *Proc. Int. Symp. Control. Rel. Bioact. Mater.*, 24:689–690 (1997).
117. J. Hirvonen and R. H. Guy. Attenuation of electroosmotic flow during transdermal iontophoresis. II. Effect of beta-blocking agents. *Proc. Int. Symp. Control. Rel. Bioact. Mater.*, 24:691–692 (1997).
118. J. Hirvonen and R. H. Guy. Iontophoretic delivery across the skin: Electroosmosis and its modulation by drug substances, *Pharm. Res.*, 14:1258–1263 (1997).
119. S. F. Taveira, A. Nomizo, and R. F. Lopez. Effect of the iontophoresis of a chitosan gel on doxorubicin skin penetration and cytotoxicity, *J. Control. Release*, 134:35–40 (2009).
120. K. D. Peck, V. Srinivasan, S. K. Li, W. I. Higuchi, and A. H. Ghanem. Quantitative description of the effect of molecular size upon electroosmotic flux enhancement during iontophoresis for a synthetic membrane and human epidermal membrane, *J. Pharm. Sci.*, 85:781–788 (1996).
121. O. W. Ackaert, J. Van Smeden, J. De Graan, D. Dijkstra, M. Danhof, and J. A. Bouwstra. Mechanistic studies of the transdermal iontophoretic delivery of 5-OH-DPAT *in vitro*, *J. Pharm. Sci.*, 99:275–285 (2010).
122. P. Glikfeld, R. S. Hinz, and R. H. Guy. Noninvasive sampling of biological fluids by iontophoresis, *Pharm. Res.*, 6:988–990 (1989).
123. P. Glikfeld, C. Cullander, R. S. Hinz, and R. H. Guy. Device for iontophoretic non-invasive sampling or delivery of substances. Assignee: University of California. 879,060 [U.S. Pat. 5,279,543]. 1994. California, USA.
124. V. Merino, Y. N. Kalia, and R. H. Guy. Transdermal therapy and diagnosis by iontophoresis, *Trends. Biotech.*, 15:288–290 (1997).
125. G. Rao, P. Glikfeld, and R. H. Guy. Reverse iontophoresis: Development of a noninvasive approach for glucose monitoring, *Pharm. Res.*, 10:1751–1755 (1993).
126. G. Rao, R. H. Guy, P. Glikfeld, W. R. Lacourse, L. Leung, J. Tamada, R. O. Potts, and N. Azimi. Reverse iontophoresis: Noninvasive glucose monitoring *in vivo* in humans, *Pharm. Res.*, 12:1869–1873 (1995).
127. R. H. Guy. A sweeter life for diabetics? *Nature Med.*, 1:1132–1133 (1995).
128. S. Numajiri, K. Sugibayashi, and Y. Morimoto. Non-invasive sampling of lactic acid ions by iontophoresis using chloride ion in the body as an internal standard, *J. Pharm. Biomed. Anal.*, 11:903–909 (1993).
129. M. B. Delgado-Charro and R. H. Guy. Transdermal reverse iontophoresis of valproate: A noninvasive method for therapeutic drug monitoring, *Pharm. Res.*, 20:1508–1513 (2003).
130. A. Sieg, R. H. Guy, and M. B. Delgado-Charro. Simultaneous extraction of urea and glucose by reverse iontophoresis *in vivo*, *Pharm. Res.*, 21:1805–1810 (2004).
131. V. Wascotte, M. B. Delgado-Charro, E. Rozet, P. Wallemacq, P. Hubert, R. H. Guy, and V. Preat. Monitoring of urea and potassium by reverse iontophoresis *in vitro*, *Pharm. Res.*, 24:1131–1137 (2007).

132. V. Wascotte, P. Caspers, J. de Sterke, M. Jadoul, R. H. Guy, and V. Preat. Assessment of the "skin reservoir" of urea by confocal Raman microspectroscopy and reverse iontophoresis *in vivo*, *Pharm. Res.*, 24:1897–1901 (2007).
133. B. Leboulanger, M. Fathi, R. H. Guy, and M. B. Delgado-Charro. Reverse iontophoresis as a noninvasive tool for lithium monitoring and pharmacokinetic profiling, *Pharm. Res.*, 21:1214–1222 (2004).
134. P. Santi and R. H. Guy. Reverse iontophoresis—Parameters determining electroosmotic flow. I. pH and ionic strength, *J. Control. Release*, 38:159–165 (1996).
135. P. Santi and R. H. Guy. Reverse iontophoresis—Parameters determining electro-osmotic flow. II. Electrode chamber formulation, *J. Control. Release*, 42:29–36 (1996).
136. P. R. Harris. Iontophoresis: Clinical research in musculoskeletal inflammatory conditions, *J. Orthop. Sports Phys. Ther.*, 4:109–112 (1982).
137. C. A. Banta. A prospective, nonrandomized study of iontophoresis, wrist splinting, and antiinflammatory medication in the treatment of early-mild carpal tunnel syndrome, *J. Occup. Med.*, 36:166–168 (1994).
138. S. W. Stralka, P. L. Head, and K. Mohr. The clinical use of iontophoresis, *Phys. Ther. Prod.*, March:48–51 (1996).
139. P. C. Panus, J. Campbell, S. B. Kulkarni, R. T. Herrick, W. R. Ravis, and A. K. Banga. Transdermal iontophoretic delivery of ketoprofen through human cadaver skin and in humans, *J. Control. Release*, 44:113–121 (1997).
140. L. E. Bertolucci. Introduction of antiinflammatory drugs by iontophoresis: Double blind study, *J. Orthop. Sports Phys. Ther.*, 4:103–108 (1982).
141. T. J. Petelenz, J. A. Buttke, C. Bonds, L. B. Lloyd, J. E. Beck, R. L. Stephen, S. C. Jacobsen, and P. Rodriguez. Iontophoresis of dexamethasone: Laboratory studies, *J. Control. Release*, 20:55–66 (1992).
142. Y. Wang, R. Thakur, Q. Fan, and B. Michniak. Transdermal iontophoresis: Combination strategies to improve transdermal iontophoretic drug delivery, *Eur. J. Pharm. Biopharm.*, 60:179–191 (2005).
143. S. Ganga, P. Ramarao, and J. Singh. Effect of Azone on the iontophoretic transdermal delivery of metoprolol tartrate through human epidermis *in vitro*, *J. Control. Release*, 42:57–64 (1996).
144. Y. Wang, Q. Fan, Y. Song, and B. Michniak. Effects of fatty acids and iontophoresis on the delivery of midodrine hydrochloride and the structure of human skin, *Pharm Res.*, 20:1612–1618 (2003).
145. M. Al Khalili, V. M. Meidan, and B. B. Michniak. Iontophoretic transdermal delivery of buspirone hydrochloride in hairless mouse skin, *AAPS. PharmSci.*, 5:E14 (2003).
146. F. Bounoure, S. M. Lahiani, M. Besnard, P. Arnaud, E. Mallet, and M. Skiba. Effect of iontophoresis and penetration enhancers on transdermal absorption of metopimazine, *J. Dermatol. Sci.*, 52:170–177 (2008).
147. K. Kigasawa, K. Kajimoto, M. Watanabe, K. Kanamura, A. Saito, and K. Kogure. *In vivo* transdermal delivery of diclofenac by ion-exchange iontophoresis with geraniol, *Biol. Pharm. Bull.*, 32:684–687 (2009).
148. J. Y. Fang, C. L. Fang, Y. B. Huang, and Y. H. Tsai. Transdermal iontophoresis of sodium nonivamide acetate. III. Combined effect of pretreatment by penetration enhancers, *Int. J. Pharm.*, 149:183–193 (1997).
149. J. Hirvonen, K. Kontturi, L. Murtomaki, P. Paronen, and A. Urtti. Transdermal iontophoresis of sotalol and salicylate—The effect of skin charge and penetration enhancers, *J. Control. Release*, 26:109–117 (1993).
150. D. F. Hager, M. J. Laubach, J. W. Sharkey, and J. R. Siverly. *In vitro* iontophoretic delivery of CQA 206-291—Influence of ethanol, *J. Control. Release*, 23:175–182 (1993).

151. S. Chesnoy, J. Doucet, D. Durand, and G. Couarraze. Effect of iontophoresis in combination with ionic enhancers on the lipid structure of the stratum corneum: An x-ray diffraction study, *Pharm. Res.*, 13:1581–1584 (1996).
152. S. N. Murthy, D. C. Waddell, H. N. Shivakumar, A. Balaji, and C. P. Bowers. Iontophoretic permselective property of human nail, *J. Dermatol. Sci.*, 46:150–152 (2007).
153. S. N. Murthy, D. E. Wiskirchen, and C. P. Bowers. Iontophoretic drug delivery across human nail, *J. Pharm. Sci.*, 96:305–311 (2007).
154. J. Dutet and M. B. Delgado-Charro. Transungual iontophoresis of lithium and sodium: Effect of pH and co-ion competition on cationic transport numbers, *J. Control. Release*, 144:168–174 (2010).
155. A. B. Nair, H. D. Kim, S. P. Davis, R. Etheredge, M. Barsness, P. M. Friden, and S. N. Murthy. An *ex vivo* toe model used to assess applicators for the iontophoretic ungual delivery of terbinafine, *Pharm. Res.*, 26:2194–2201 (2009).
156. B. Amichai, R. Mosckovitz, H. Trau, O. Sholto, S. Ben Yaakov, M. Royz, D. Barak, B. Nitzan, and A. Shemer. Iontophoretic terbinafine HCL 1.0% delivery across porcine and human nails, *Mycopathologia*, 169:343–349 (2010).
157. V. Sachdeva, H. D. Kim, P. M. Friden, and A. K. Banga. Iontophoresis mediated *in vivo* intradermal delivery of terbinafine hydrochloride, *Int. J. Pharm.*, 393:112–118 (2010).
158. V. Sachdeva, S. Siddoju, Y. Y. Yu, H. D. Kim, P. M. Friden, and A. K. Banga. Transdermal iontophoretic delivery of terbinafine hydrochloride: Quantitation of drug levels in stratum corneum and underlying skin, *Int. J. Pharm.*, 388:24–31 (2010).
159. K. A. Smith, J. Hao, and S. K. Li. Influence of pH on transungual passive and iontophoretic transport, *J. Pharm. Sci.*, [Epub] (2009).
160. K. A. Smith, J. Hao, and S. K. Li. Effects of ionic strength on passive and iontophoretic transport of cationic permeant across human nail, *Pharm. Res.*, 26:1446–1455 (2009).
161. J. Hao, K. A. Smith, and S. K. Li. Time-dependent electrical properties of human nail upon hydration *in vivo*, *J. Pharm. Sci.*, 99:107–118 (2010).
162. B. Amichai, B. Nitzan, R. Mosckovitz, and A. Shemer. Iontophoretic delivery of terbinafine in onychomycosis: A preliminary study, *Br. J. Dermatol.*, 162:46–50 (2010).
163. J. Dutet and M. B. Delgado-Charro. *In vivo* transungual iontophoresis: Effect of DC current application on ionic transport and on transonychial water loss, *J. Control. Release*, 140:117–125 (2009).
164. Transdermal Nicotine Study Group. Transdermal nicotine for smoking cessation: Six-month results from two multicenter controlled clinical trials, *JAMA*, 266:3133–3138 (1991).
165. M. C. Fiore, S. S. Smith, D. E. Jorenby, and T. B. Baker. The effectiveness of the nicotine patch for smoking cessation—A meta-analysis, *JAMA*, 271:1940–1947 (1994).
166. G. M. Kochak, J. X. Sun, R. L. Choi, and A. J. Piraino. Pharmacokinetic disposition of multiple-dose transdermal nicotine in healthy adult smokers, *Pharm. Res.*, 9:1451–1455 (1992).
167. R. D. Prather, T. G. Tu, C. N. Rolf, and J. Gorsline. Nicotine pharmacokinetics of Nicoderm® (Nicotine transdermal system) in women and obese men compared with normal-sized men, *J. Clin. Pharmacol.*, 33:644–649 (1993).
168. S. S. Lin, H. Ho, and Y. W. Chien. Development of a new nicotine transdermal delivery system—*In vitro* kinetics studies and clinical pharmacokinetic evaluations in two ethnic groups, *J. Control. Release*, 26:175–193 (1993).
169. J. D. Lane, E. C. Westman, G. V. Ripka, J. L. Wu, C. C. Chiang, and J. E. Rose. Pharmacokinetics of a transdermal nicotine patch compared to nicotine gum, *Drug Dev. Ind. Pharm.*, 19:1999–2010 (1993).
170. H. Ho and Y. W. Chien. Kinetic evaluation of transdermal nicotine delivery systems, *Drug Develop. Ind. Pharm.*, 19:295–313 (1993).

171. S. S. Lin, Y. W. Chien, W. C. Huang, C. H. Li, C. L. Chueh, R. R. L. Chen, T. M. Hsu, T. S. Jiang, J. L. Wu, and K. H. Valia. Transdermal nicotine delivery systems—Multi-institutional cooperative bioequivalence studies, *Drug Dev. Ind. Pharm.*, 19:2765–2793 (1993).

172. B. J. Aungst. Nicotine skin penetration characteristics using aqueous and non-aqueous vehicles, anionic polymers, and silicone matrices, *Drug Dev. Ind. Pharm.*, 14:1481–1494 (1988).

173. R. M. Brand and R. H. Guy. Iontophoresis of nicotine *in vitro*: Pulsatile drug delivery across the skin? *J. Control. Release*, 33:285–292 (1995).

174. S. D. Roy and G. L. Flynn. Transdermal delivery of narcotic analgesics: pH, anatomical, and subject influences on cutaneous permeability of fentanyl and sufentanil, *Pharm. Res.*, 7:842–847 (1990).

175. S. D. Roy and G. L. Flynn. Solubility and related physicochemical properties of narcotic analgesics, *Pharm. Res.*, 5:580–586 (1988).

176. H. E. Junginger. Iontophoretic delivery of apomorphine: from in-vitro modelling to the Parkinson patient, *Adv. Drug Deliv. Rev.*, 54 Suppl 1:S57–S75 (2002).

177. G. L. Li, R. van der Geest, L. Chanet, E. van Zanten, M. Danhof, and J. A. Bouwstra. *In vitro* iontophoresis of R-apomorphine across human stratum corneum. Structure-transport relationship of penetration enhancement, *J. Control. Release*, 84:49–57 (2002).

178. G. L. Li, M. Danhof, P. M. Frederik, and J. A. Bouwstra. Pretreatment with a water-based surfactant formulation affects transdermal iontophoretic delivery of R-apomorphine *in vitro*, *Pharm. Res.*, 20:653–659 (2003).

179. A. M. de Graaff, G. L. Li, A. C. van Aelst, and J. A. Bouwstra. Combined chemical and electrical enhancement modulates stratum corneum structure, *J. Control. Release*, 90:49–58 (2003).

180. G. L. Li, T. J. Van Steeg, H. Putter, J. van der Spek, S. Pavel, M. Danhof, and J. A. Bouwstra. Cutaneous side-effects of transdermal iontophoresis with and without surfactant pretreatment: A single-blinded, randomized controlled trial, *Br. J. Dermatol.*, 153:404–412 (2005).

181. G. L. Li, J. J. de Vries, T. J. Van Steeg, H. van den Bussche, H. J. Maas, H. J. Reeuwijk, M. Danhof, J. A. Bouwstra, and T. van Laar. Transdermal iontophoretic delivery of apomorphine in patients improved by surfactant formulation pretreatment, *J. Control. Release*, 101:199–208 (2005).

182. A. K. Nugroho, G. L. Li, M. Danhof, and J. A. Bouwstra. Transdermal iontophoresis of rotigotine across human stratum corneum *in vitro*: Influence of pH and NaCl concentration, *Pharm. Res.*, 21:844–850 (2004).

183. A. K. Nugroho, S. G. Romeijn, R. Zwier, J. B. de Vries, D. Dijkstra, H. Wikstrom, O. Della-Pasqua, M. Danhof, and J. A. Bouwstra. Pharmacokinetics and pharmacodynamics analysis of transdermal iontophoresis of 5-OH-DPAT in rats: *In vitro–in vivo* correlation, *J. Pharm. Sci.*, 95:1570–1585 (2006).

184. O. W. Ackaert, J. De Graan, R. Capancioni, D. Dijkstra, M. Danhof, and J. A. Bouwstra. Transdermal iontophoretic delivery of a novel series of dopamine agonists *in vitro*: Physicochemical considerations, *J. Pharm. Pharmacol.*, 62:709–720 (2010).

185. A. K. Nugroho, L. Li, D. Dijkstra, H. Wikstrom, M. Danhof, and J. A. Bouwstra. Transdermal iontophoresis of the dopamine agonist 5-OH-DPAT in human skin *in vitro*, *J. Control. Release*, 103:393–403 (2005).

186. A. Luzardo-Alvarez, M. B. Delgado-Charro, and J. Blanco-Mendez. *In vivo* iontophoretic administration of ropinirole hydrochloride, *J. Pharm. Sci.*, 92:2441–2448 (2003).

187. S. Y. Oh, S. Y. Jeong, T. G. Park, and J. H. Lee. Enhanced transdermal delivery of AZT (Zidovudine) using iontophoresis and penetration enhancer, *J. Control. Release*, 51:161–168 (1998).

188. S. N. Murthy and S. R. Vaka. Irontophoresis: Transdermal delivery of iron by iontophoresis, *J. Pharm. Sci.*, 98:2670–2676 (2009).
189. J. Y. Fang, P. C. Wu, Y. B. Huang, and Y. H. Tsai. *In vivo* percutaneous absorption of capsaicin, nonivamide and sodium nonivamide acetate from ointment bases: Pharmacokinetic analysis in rabbits, *Int. J. Pharm.*, 128:169–177 (1996).
190. J. Y. Fang, P. C. Wu, Y. B. Huang, and Y. H. Tsai. *In vivo* percutaneous absorption of capsaicin, nonivamide and sodium nonivamide acetate from ointment bases: Skin erythema test and non-invasive surface recovery technique in humans, *Int. J. Pharm.*, 131:143–151 (1996).
191. P. C. Wu, J. Y. Fang, Y. B. Huang, and Y. H. Tsai. Development and evaluation of transdermal patches of nonivamide and sodium nonivamide acetate, *Pharmazie.*, 52:135–138 (1997).
192. J. Y. Fang, Y. B. Huang, P. C. Wu, and Y. H. Tsai. Transdermal iontophoresis of sodium nonivamide acetate. I. Consideration of electrical and chemical factors, *Int. J. Pharm.*, 143:47–58 (1996).
193. J. Y. Fang, Y. B. Huang, P. C. Wu, and Y. H. Tsai. Transdermal iontophoresis of sodium nonivamide acetate. II. Optimization and evaluation on solutions and gels, *Int. J. Pharm.*, 145:175–186 (1996).
194. P. Singh, S. Boniello, P. Liu, and S. Dinh. Iontophoretic transdermal delivery of methylphenidate hydrochloride, *Pharm. Res.*, 14:S-309 (1997).
195. P. Singh, S. Boniello, P. C. Liu, and S. Dinh. Transdermal iontophoretic delivery of methylphenidate HCl *in vitro*, *Int. J. Pharm.*, 178:121–128 (1999).
196. W. Liu, M. Hu, W. Liu, C. Xue, H. Xu, and X. Yang. Investigation of the carbopol gel of solid lipid nanoparticles for the transdermal iontophoretic delivery of triamcinolone acetonide acetate, *Int. J. Pharm.*, 364:135–141 (2008).
197. M. Barza, C. Peckman, and J. Baum. Transscleral iontophoresis of gentamicin in monkeys, *Invest. Ophthalmol. Vis. Sci.*, 28:1033–1036 (1987).
198. M. Barza, C. Peckman, and J. Baum. Transscleral iontophoresis of cefazolin, ticarcillin and gentamicin in the rabbit, *Ophthalmology*, 93:133–139 (1986).
199. D. M. Maurice. Iontophoresis of fluorescein into the posterior segment of the rabbit eye, *Ophthalmology*, 93:128–132 (1986).
200. D. Sarraf and D. A. Lee. The role of iontophoresis in ocular drug delivery, *J. Ocular Pharmacol.*, 10:69–81 (1994).
201. E. Eljarrat-Binstock, F. Raiskup, J. Frucht-Pery, and A. J. Domb. Transcorneal and transscleral iontophoresis of dexamethasone phosphate using drug loaded hydrogel, *J. Control. Release*, 106:386–390 (2005).
202. E. Eljarrat-Binstock and A. J. Domb. Iontophoresis: A non-invasive ocular drug delivery, *J. Control. Release*, 110:479–489 (2006).
203. S. A. Molokhia, E. K. Jeong, W. I. Higuchi, and S. K. Li. Transscleral iontophoretic and intravitreal delivery of a macromolecule: Study of ocular distribution *in vivo* and post-mortem with MRI, *Exp. Eye Res.*, 88:418–425 (2009).
204. S. A. Molokhia, E. K. Jeong, W. I. Higuchi, and S. K. Li. Examination of barriers and barrier alteration in transscleral iontophoresis, *J. Pharm. Sci.*, 97:831–844 (2008).
205. S. Nicoli, G. Ferrari, M. Quarta, C. Macaluso, and P. Santi. *In vitro* transscleral iontophoresis of high molecular weight neutral compounds, *Eur. J. Pharm. Sci.*, 36:486–492 (2009).
206. M. E. Myles, D. M. Neumann, and J. M. Hill. Recent progress in ocular drug delivery for posterior segment disease: Emphasis on transscleral iontophoresis, *Adv. Drug Deliv. Rev.*, 57:2063–2079 (2005).
207. S. A. Molokhia, E. K. Jeong, W. I. Higuchi, and S. K. Li. Examination of penetration routes and distribution of ionic permeants during and after transscleral iontophoresis with magnetic resonance imaging, *Int. J. Pharm.*, 335:46–53 (2007).

208. E. N. Lerner, E. H. van Zanten, and G. R. Stewart. Enhanced delivery of octreotide to the brain via transnasal iontophoretic administration, *J. Drug Target.*, 12:273–280 (2004).
209. K. M. Brough, D. M. Anderson, J. Love, and P. R. Overman. The effectiveness of iontophoresis in reducing dentin hypersensitivity, *J. Am. Dent. Assoc.*, 111:761–765 (1985).
210. L. P. Gangarosa. Fluoride iontophoresis for tooth desensitization, *J. Am. Dent. Assoc.*, 112:808–810 (1986).
211. J. C. Orsini, F. C. Barone, D. L. Armstrong, and M. J. Wayner. Direct effects of androgens on lateral hypothalamic neuronal activity in the male rat I. A microiontophoretic study, *Brain Res. Bull.*, 15:293–297 (1985).
212. R. A. Warren and R. W. Dykes. Transient and long-lasting effects of iontophoretically administered norepinephrine on somatosensory cortical neurons in halothane-anesthetized cats, *Can. J. Physiol. Pharmacol.*, 74:38–57 (1996).
213. M. Skydsgaard and J. Hounsgaard. Multiple actions of iontophoretically applied serotonin on motorneurones in the turtle spinal cord *in vitro*, *Acta Physiol. Scand.*, 158:301–310 (1996).
214. S. M. Roychowdhury and M. M. Heinricher. Effects of iontophoretically applied serotonin on three classes of physiologically characterized putative pain modulating neurons in the rostral ventromedial medulla of lightly anesthetized rats, *Neurosci. Lett.*, 226:136–138 (1997).
215. T. W. Stone. Responses of blood vessels to various amines applied by microiontophoresis, *J. Pharm. Pharmacol.*, 24:318–323 (1972).
216. J. J. Fu and J. F. Lorden. An easily constructed carbon fiber recording and microiontophoresis assembly, *J. Neurosci. Methods*, 68:247–251 (1996).
217. V. Labhasetwar and R. J. Levy. Novel delivery of antiarrhythmic agents, *Clin. Pharmacokinet.*, 29:1–5 (1995).
218. Avitall, Boaz. Myocardial iontophoresis. Assignee: Avitall, Boaz. 539,611 [U.S. Pat. 5,087,243]. 1992. Wisconsin, USA.
219. V. Labhasetwar, T. Underwood, S. P. Schwendeman, and R. J. Levy. Iontophoresis for modulation of cardiac drug delivery in dogs, *Proc. Natl. Acad. Sci. USA*, 92:2612–2616 (1995).
220. S. P. Schwendeman, V. Labhasetwar, and R. J. Levy. Model features of a cardiac iontophoretic drug delivery implant, *Pharm. Res.*, 12:790–795 (1995).
221. S. P. Schwendeman, G. L. Amidon, V. Labhasetwar, and R. J. Levy. Modulated drug release using iontophoresis through heterogeneous cation-exchange membranes. 2. Influence of cation-exchanger content on membrane resistance and characteristic times, *J. Pharm. Sci.*, 83:1482–1494 (1994).
222. R. Goldman and S. Pollack. Electric fields and proliferation in a chronic wound model, *Bioelectromagnetics*, 17:450–457 (1996).
223. K. Cheng, P. P. Tarjan, and P. M. Mertz. Conductivities of pig dermis and subcutaneous fat measured with rectangular pulse electrical current, *Bioelectromagnetics*, 17:458–466 (1996).
224. J. B. Sloan and K. Soltani. Iontophoresis in dermatology, *J. Am. Acad. Dermatol.*, 15:671–684 (1986).
225. D. G. Kassan, A. M. Lynch, and M. J. Stiller. Physical enhancement of dermatologic drug delivery: Iontophoresis and phonophoresis, *J. Am. Acad. Dermatol.*, 34:657–666 (1996).
226. C. T. Costello and A. H. Jeske. Iontophoresis: Applications in transdermal medication delivery, *Phys. Ther.*, 75:554–563 (1995).
227. A. C. Lauer, L. M. Lieb, C. Ramachandran, G. L. Flynn, and N. D. Weiner. Transfollicular drug delivery, *Pharm. Res.*, 12:179–186 (1995).
228. B. Illel. Formulation for transfollicular drug administration: Some recent advances, *Crit. Rev. Ther. Drug Carr. Syst.*, 14:207–219 (1997).
229. J. E. Riviere, B. Sage, and P. L. Williams. Effects of vasoactive drugs on transdermal lidocaine iontophoresis, *J. Pharm. Sci.*, 80:615–620 (1991).

230. S. M. Sammeta and S. N. Murthy. "ChilDrive": A technique of combining regional cuta-
neous hypothermia with iontophoresis for the delivery of drugs to synovial fluid, *Pharm. Res.*, 26:2535–2540 (2009).
231. J. Cazares-Delgadillo, C. Balaguer-Fernandez, A. Calatayud-Pascual, A. Ganem-Rondero, D. Quintanar-Guerrero, A. C. Lopez-Castellano, V. Merino, and Y. N. Kalia. Transdermal iontophoresis of dexamethasone sodium phosphate *in vitro* and *in vivo*: Effect of experimental parameters and skin type on drug stability and transport kinetics, *Eur. J. Pharm. Biopharm.*, 75:173–178 (2010).
232. R. H. Bogner and A. K. Banga. Iontophoresis and phonophoresis, *US Pharmacist*, 19:H10–H26 (1994).
233. L. V. Allen. Dexamethasone iontophoresis, *US Pharmacist*, November:86 (1992).
234. J. P. Sylvestre, R. H. Guy, and M. B. Delgado-Charro. *In vitro* optimization of dexamethasone phosphate delivery by iontophoresis, *Phys. Ther.*, 88:1177–1185 (2008).
235. J. P. Sylvestre, C. Diaz-Marin, M. B. Delgado-Charro, and R. H. Guy. Iontophoresis of dexamethasone phosphate: Competition with chloride ions, *J. Control. Release*, 131:41–46 (2008).
236. J. M. Glass, R. L. Stephen, and S. C. Jacobson. The quantity and distribution of radiolabeled dexamethasone delivered to tissue by iontophoresis, *Int. J. Dermatol.*, 19:519–525 (1980).
237. S. M. Hasson, C. L. Wible, M. Reich, W. S. Barnes, and J. H. Williams. Dexamethasone iontophoresis: Effect on delayed muscle soreness and muscle function, *Can. J. Spt. Sci.*, 17:8–13 (1992).
238. S. H. Hasson, G. H. Henderson, J. C. Daniels, and D. A. Schieb. Exercise training and dexamethasone iontophoresis in rheumatoid arthritis, *Physiotherapy Canada*, 43:11–14 (1991).
239. L. C. Li, R. A. Scudds, C. S. Heck, and M. Harth. The efficacy of dexamethasone iontophoresis for the treatment of rheumatoid arthritic knees: A pilot study, *Arthritis Care Res.*, 9:126–132 (1996).
240. D. Doeglas, T. Suurmeijer, B. Krol, R. Sanderman, M. Vanleeuwen, and M. Vanrijswijk. Work disability in early rheumatoid arthritis, *Ann. Rheum. Dis.*, 54:455–460 (1995).
241. P. T. Dawes and P. D. Fowler. Treatment of early rheumatoid arthritis: A review of current and future concepts and therapy, *Clin. Exp. Rheumatol.*, 13:381–394 (1995).
242. H. Liss and D. Liss. Lateral epicondylitis, *New Jersey Rehab*, Nov.:8–11 (1995).
243. G. L. Pellecchia, H. Hamel, and P. Behnke. Treatment of infrapatellar tendinitis: A combination of modalities and transverse friction message versus iontophoresis, *J. Sport Rehab.*, 3:135–145 (1994).
244. K. I. Reid, R. A. Dionne, L. Sicard-Rosenbaum, D. Lord, and R. A. Dubner. Evaluation of iontophoretically applied dexamethasone for painful pathologic temporomandibular joints, *Oral Surg. Oral Med. Oral Pathol.*, 77:605–609 (1994).
245. A. Chantraine, J. P. Ludy, and D. Berger. Is cortisone iontophoresis possible? *Arch. Phys. Med. Rehabil.*, 67:38–40 (1986).
246. U. Tauber. Dermatocorticosteroids: Structure, activity, pharmacokinetics, *Eur. J. Dermatol.*, 4:419–429 (1994).
247. L. C. Fuhrman, B. B. Michniak, C. R. Behl, and A. W. Malick. Effect of novel penetration enhancers on the transdermal delivery of hydrocortisone: An *in vitro* species comparison, *J. Control. Release*, 45:199–206 (1997).
248. B. B. Michniak, M. R. Player, J. M. Chapman, and J. W. Sowell. Azone analogues as penetration enhancers: Effect of different vehicles on hydrocortisone acetate skin permeation and retention, *J. Cont. Rel.*, 32:147–154 (1994).
249. Y. Wang, L. V. Allen, L. C. Li, and Y. H. Tu. Iontophoresis of hydrocortisone across hairless mouse skin: Investigation of skin alteration, *J. Pharm. Sci.*, 82:1140–1144 (1993).

250. S. C. Seth, L. V. Allen, and P. Pinnamaraju. Stability of hydrocortisone salts during ion-tophoresis, *Int. J. Pharm.*, 106:7–14 (1994).
251. M. P. James, R. M. Graham, and J. English. Percutaneous iontophoresis of predniso-lone—A pharmacokinetic study, *Clin. Exp. Dermatol.*, 11:54–61 (1986).
252. S. H. Rothfeld and W. Murray. The treatment of Peyronie's disease by iontophoresis of C_{21} esterified glucocorticoids, *J. Urology*, 97:874–875 (1967).
253. R. L. Dobson. Treatment of hyperhidrosis, *Arch. Dermatol.*, 123:883–884 (1987).
254. F. Levit. Treatment of hyperhidrosis by tap water iontophoresis, *Cutis*, 26:192–194 (1980).
255. L. P. Stolman. Treatment of excess sweating of the palms by iontophoresis, *Arch. Dermatol.*, 123:893–896 (1987).
256. T. Schlereth, M. Dieterich, and F. Birklein. Hyperhidrosis—Causes and treatment of enhanced sweating, *Dtsch. Arztebl. Int.*, 106:32–37 (2009).
257. S. N. Shrivastava and G. Singh. Tap water iontophoresis in palmoplantar hyperhidrosis, *Br. J. Dermatol.*, 96:189–195 (1977).
258. M. L. Elgart and G. Fuchs. Tapwater iontophoresis in the treatment of hyperhidrosis, *Int. J. Dermatol.*, 26:194–197 (1987).
259. D. L. Akins, J. L. Meisenheimer, and R. L. Dobson. Efficacy of the drionic unit in the treatment of hyperhidrosis, *J. Am. Acad. Dermatol.*, 16:828–832 (1987).
260. K. Grice, H. Sattar, and H. Baker. Treatment of idiopathic hyperhidrosis with iontopho-resis of tap water and poldine methosulfate, *Br. J. Dermatol.*, 86:72–78 (1972).
261. E. Abell and K. Morgan. The treatment of idiopathic hyperhidrosis by glycopyrronium bromide and tap water iontophoresis, *Br. J. Dermatol.*, 91:87–91 (1974).
262. G. M. Kavanagh, C. Oh, and K. Shams. BOTOX delivery by iontophoresis, *Br. J. Dermatol.*, 151:1093–1095 (2004).
263. W. H. Yeung, J. Palmer, D. Schidlow, M. R. Bye, and N. N. Huang. Evaluation of a paper-patch test for sweat chloride determination, *Clin. Pediatrics*, 23:603–607 (1984).
264. W. J. Warwick, L. G. Hansen, and M. E. Werness. Quantitation of chloride in sweat with the cystic fibrosis indicator system, *Clin. Chem.*, 36:96–98 (1990).
265. W. J. Warwick, N. N. Huang, W. W. Waring, A. G. Cherian, I. Brown, E. Stejskal-Lorenz, W. H. Yeung, G. Duhon, J. G. Hill, and D. Strominger. Evaluation of a cystic fibrosis screening system incorporating a minature sweat stimulator and disposable chloride sensor, *Clin. Chem.*, 32:850–853 (1986).
266. Wescor, http://www.wescor. com (accessed September 4, 2010).
267. J. M. Rattenbury and E. Worthy. Is the sweat test safe? Some instances of burns received during pilocarpine iontophoresis, *Ann. Clin. Biochem.*, 33:456–458 (1996).
268. S. Thysman, A. Jadoul, T. Leroy, D. Vanneste, and V. Preat. Laser doppler evaluation of skin reaction in volunteers after histamine iontophoresis, *J. Cont. Rel.*, 36:215–219 (1995).
269. D. Van Neste, D. T. Leroy, B. deBrouwer, and J. P. Rihoux. Histamine-induced skin reac-tions using iontophoresis and H_1- blockade, *Inflamm. Research.*, 45:S48–S49 (1996).
270. U. Darsow, J. Ring, E. Scharein, and B. Bromm. Correlations between histamine-induced wheal, flare and itch, *Arch. Dermatol. Res.*, 288:436–441 (1996).
271. R. Groning. Electrophoretically controlled dermal or transdermal application systems with electronic indicators, *Int. J. Pharm.*, 36:37–40 (1987).
272. J. M. Hill, L. P. Gangarosa, and N. H. Park. Iontophoretic application of antiviral che-motherapeutic agents, *Ann. N. Y. Acad. Sci.*, 284:604 (1977).
273. L. P. Gangarosa, N. H. Park, and J. M. Hill. Iontophoretic assistance of 5-iodo 2′-deoxyu-ridine penetration into neonatal mouse skin and effects on DNA synthesis, *Proc. Soc. Exp. Biol. Med.*, 154:439–443 (1977).

274. L. P. Gangarosa, H. W. Merchant, N. H. Park, and J. M. Hill. Iontophoretic application of idoxuridine for recurrent herpes labialis: Report of preliminary clinical findings, *Methods Find. Exp. Clin. Pharmacol.*, 1:105–109 (1979).
275. N. H. Park, L. P. Gangarosa, B. S. Kwon, and J. M. Hill. Iontophoretic application of adenosine arabinoside monophosphate to herpes simplex virus type 1 infected hairless mouse skin, *Antimicrob. Agents Chemother.*, 14:605–608 (1978).
276. B. S. Kwon, J. M. Hill, C. Wiggins, C. Tuggle, and L. P. Gangarosa. Iontophoretic application of adenosine arabinoside monophosphate for the treatment of herpes simplex virus type 2 infections in hairless mice, *J. Infect. Dis.*, 140:1014 (1979).
277. M. Boxhall and J. Frost. Iontophoresis and herpes labialis, *Med. J. Aust.*, 140:686–687 (1984).
278. J. Henley-Cohn and J. N. Hausfeld. Iontophoretic treatment of oral herpes, *Laryngoscope*, 94:118–121 (1984).
279. P. R. Layman, E. Argyras, and C. J. Glynn. Iontophoresis of vincristine versus saline in post-herpetic neuralgia. A controlled trial, *Pain*, 25:165–170 (1986).
280. K. J. Smith, J. L. Konzelman, F. A. Lombardo, H. G. Skelton, T. T. Holland, J. Yeager, K. F. Wagner, C. N. Oster, and R. Chung. Iontophoresis of vinblastine into normal skin and for treatment of Kaposi's sarcoma in human immunodeficiency virus-positive patients, *Arch. Dermatol.*, 128:1365–1370 (1992).
281. J. L. Haynes. Method for obtaining blood using iontophoresis. Assignee: Becton Dickinson. 710,420 [U.S. Pat. 5,131,403]. 1992. New Jersey, USA.
282. L. P. Woodson. Biofilm reduction method. Assignee: MN Mining & Mfg. 257,877 [U.S. Pat. 5,462,644]. 1995. Minnesota, USA.
283. J. Jass, J. W. Costerton, and H. M. Lappin-Scott. The effect of electrical currents and tobramycin on *Pseudomonas aeruginosa* biofilms, *J. Ind. Microb.*, 15:234–242 (1995).
284. A. S. Rapperport, D. L. Larson, D. F. Henges, J. B. Lynch, T. G. Blocker, and R. S. Lewis. Iontophoresis: A method of antibiotic administration in the burn patient, *Plastic Reconstruc. Surg.*, 36:547–552 (1965).
285. N. T. LaForest and C. Confrancesco. Antibiotic iontophoresis in the treatment of ear chronditis, *Phys. Ther.*, 58:32–34 (1978).
286. R. F. Greminger, R. A. Elliott, and A. Rapperport. Antibiotic iontophoresis for the management of burned ear chondritis, *Plastic Reconst. Surg.*, 66:356–359 (1980).
287. R. A. Macaluso and T. L. Kennedy. Antibiotic iontophoresis in the treatment of burn perichondritis of the rabbit ear, *Otolaryngol. Head Neck Surg.*, 100:568–572 (1989).
288. W. Rigano, M. Yanik, F. A. Barone, G. Baibak, and C. Cislo. Antibiotic iontophoresis in the management of burned ears, *J. Burn Care Rehabil.*, 13:407–409 (1992).
289. M. N. Berliner. Skin microcirculation during tapwater iontophoresis in humans: Cathode stimulates more than anode, *Microvascular. Res.*, 54:74–80 (1997).
290. M. A. Ashburn, R. L. Stephen, E. Ackerman, T. J. Petelenz, B. Hare, N. L. Pace, and A. A. Hofman. Iontophoretic delivery of morphine for postoperative analgesia, *J. Pain Symptom Mgmt.*, 7:27–33 (1992).
291. R. L. Stephen, C. Bonezzi, C. Rossi, and S. Eruzzi. Morphine formulations for use by electromotive administration. 276,613 [U.S. Pat. 5,607,940]. 1997. Utah, USA.
292. M. A. Ashburn, J. Streisand, J. Zhang, G. Love, M. Rowin, S. Niu, J. K. Kievit, J. R. Kroep, and M. J. Mertens. The iontophoresis of fentanyl citrate in humans, *Anesthesiology*, 82:1146–1153 (1995).
293. C. W. Poulos, G. M. Brenner, L. V. Allen, V. A. Prabhu, and P. L. Huerta. Method for stimulating hair growth with cationic derivative of minoxidil using therapeutic iontophoresis. 272,880 [U.S. Pat. 5,466,695]. 1995. Oklahoma, USA.
294. R. L. Stephen, T. J. Petelenz, and S. C. Jacobsen. Method of iontophoretically treating acne, furuncles and like skin disorders. Assignee: Iomed. 350,227 [U.S. Pat. 4,979,938]. 1990. Utah, USA.

295. J. Kahn. A case report: Lithium iontophoresis for gouty arthritis, *J. Orthop. Sports Phys. Ther.*, 4:113–114 (1982).

296. M. W. Cornwall. Zinc iontophoresis to treat ischemic skin ulcers, *Phys. Ther.*, 61:359–360 (1981).

297. J. Kahn. Acetic acid iontophoresis for calcium deposits, *Phys. Ther.*, 57:658–659 (1977).

298. D. L. Wieder. Treatment of traumatic myositis ossificans with acetic acid iontophoresis, *Phys. Ther.*, 72:133–137 (1992).

299. B. K. Chang, T. H. Guthrie, K. Hayakawa, and L. P. Gangarosa. A pilot study of iontophoretic cisplatin chemotherapy of basal and squamous cell carcinomas of the skin, *Arch. Dermatol.*, 129:425–427 (1993).

300. B. Csillik, E. Knyihar-Csillik, and A. Szucs. Treatment of chronic pain syndromes with iontophoresis of vinca alkaloids to the skin of patients, *Neurosci. Lett.*, 31:87–90 (1982).

301. A. Satyanand, A. K. Saxena, and A. Agarwal. Silver iontophoresis in chronic osteomyelitis, *J. Indian Med. Assoc.*, 84:135–136 (1986).

302. J. E. Wahlberg. Skin clearance of iontophoretically administered chromium (^{51}Cr) and sodium (^{22}Na) ions in the guinea pig, *Acta Dermatovener (Stockholm)*, 50:255–262 (1970).

303. F. J. M. Puttemans, D. L. Massart, F. Giles, P. C. Lievens, and M. H. Jonckeer. Iontophoresis: Mechanism of action studied by potentiometry and X-ray fluorescence, *Arch. Phys. Med. Rehabil.*, 63:176–180 (1982).

304. M. Tannenbaum. Iodine iontophoresis in reducing scar tissue, *Phys. Ther.*, 60:792 (1980).

5 Skin Electroporation and Its Applications

5.1 INTRODUCTION

Electroporation is an electric enhancement technique that involves application of high-voltage pulses for very short durations of time. While iontophoresis (see Chapter 4) involves the use of relatively low transdermal voltages (typically less than 20 V), electroporation of skin takes place at high transdermal voltages (about 100 V or more). Iontophoresis enhances flux by acting primarily on the drug, and electroporation acts on the skin with some driving force on the drug during a pulse (1). However, iontophoresis will have secondary effects on the skin just like electroporation would apply direct electromotive force on the drug during the brief pulse period. This may be particularly true for the thin cell lining of the sweat ducts which might be electroporated with low voltages of the order used in iontophoresis. Electroporation can expand the scope of iontophoresis to deliver peptides in greater quantities or enable delivery of larger molecules than what can be delivered by iontophoresis. Electroporation, sometimes also termed *electropermeabilization*, refers to the dramatic changes observed in cell membranes or artificial planar bilayer membranes upon application of large transmembrane voltages. These changes involve structural rearrangement and conductance changes leading to temporary loss of semipermeability of cell membranes, and the observations are consistent with the formation of pores. This technique is best known as a physical transfection method in which cells are exposed to a brief electrical pulse, thereby opening pores in the cell membrane, allowing DNA or other macromolecules to enter the cell (2,3). During this process, the transmembrane potential produces a current that passes through the membrane defects and leads to a reversible increase in its permeability by the formation of aqueous pathways that are straight-through with radii of a few nanometers; this process is more energetically favorable than long, tortuous pathways around corneocytes. Even though the exact mechanism for electroporation is not clear, the pore mechanism is generally believed to be the case (4), as changes in the behavior of membranes seen following electroporation are consistent with the theory of pore formation. These include changes in electrical behavior (e.g., changes in membrane conductance are often dramatic and are believed to be due to ionic conduction through transient aqueous pores). However, these pores have not been visualized by any microscopic techniques, presumably due to factors such as their small size and transient nature. Electroporation is considered to be a nonthermal phenomena because pore formation occurs by membrane rearrangement much before any significant temperature rise takes place in the pulsing medium (5).

Electroporation as a science has about 35 years of history, though its use for gene transfer to tissues or its investigation for intradermal or transdermal drug delivery only began in the 1990s (1,6–10). Skin electroporation for drug delivery was explored by several investigators over the next 10 years, but interest in this technology seems to have decreased as it is being realized that the technique does not lead to severalfold enhancements in drug delivery unless high-voltage pulses are used at an intensity that may not be acceptable for routine drug delivery applications, and such high-voltage pulses can also result in involuntary muscle contractions (11). However, interest has increased for application of electroporation for delivery of DNA vaccines and for electrochemotherapy. The transition of electroporation from basic research to clinical trials has been recently reviewed (12). A detailed discussion of these applications is outside the scope of this book, but electrochemotherapy applications will be briefly discussed in this chapter, and application of electroporation to DNA vaccination is briefly discussed in Chapter 9.

As discussed, the technique of electroporation is normally used on the unilamellar phospholipid bilayers of cell membranes. However, it was demonstrated that electroporation of skin is feasible, even though the stratum corneum contains multilamellar, intercellular lipid bilayers with few phospholipids (1). The electrical behavior of human epidermal membrane as a function of the magnitude and duration of applied voltage closely parallels the electrical breakdown/recovery of bilayer membranes seen during electroporation (13). The approximately 100 multilamellar bilayers of the stratum corneum need about 100 V pulses for electroporation, or about 1 V per bilayer (14). Typically, high-voltage pulses of 50 to 500 V applied for micro- to millisecond duration have been used for skin electroporation (15). The stratum corneum does not contain any living cells, but it can still be permeabilized by an electric pulse because electroporation is a physical process based on electrostatic interactions and thermal fluctuations within fluid membranes and no active transport processes are involved. It was suggested that electroporation of skin causes the generation of localized transport regions in the stratum corneum. In a study using charged fluorescent molecules, it was shown that they could not be delivered across snake skin by iontophoresis, because snake skin has no preexisting pathways in the form of hair follicles and sweat ducts. In contrast, electroporation was able to transport these molecules across snake skin by generating new transport pathways (16). A series of experiments using two molecules of similar size but different charges (calcein, 623 Da, charge −4; and sulforhodamine, 607 Da, charge −1) indicated that their electrical behavior and transport was consistent with the hypothesis that aqueous pathways are caused in skin by high-voltage pulsing (17). Transport of calcein through human stratum corneum during electroporation was shown to occur through intercellular and transcellular pathways (18). There is considerable indirect evidence that high-voltage pulses cause changes in the skin structure (19,20). As discussed, the use of electroporation in conjunction with iontophoresis can expand the scope of transdermal delivery to larger molecules such as therapeutic proteins and oligonucleotides. Electroporation has been shown to significantly and reversibly increase the flux of luteinizing hormone-releasing hormone (LHRH) through human skin. The application of a single pulse prior to iontophoresis increased the flux 5 to 10 times over that achieved by iontophoresis alone (21). Iontophoresis can be used to provide baseline levels, and electroporation pulses can potentially be applied to provide rapid boluses. This has been shown to be feasible for transdermal transport of calcein across human epidermis. More complex delivery schedules were also achieved by changing pulse voltage (22). Thus, the combined use of iontophoresis and electroporation is likely to yield useful and interesting results that will intensify ongoing efforts to explore electroporation as a means of transdermal drug delivery. One advantage of giving an electroporation pulse in conjunction with iontophoresis for proteins can be that the size of the protein may become less of a limitation for its delivery across or into skin, because electroporation can potentially deliver drugs of higher molecular weight as compared to iontophoresis.

5.2   ELECTROPERMEABILIZATION OF THE SKIN

Electrical properties of skin were discussed in Section 4.2. Above a certain threshold value, resistance of skin is progressively reduced. A retrospective investigation using data from several studies showed that resistance of skin depends most strongly on voltage, followed by current density and power density. Because electroporation involves the use of very short pulses, much higher voltages (as compared to ionto-phoresis) can be used. The minimum current required to be applied to the skin to evoke a sensation is termed *perception threshold*, and the minimum current to cause a painful sensation is called *pain threshold*. The latter is more variable as it depends on both physical and sociological factors and is typically 3 to 25 times greater than the former. Both perception and pain thresholds do not scale directly with either current (I) or current density (I_j) but exhibit more complex functionalities. Power-function least-squares fit of the data resulted in the following relationships for perception thresholds (23,24):

$$I = 0.00042 \, A^{0.21} \; (r^2 = 0.47)$$

$$I_j = 0.00042 \, A^{-0.79} \; (r^2 = 0.93)$$

and the following for pain thresholds:

$$I = 0.0033 \, A^{0.33} \; (r^2 = 0.45)$$

$$I_j = 0.0033 \, A^{-0.67} \; (r^2 = 0.77)$$

where A is the electrical contact area.

For single pulses of about 90 V across human stratum corneum, recovery of skin resistance has been shown to be almost complete, returning to about 90% of the prepulse value. The recovery consisted of four phases, with about 60% recovery in phase I (20 ms), 40% to 70% in Phase II (0.4 to 0.8 seconds) and 60% to 90% in Phase III (10 seconds). This is followed by a slow phase of recovery over a minute to a few hours. For higher-voltage pulses (>130 V), recovery was typically less than 50% (25). The recovery time for these conditions may be quicker if full-thickness skin is used. Using excised full-thickness porcine skin, the recovery was shown to take place almost instantaneously or within the time required to switch to measuring instrument. This was claimed to be due to the stratum corneum being still attached to epidermis and underlying tissue, and being relaxed unlike the former study where the stratum corneum was heat stripped, hydrated, and mounted in a chamber. The permeabilization was found to depend on the electrical exposure dose that is the product of the pulse voltage and cumulative pulsing exposure time. Skin resistance was observed to drop to 20% of its prepulsing value when pulsed beyond a critical dosage of 0.4 V-s but recovered rapidly. When the dose exceeded 200 V-s, the recovery was slow and incomplete (26).

As mentioned earlier, enhancement due to electroporation is believed to be due to its effects on the skin, with contribution of the direct electrophoretic effects on the drug relatively less. For instance, it was shown that application of continuous low voltage resulted in a calcein flux three orders of magnitude smaller than pulsing at high voltage under "electrophoretically equivalent" conditions, suggesting that structural changes induced in the skin by pulsing contribute more significantly than the direct electrophoretic force acting on the drug (1). Calcein, being a fluorescent molecule (with excitation/emission wavelengths of 495 nm/515 nm), has also allowed measurement of rapid kinetics of transdermal transport. The measured fluorescence is a function of the rate of transdermal transport, but because continuous measurements are being made, the reading may also be a function of mixing and pumping times of the measurement systems. After deconvoluting the signal, it was suggested that the transport due to each pulse occurred over a time scale on the order of 10 seconds or less (27). It was shown that calcein flux reached steady-state levels within minutes and then decreased below the detection limit within seconds following the termination of a pulse. While flux was dependent on both pulse voltage and pulse rate, the steady-state lag time and onset time were found to depend only on the pulse rate (22). In contrast, flux of a macromolecule, heparin, remained partially elevated for several hours after high-voltage pulsing (28).

An enhancer effect for transdermal transport has been shown by high-voltage pulsing in the presence of macromolecules. Heparin has been shown to alter its own transport during electroporation (28) as well as the transport of other molecules (29,30). It has been hypothesized that heparin or other linear macromolecules (such as DNA) can enter an aqueous pathway between corneocytes, get trapped, and keep the pathways open. This resulted in a persistent low postpulse electrical resistance of the skin in the presence of heparin ($-300 \ \Omega$). In the absence of heparin, the postpulse recovery was from the initial postpulse $400 \ \Omega$ to about $800 \ \Omega$ within 2 hours (prepulse skin resistance was about 82 kΩ). As a result of these open pathways, transport of sulforhodamine (charge -1) was increased in the presence of heparin. However, the transport of highly charged (charge -4) calcein was decreased, presumably because of the electrostatic repulsion between the highly negatively charged calcein and the negatively charged heparin trapped in the aqueous pathways (29). Therefore, use of pathway-enlarging molecules to enhance macromolecule transport by skin electroporation was proposed (31). Several macromolecules such as dextran sulfate, neutral dextran, and polylysine have been shown to have an enhancer effect on the electroporation-induced transdermal delivery of mannitol across freshly excised abdominal hairless rat skin. Macromolecules with a greater charge and size were found to be more effective enhancers. The effect of macromolecules added at the time of pulsing lasted for several hours even after pulsing. It was suggested that these macromolecules do not disrupt lipids but rather stabilize the transient disruptions already induced by electroporation of skin due to their flexible linear structures entering the skin. This hypothesis was supported by the observation that the enhancer effect was seen only for electroporation, and not for passive diffusion or iontophoresis. Small ions were also used but provided no enhancement for electroporation (30).

5.3 ELECTROPORATION EQUIPMENT

Some of the electroporation equipment is manufactured for use in recombinant DNA technology–type applications but has been used as such or adapted for transdermal studies. Commercially available equipment comes with a chamber that typically is a sealable cylinder with an electrode plate at each end. These chambers are typically disposable, presterilized cuvettes with molded-in aluminum electrodes and come in different gap sizes, depending on whether bacteria or mammalian cells need to be electroporated. These chambers are not required for electroporation of tissues and will not be discussed here. In such studies, the nominal field strength, E, is related to the applied voltage, V, and the interelectrode gap, d ($E = V/d$). Field strengths vary from a few hundred V/cm for mammalian cells to many kV/cm for bacteria. For many drug delivery studies, this equipment has been used and directly connected to electrodes placed on the skin, bypassing the chamber. An example of such equipment is the ECM line of square wave and exponential decay electroporators from the BTX Molecular Delivery Systems Division of Harvard Biosciences (Holliston, Massachusetts). These electroporators can switch between a high- and a low-voltage mode. The high-voltage mode is for bacterial transformation, but the low-voltage mode is designed for plant or mammalian transfection and can be used for skin electroporation. These instruments have several capacitors and resistors allowing for wide choice in setting the pulse length.

In typical electroporation equipment, a capacitor is connected to a power supply to build up an electric charge and is then discharged after isolating from its charging source. The current and pulse length obtained depend on the speed at which the stored energy is released, which, in turn, can be controlled by using resistors. The voltage output from electroporation equipment may be in the form of exponential decay or square wave pulses. The exponential decay waveform represents complete discharge of the capacitor into a resistor, and the square wave pulse is often produced by a partial discharge of a large capacitor. For an exponential waveform generator, the voltage of a capacitor, C, discharging into a resistor, R, follows an exponential decay law:

$$V = V_0 \exp (-t/RC)$$

Pulse length will be characterized by the $1/e$ time constant, which is the time required for the initial voltage to decay to $1/e \approx$ one-third of the initial value. This time constant is a product of R and C, where C is the capacitance of the storage capacitor in the generator, and R is the total resistance into which the capacitor discharges. Both waveforms have been successfully used for electroporation of skin (32). Waveforms are determined by the principles of electrical engineering used by the pulser which generally have to be designed for one waveform only. Unlike capacitive discharge devices, a microprocessor-controlled logic-driven unit can produce reproducible direct current (DC) pulse lengths independent of the conductivity of the solutions bathing the cells.

During the last few years, constant current electroporation has been used in many studies by monitoring the resistance of the tissue and using a constant amperage for electroporation (33–39). A recent mechanistic study with constant current electroporation of planar lipid bilayer membranes reported that charge accumulates on the membrane over several seconds, and metastable nanopores are observed just before conditions at which electroporation occurs (39). In recent years, electroporation devices especially designed for electrochemotherapy or DNA vaccination applications have been designed and introduced in the marketplace. Inovio Pharmaceuticals (San Diego, California), formerly known as Genetronics, produces a range of electroporation equipment for DNA vaccination. The CELLECTRA® DNA delivery system from Inovio has been used for DNA immunizations. The device can adjust voltage based on tissue resistance to maintain constant current and a square wave during electroporation (38,40). This portable device is designed to deliver DNA vaccines intramuscularly and intradermally. Inovio also produces a MedPulser® DNA delivery system for immunization into muscle cells and an Elgen® DNA delivery system for use with muscle or skin. The Elgen device consists of a needle-electrode applicator and a pulse generator. DNA is injected from the needles followed by application of low-voltage electric pulses to colocalize the DNA and the electric field (41). Pulses are delivered using the same needles that are used to inject the DNA, thereby optimizing delivery efficiency (42,43). Through mergers over the last few years, ADViSYS Inc. and VGX Pharmaceuticals are now part of Inovio. Ichor Medical Systems (San Diego, California) makes a TriGrid Delivery System that has a pulse stimulator, an integrated applicator, and an application cartridge enclosing four electrodes arranged in two interlocking triangles around a central injection site (44). CytoPulse Sciences, Inc. (Glen Burnie, Maryland) makes electroporation devices designed for skin rather than muscle electroporation. They market the Derma Vax™ and Easy Vax™ systems, with the former primarily targeting the dermal layer of the skin and the latter primarily targeting the epidermal layer of the skin. The company uses its Pulse Agile® technology with a sequence of short, high-intensity pulses to porate the cell membranes followed by long, low-intensity pulses to further drive the DNA into cells via electrophoresis. Electroporation devices are also available from vendors such as Bio-Rad (Hercules, California), Invitrogen (Carlsbad, California), and Wolf Labs (York, United Kingdom). Miniaturization of this technology to develop wearable patches is currently not feasible for electroporation due to the need for a capacitor.

5.4 FACTORS AFFECTING DELIVERY BY ELECTROPORATION

For electroporation of cells, the most important factors are the strength of the electric field (V/cm) and the duration of time (pulse length) for which this electric field

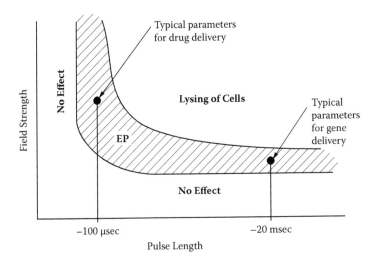

FIGURE 5.1 Relationship between field strength and pulse length as it applies to successful electroporation of cells. The shaded area depicts effective electroporation conditions, allowing drug and gene delivery into cells. (Reproduced with permission from G. Widera and D. Rabussay, *Drug Del. Technol.*, 2:34–40, 2002 [45].)

is applied (45). Electroporation of cells is dependent on the lipid bilayer structure but not so much on the cell-specific characteristics that can therefore enable the prediction of molecular uptake of several different molecules in several different cell types (46). Figure 5.1 represents these parameters schematically. If the field strength is too low or the pulse length is too short, breakdown transmembrane potential is not reached. If it is too high, cells may undergo lyses. In fact, it was suggested that cell death by irreversible electroporation can be used in medicine as a tissue ablation tool by surgeons (47). When optimum field strength of sufficient duration is applied to the cells, pores are created in the membranes (shaded area of Figure 5.1). Field strength and pulse length can compensate each other in the effective poration (shaded) area (45).

Key variables in the enhancement of percutaneous absorption of drugs by electroporation are the selection of the pulse voltage, pulse duration, number of pulses, and spacing between pulses. The goal is to maximize drug delivery using a protocol that will be acceptable for *in vivo* human studies. For *in vitro* experiments, significant voltage drop occurs within donor and receptor solutions, so that applied voltages need to be significantly higher (22). The application of high-voltage electric pulses may be accompanied by temperature changes in the pulsing medium due to electrical power dissipation or even interfacial electrochemical heat production (48). Thus, results obtained under different conditions should be interpreted cautiously. The use of short pulses adding up to the duration of a longer single pulse will help to minimize the heating. Because the bulk of research with electroporation has been done on cells and not tissues, literature relating to the former can be important to shed some light on the factors involved. A quantitative dependence on the percentage of porated cells and the parameters of the pulse such as shape, amplitude, and duration has been derived, and the expressions compared well with experimental results on the electroporation of human erythrocytes (4). The voltage used to electroporate cells is typically very high. For studies with skin, a typical applied voltage used is 100 to 300 V (32), though voltages as high as 1000 V have been used during *in vitro* transdermal studies. The electric field represents the voltage (V) that is applied across the electrode gap (cm). For instance, if a pair of electrodes is put across a 5-mm tumor and a voltage of 500 V is applied, then the strength of the electric field is 500 V/0.5 cm = 1000 V/cm or 1 kVcm (49).

For the transport of calcein (623 Da) across human epidermis, it was found that flux increased as a strong function of voltage up to 100 V but only weakly above this value. Transdermal flux of calcein was enhanced by up to four orders of magnitude by pulsing for 1 hour. After the pulsing was stopped, flux decreased by about 90% within 30 minutes and by >99% within 1 or 2 hours, indicating reversibility when the voltage was at or below approximately 100 V (1). It was suggested that a large number of high-voltage, short-duration pulses may permeabilize the skin efficiently but may be less efficient than a small number of low-voltage, long-duration pulses to obtain an electrophoretic movement of drug (32). A long-duration pulse may also be associated with increased skin irritation. In a study on the transport of metoprolol, a linear correlation was observed between pulse voltage (24 to 450 V; pulse time 620 msec) and cumulative amount transported after 4 hours. A linear correlation was also observed with pulse time (80 to 710 ms) as well at 100 V (50). For studies with calcein, the transport increased almost linearly with transdermal voltage above a threshold of about 80 V and then leveled off at higher voltages (>250 V) or shorter spacing (<10 s) between pulses (51). If desired, the voltage traces such as pulsing voltage or transdermal voltage may be acquired and stored on a digital oscilloscope (52–54). The current flowing through the skin during the first pulse will be different from subsequent pulses because the resistance of the skin will drop.

The electrodes used for pulsing in transdermal drug delivery studies have mostly been silver/silver chloride. The use of Ag/AgCl electrodes in iontophoresis was discussed in Section 4.6. The material should be such that it can accommodate a high instantaneous charge density and should not form harmful electrochemical products. Homogeneously mixed Ag/AgCl electrodes may be best, because Ag wires electrochemically plated with AgCl may be susceptible to the detachment of the outer layer of coating during pulsing. Wire electrodes have been used for *in vitro* studies, but more elaborate designs are required for *in vivo* studies. One possible design for *in vivo* tissue electroporation was described in the literature (55). A combination of a stainless steel cathode and a silver anode has been used to provide an inexpensive alternative applicable to the particular situation. A second pair of inner electrodes located closer to the skin were made of Ag/AgCl and were used for electrical measurements due to their low electrode polarization and phase stability (17). The effect of competitive ions and electrode polarity may be less significant in electroporation if passive permeation of the drug through electropermeabilized skin represents significant percentage of flux. Delivery of fentanyl following electroporation of hairless rats was not affected significantly by a 10-fold change in the buffer concentration or by reversal of electrode polarity (32). Drug delivery by electroporation may be hindered by the presence of high concentrations (>15% v/v) of cosolvents (such as propylene glycol) in the formulation. This could be due to a decrease in the conductivity of the drug solution (56).

Factors identified for optimal gene delivery to achieve maximum gene expression with constant current electroporation are amperage, pulse length, number of pulses, and lag time. Furthermore, these factors have to be optimized for each species due to the differences in tissue resistance. For mice, the best results were obtained at 0.1 Amps with 20 msec pulses applied with no lag time, while pulses of 0.4 to 0.6 Amps were used to regulate gene expression in pigs (33,34). For delivery of plasmid DNA by *in vivo* electroporation, optimal pulse lengths should be in the range of 0.1 to 50 ms, and effective field strengths need to be in the range of 200 to 1800 V/cm (57). For electrochemotherapy, positioning and geometry of the electrodes is also very important to achieve uniform electric field across the entire tumor for effective electropermeabilization (58). Similarly, the shape of the electrodes is important (59). Optimal electroporation factors may differ depending on the intended application—drug delivery, DNA vaccination, or long-term expression of genes following delivery of DNA (45,60). For delivery to the skin, shorter electrodes are required. Microneedle arrays that can potentially be used to electroporate the epidermal layers of the skin were reported (61,62).

5.5   SKIN TOXICOLOGY OF ELECTROPORATION

Clinical precedent for safely applying electrical pulses of hundreds of volts to the skin exists with the use of techniques such as transcutaneous electrical nerve stimulation (TENS) (1). TENS is commonly used to treat chronic conditions such as low back pain, arthritis, and pain caused by a variety of neurologic disorders (23). Its use for acute conditions such as reduced pain during labor (63), delayed onset muscle soreness in humans (64), sore neck muscles (65), and other uses (66,67) is under investigation. The range of sensations evoked by TENS have varied from tactile (touch, vibration, etc.) to pricking pain and itch. Very small changes in stimulus parameters are needed to convert tactile sensations to pain and vice versa, though thermal sensations are rarely reported (68). Skin irritation resulting from electroporation has generally been considered to be mild and temporary (69). Nevertheless, electroporation of skin or tissue does have safety risks if application parameters are not carefully controlled. In a study with 20 impotent patients, the tolerability of voltages from 50 to 80 V, delivered as a single 3-ms pulse to the shaft and part of the glans of the penis was tested. Treatment was generally well tolerated but did cause significant distress to 25% of the patients, though a sensitive area of the body was used in this study (70). In recent years, interest in using electroporation for drug delivery applications seems to have declined, and instead, many researchers have started to focus on the use of microneedles for drug and vaccine delivery applications. However, electroporation continues to be extensively investigated for its applications for electrochemotherapy and DNA vaccination (see Section 9.7). It was suggested that constant current electroporation will be safer because it will avoid uncontrolled variations or surges in current due to variations in tissue resistance and maintain a square wave function. A constant voltage device will not be able to maintain the amperage in such cases, possibly producing current variations that may damage the tissue (34). Muscle contractions are typically observed with electropora-tion pulses but can be minimized by changing the electrode configuration. These contractions may not be acceptable for routine drug delivery applications but may be acceptable for electrochemotherapy, which is usually done under local or general anesthesia as palliative treatment of cutaneous metastases in patients with stage III melanoma. Erythema and edema may also be observed at the site of treated lesions in some patients, and marks from electrodes inserted into tumors may be seen, but all these effects disappear within a few days to a month (71).

Electroporation of the back skin of anesthetized hairless rats has been reported to result in direct stimulation of motor nerves. As each pulse was given, the hind legs of the rat were observed to kick, and the intensity of kicking varied with the applied voltage and electrode position (55). For electrochemotherapy on terminally ill patients (see Section 5.6), electric fields as high as 1.3 kV/cm have been used after 1% lidocaine was injected around each nodule. Muscle contractions were observed and mild pain was felt during each pulse which then subsided at the end of each pulse (53). Toxicology of exposure to currents or voltages much higher than those used in electroporation may also give some insights to understand mechanistic toxico-logical details. Common electrical injuries have been proposed to involve extensive electroporation of muscle cells. In cardiac procedures including electric shock, elec-troporation of muscle cells and the accompanying Ca^{++} leakage is a source of clini-cal complications (6). Poloxamer 188 has been found to adsorb to lipid bilayers and modify the surface properties. It thus decreases their susceptibility to electropora-tion and may contribute toward decreasing damage sustained by cells inadvertently exposed to an electric shock (72). Repeated electrical stress will have an adverse effect on epidermal mitosis (73).

As the stratum corneum has a much higher electrical resistance than other parts of the skin, an electric field applied to the skin will concentrate in the nonviable stratum corneum to induce electroporation. In contrast, the field will be much lower in the viable tissues, thereby protecting the already permeable viable parts of the skin (part of epidermis and all of dermis) and deeper tissues. The reversibility of permeation following electroporation suggests that the technique is not damaging to the skin. As discussed, diagnostic and therapeutic applications such as TENS safely apply electric pulses to skin with voltages up to hundreds of volts and durations up to milliseconds. Long-term toxicological implications of the use of this technique, sup-ported by histological examinations, are currently underway. As the promise of this technique is demonstrated, such studies will be conducted by several groups active in this area before U.S. Food and Drug Administration (FDA) approval for com-mercialization is sought. Skin toxicology of electroporation relative to iontophoresis has been studied using 14 pigs. In the first study using eight pigs, exponential volt-age pulses were applied followed by constant current anodal iontophoresis. Pulses of 0, 250, 500, and 1000 V were applied followed by iontophoresis of 0, 0.2, and 2.0 mA/cm² for 30 minutes or 10 mA/cm² for 10 minutes, and any resulting erythema, edema, and petechiae were observed. Results of gross evaluation immediately after or 4 hours after treatment indicated that erythema increases immediately after treat-ment with increasing pulse voltage but was absent or minimal after 4 hours. It was reported that it disappeared or reduced within 5 minutes, and pulse voltage had no effect on edema or petechiae. Erythema, edema, and petechiae all increased with increasing current, though the application of a pulse did not increase the irritation induced by iontophoresis. These changes were comparable to those seen with iontophoresis alone. In the second study with six pigs, it was found that at both the gross and light microscopic levels, electroporation does not result in any skin changes not previously seen with iontophoresis alone (74). It should be noted that very high electric fields can cause permanent permeabilization of cell membranes and therefore cell death, in a process known as irreversible electroporation (47,57).

5.6 ELECTROPORATION FOR CANCER CHEMOTHERAPY

A novel form of cancer treatment, called electrochemotherapy, has utilized a combination of electroporation and chemotherapeutic agents (49,58,71,75–80). The technique typically involves systemic administration or intratumoral injection of anticancer drugs followed by delivery of electric field pulses at the site of the tumor nodules. The rationale for this approach is that many cancer drugs are very poorly permeable into the tumor cells, and pulsing the tumor site will increase the uptake of the drug from the systemic circulation (or from the tumor) into the tumor cells via the pores created in their membranes (81). The drugs used are typically bleomycin or cisplatin. Significant progress has been made with electrochemotherapy in recent years to the point that European Standard Operating Procedures of Electrochemotherapy (ESOPE) have been proposed (78), and an electroporation device especially designed for electrochemotherapy, the Cliniporator™ device (IGEA Ltd., Modena, Italy) (Figure 5.2) (71), has received marketing approval in the European Union. This device generates square wave pulses of variable amplitude with delivery frequencies of 1 to 5000 Hz. The procedure can be performed in about 30 minutes with very minimal training required by the medical staff to carry out the procedure. For larger lesions, repeated treatments may be needed.

In one of the early studies with electrochemotherapy, bleomycin was delivered to a total of 18 nodules in six patients. Three of the patients had malignant melanoma, two had basal cell carcinoma, and one had metastatic adenocarcinoma. Eight 99-μsec pulses of 1.3 kV/cm were administered to the tumors, 5 to 15 minutes after intravenous administration of bleomycin to the patients. Square wave pulses were delivered through electrodes placed on both sides of the protruding tumors. Five of the six patients responded positively to the treatment, with responses (partial to complete regression) seen in 13 of the 14 nodules. Some nodules were left untreated as positive controls. The patient who failed to respond had severe peripheral vascular disease that may have compromised the delivery of drug to the tumor (53). To avoid systemic drug delivery for localized therapy, bleomycin can be injected directly into the tumors for electrochemotherapy. Studies in mice have shown that intratumoral injection of bleomycin in combination with electric pulses is effective (82). The parallel plate electrodes used in the human study described earlier can only treat tumors at the level of the skin but cannot be used on deep-seated tumor or internal organs. Furthermore, they are difficult to fit and hold in place, and the electric fields may not penetrate the deepest regions of cutaneous and subcutaneous tumors. Furthermore, burns may occur where the electrodes touch the skin

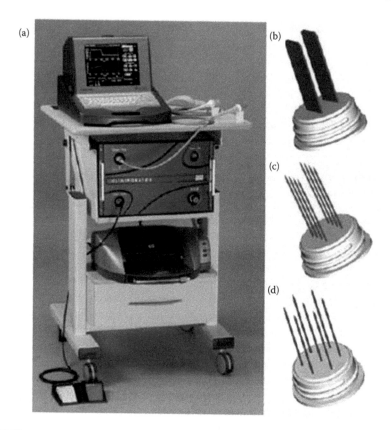

FIGURE 5.2 The electric pulse delivery unit, (a) Cliniporator™ from IGEA (Modena, Italy) and the electrodes used with the unit; (b) type I plate electrodes; (c) needle electrodes type II and (d) type III. (With kind permission from Springer Science+Business Media: *Ann. Surg. Oncol.*, Electrochemotherapy with Intravenous Bleomycin in the Local Treatment of Skin Melanoma Metastases, 15:2215–2222, 2008, P. Quaglino et al. [71]. Copyright 2008.) (**See color insert.**)

(77). Improvements in electrode design have thus been investigated (83). A needle array arrangement composed of six 28-gauge stainless steel acupuncture needles spaced at 60° intervals around a 1-cm diameter circle and extending 1 cm from the electrode body was found to be useful. These needles can be inserted into the tumor to a depth of 0.5 cm. Each needle has an independent electrical connection so that a pulse sequence to energize individual needles can be developed. The design resulted in a 50% improvement in response rate compared to the standard parallel plate electrodes (84). The ESOPE study referred to earlier (78) used plate electrodes for superficial tumor nodules but needle electrodes for deeper-seated tumors. Two types of needle electrodes were used: a needle row for small tumor nodules and a hexagonal centered configuration for larger nodules (Figure 5.2). Depending on the tumor and type of electrode being utilized, amplitudes were determined to provide high electric field throughout the tumor. For plate (Type I) electrodes, eight pulses of 1300 V/cm were used, and for the needle row (Type II) electrodes, eight pulses of 1000 V/cm were used. The hexagonal centered configuration (Type III) delivered 96 pulses of 1000 V/cm amplitude. The pulses were of 100 µs duration. Four cancer centers participated in this two-year study with 41 patients, and a response rate of 85% was achieved on tumor nodules treated with electrochemotherapy. The response rate was not significantly affected by the drug (bleomycin or cisplatin) or the route (intravenous or intratumoral injection) of administration, suggesting that either drug can be used. Cisplatin is usually given intratumorally, and bleomycin can be given by either route. All patients had progressive cutaneous and subcutaneous metastases of any histologically proven cancer, and the response rate was not significantly dependent on tumor histology. Low-level or no muscle contractions were reported for 78% of the patients. Muscle contractions were stronger for Type I electrodes and least for Type III electrodes. In a subsequent study in 14 patients with Cliniporator™ pulse generator used as per ESOPE guidelines, a response rate of 93% was achieved after the first electrochemotherapy treatment, and complete regression was observed in 50% of patients. A lower response rate was observed in lesions that are >1 cm^2 (71).

5.7   DELIVERY OF DRUGS BY ELECTROPORATION

Electroporation can potentially enable transdermal delivery of both small molecules and macromolecules such as proteins and gene-based drugs (85). In addition to the model compounds calcein, sulforhodamine, and caffeine (17,86), other drugs that have been investigated for transdermal delivery via electroporation include buprenorphine (87), calcitonin (88), fentanyl (32,89), metoprolol (50), parathyroid hormone (88), tacrine (90), and timolol (91). Some other drugs investigated include dextran, insulin, interleukin, methotrexate, and terazosin, and these studies were recently reviewed (15).

Results with many of these drugs have been less dramatic than those seen with model compounds like calcein. Studies with model compounds have given excellent mechanistic insights, but the magnitude of flux enhancement observed may not happen for all drugs; therefore, each drug needs to be studied as a separate case. This is likely to be especially true for drugs that have some passive flux. In an *in vitro* study with human skin, it was projected that a flux in the therapeutic range (>50 µg/cm^2/hour) can be achieved for timolol (91). For the delivery of the nonsteroidal anti-inflammatory drug flurbiprofen, both iontophoresis and electroporation were effective in rapid delivery of the drug into the systemic circulation of male Wistar rats. The same total charge (1.3 C) and duration of treatment were used for both techniques. The average (± SEM) area under the flurbiprofen plasma concentration–time curve (AUC) was 87.1 ± 20.5 µg.h/ml by iontophoresis, 110.5 ± 18.0 µg.h/ml by electroporation, and 91.7 ± 4.3 µg.h/ml following IV bolus (92). Delivery of genes by electroporation is discussed separately in Chapter 9. Delivery of cyclosporine A by electroporation was also reported. The drug is useful for treatment of psoriasis but is toxic if given systemically and no topical formulations currently exist. By using a single pulse at a field strength of 200 V/cm and a 10-msec pulse interval, delivery through rat skin was enhanced by a factor of 8.5 over passive diffusion (93).

Meander electrodes, which consist of an array of interweaving electrode fingers, may provide an electrode configuration suitable for including in a patch or device. The use of meander electrodes was reported in some studies (70,94) done by Genetronics which is now Inovio Pharmaceuticals (San Diego, California). Use of these electrodes to deliver particulates into skin in a process known as electroincorporation has been suggested, but these studies have generally been inconclusive. Electrodes shaped like calipers are also available and have been used to pulse back skin folds of hairless mice. Particles of 0.2, 4, and 45 μm size were shown to be imbedded in hairless mouse skin when pulsed with three exponential decay pulses of an amplitude of 120 V and a pulse length of 1.2 ms (95). Pressure-mediated electroincorporation was reported to deliver Lupron Depot® (Leuprolide acetate) microspheres into hairless mouse skin and into human skin xenografted on immunodeficient nude mice (96). Similarly, gold particles have been used to enhance the transdermal delivery of caffeine through full-thickness human cadaver skin via pressure-mediated electroincorporation. In this study, 12 exponential decay pulses (120 V, 8 msec) were administered using meander electrodes (97). Use of electroporation for transcutaneous extraction of drugs from the dermal tissue has also been suggested. This could serve as a noninvasive alternative to phlebotomy and microdialysis, and the feasibility of determining salicylic acid as a model drug in dermal extracellular fluid has been demonstrated (98). Electroporation has also been used in combination with various passive and physical enhancement strategies (15). If iontophoresis and electroporation are compared at the same total charge delivered, electroporation appears to be better based on some literature reports, but the safety of these two protocols is not likely to be the same, with iontophoretic conditions being more acceptable. In a study that compared the transport efficiency of electroporation and iontophoresis, transport numbers were in the same range and were a function of voltage and current. They did not show any dependence on pulse length, rate, energy, waveform, or total charge transferred. However, the fraction of area of skin available for transport was larger during high-voltage pulsing than during iontophoresis (54). We used a full factorial design approach to study the effect of iontophoresis and electroporation and their interaction mechanisms on delivery of tacrine hydrochloride *in vitro* across intact and stripped rat skin (90). Electrolytes, especially those containing $CaCl_2$, $MgCl_2$, or $CaBr_2$ were reported to further enhance electroporation-mediated transdermal delivery of calcein, presumably due to their disruptive effects on the stratum corneum (86). Unlike iontophoresis, very few human studies have been done with electroporation for drug delivery applications. In one reported double-blind study, garlic juice was used as a simple model compound to demonstrate the feasibility of using electroporation for transdermal delivery. Six pulses (100 V, 10 ms) were applied on the volar side of each arm, with some pressure applied during and after pulsing. The intensity of tongue sensation and taste was scored by the volunteers, and it was shown that pulses and some pressure were required to obtain a positive response (97). Electric fields developed in the tissue during pulsing will have to be characterized. These fields have typically been modeled by computer technology, but direct measurement of the field has also been reported because the conductivities in the body are very different and complicated (99). Thus, the use of electroporation pulses to permeabilize skin, followed by low-current iontophoresis could be a very

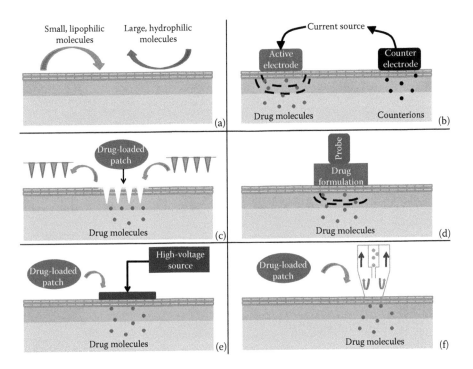

FIGURE 1.1 Skin is not permeable to hydrophilic molecules (a), but such delivery can be enabled by enhancement technologies such as iontophoresis (b), microneedles (c), sonophoresis (d), electroporation (e), or microdermabrasion (f).

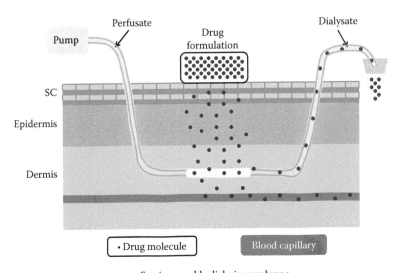

FIGURE 2.2 Linear microdialysis probe inserted into the dermis with the dialysis membrane of the probe located under the patch having the drug formulation.

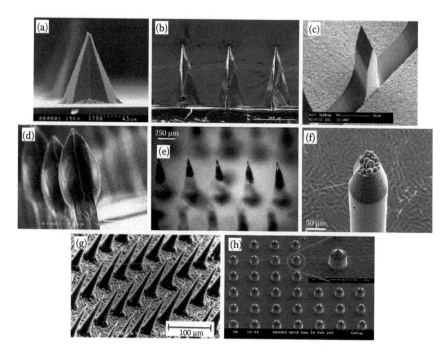

FIGURE 3.1 Different types of solid microneedles. (a) A single, short silicon microneedle; (b) maltose microneedles; (c) polycarbonate microneedles; (d) coated microneedles; (e) tapered-cone PLGA microneedles; (f) biodegradable microneedles with encapsulated microparticles; (g) an array of conical-shaped silicon microneedles; and (h) a single-crystal silicon microneedle array.

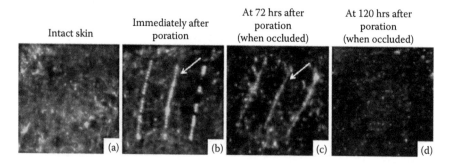

FIGURE 3.5 Calcein imaging studies to show pore closure in hairless rat skin. Fluorescent images showing (a) absence of microchannels before poration; presence of microchannels (b) immediately after poration; (c) at 72 hours after poration; and (d) at 120 hours after poration, under occluded conditions [48].

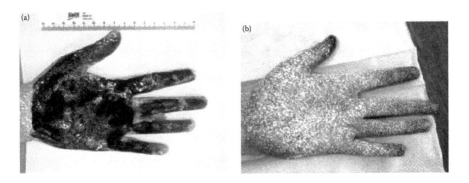

FIGURE 4.2 A starch–iodine test to show hyperhidrotic area in patient's palm (a) before and (b) 2 weeks post-BOTOX® iontophoresis [262].

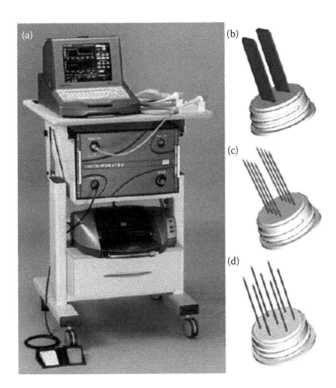

FIGURE 5.2 The electric pulse delivery unit, (a) Cliniporator™ from IGEA (Modena, Italy) and the electrodes used with the unit; (b) type I plate electrodes; (c) needle electrodes type II and (d) type III [71].

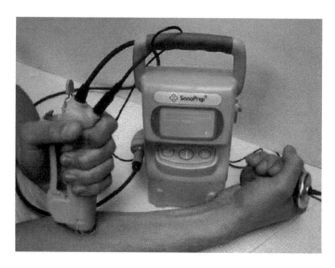

FIGURE 6.1 Sonication of the subject's volar forearm by the SonoPrep® device. The hand piece interacts with the main console of the device to determine the duration of sonication by measuring the drop in skin resistance [49].

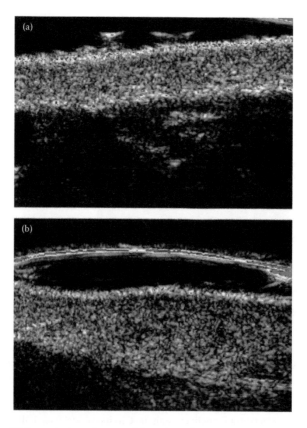

FIGURE 6.2 High-resolution ultrasound imaging of skin thickness (a) before treatment, and (b) after 30 minutes of histamine application by sonophoresis [45].

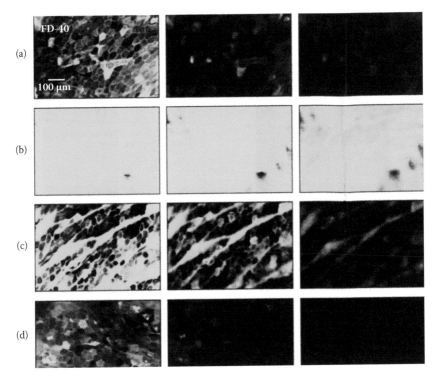

FIGURE 6.4 Delivery of FITC-labeled dextran (MW 38 kDa) into hairless rat skin as imaged by confocal microscopy up to a depth of 20 μm by (a) passive diffusion over 12 hours or by sonication at 41 kHz (b) directly under the horn (c) in boundary area; or (d) in area outside the horn [39].

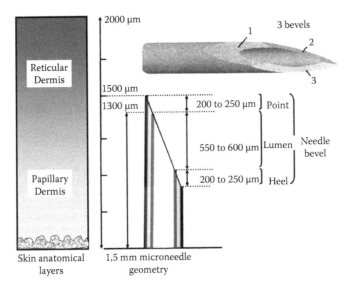

FIGURE 9.1 Microneedle geometry of a hollow 1.5-mm microneedle for intradermal injection showing how the length of the microneedle and the bevel relate to anatomical placement in skin [26].

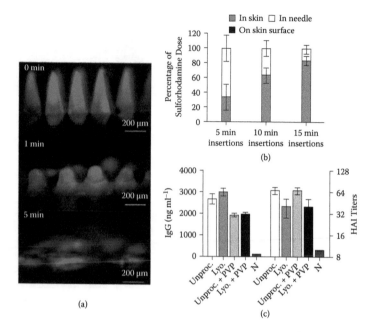

FIGURE 9.3 Dissolution of polymeric microneedles (a) *ex vivo* in pig skin or (b) *in vivo* in mice skin, and (c) effect of processing (lyophilization) or formulation (PVP) on immunogenicity in mice following intramuscular administration [57].

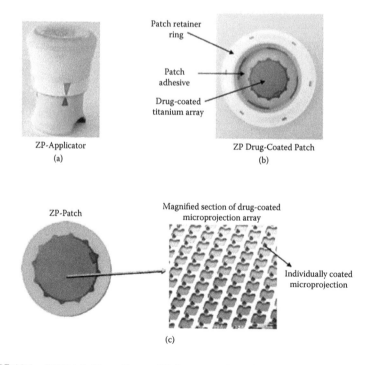

FIGURE 10.1 ZP Patch (a) applicator, (b) drug-coated patch, and (c) microprojection array [67].

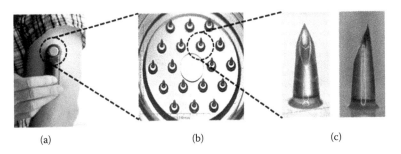

(a)   (b)   (c)

FIGURE 10.2   3M's hollow microstructured transdermal system (hMTS) integrated device for drug delivery. (a) Subject demonstrating the use of wearable hMTS integrated device. It consists of an applicator, drug reservoir, and delivery system, all in a single, compact device; (b) hMTS array consisting of 18 "mini-hypodermic" structures distributed over 1 cm^2 area; (c) front and side views of single polymeric mini-hypodermic needle (approximately 900 μm in length). (Reproduced with permission from 3M, St. Paul, Minnesota.)

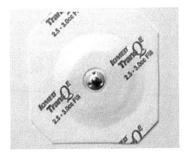
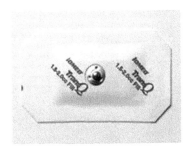

FIGURE 10.3   Iontophoresis device and electrodes. (Reproduced with permission from Chattanooga Group [DJO Global], Vista, California, http://wwww.chattgroup.com.)

Polymer thick
film electrode

Hydrogel/drug

Selective barrier
membrane

FIGURE 10.4 IsisIQ™ patch with polymer thick-film electrodes that transport drug from a hydrogel across a selective barrier membrane. (Reproduced with permission from Biopolymer, Providence, Rhode Island, http://www.isisbiopolymers.com.)

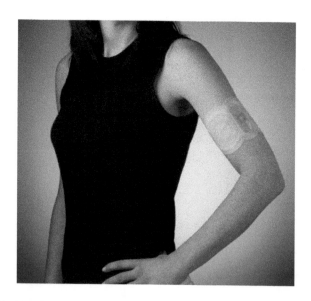

FIGURE 10.5 Developmental Zelrix patch (NuPathe Inc., Conshohocken, Pennsylvania) as worn by a patient [106].

useful technique to achieve high flux levels of a drug without the skin irritation. Also, iontophoresis alone will not be able to deliver large molecular weight drugs, even if high current is used. The use of electroporation in combination with ultrasound has also been investigated, and it was found that application of ultrasound reduces the threshold voltage required to facilitate transdermal drug transport in the presence of electric fields (100).

Some studies with specific drug molecules are discussed in the following paragraphs. Use of electroporation and other physical enhancement strategies for delivery of buprenorphine, calcitonin, fentanyl, and parathyroid hormone is discussed in Chapters 7 and 10 to make the reader aware of the rationale of selection of the most appropriate enhancement technology for a specific drug molecule based on the therapeutic rationale, physicochemical properties of the drug molecule, economic considerations, and the status of commercial development of the technology.

5.7.1 LUTEINIZING HORMONE-RELEASING HORMONE

As discussed, iontophoretic delivery of luteinizing hormone-releasing hormone (LHRH) across epidermis separated from human cadaver skin has been shown to be significantly enhanced following a single electroporation pulse. The pulse was an exponentially decaying type with an initial amplitude of 1000 V and a time constant of 5 ± 1 msec. At a current of 0.5 mA/cm^2, flux was 0.27 ± 0.08 without the pulse and increased to 1.62 ± 0.05 µg/hour/cm^2 with the pulse (21). Usefulness of electroporation to enhance the iontophoretic flux of LHRH has been verified using the isolated perfused porcine skin flap (IPPSF) model, a model that closely resembles human clinical use (see Section 2.2.3). It was found that application of a single pulse (500 V, 5 ms) immediately prior to 30 minutes of iontophoresis increased LHRH concentration in the IPPSF perfusate nearly twofold, while application of a pulse every 10 minutes resulted in a threefold increase (74). By using repeated applications of the pulse/iontophoresis protocol, it was shown that electroporation is able to repeatedly enhance LHRH transport in a pulsatile manner relative to iontophoresis, thus allowing pulsatile delivery of therapeutic peptides.

5.7.2 HEPARIN

Heparin is a class of molecules (MW 5000 to 30,000 Da) with smaller molecules having weaker anticoagulation activity. It is used clinically for anticoagulation and prophylaxis of thromboembolism. It is not bioavailable if used orally, and continuous parenteral infusion is required for full-dose therapy. Thus, transdermal delivery would be useful which also has applications for topical therapy of vascular permeability diseases, superficial thromboses, and inflammatory and arthritic conditions. Due to its high molecular weight, skin permeation needs to be enhanced by a suitable mechanism such as penetration enhancers (101) or electroporation. Transdermal transport across human skin, *in vitro*, has been shown to be feasible, delivering therapeutic rates (100 to 500 µg/cm^2-hour) by application of short (pulse length 1.9 ms), high-voltage (150 to 350 V) pulses at the rate of 12 pulses/minute for 1 hour. Heparin retained its biological activity as it was transported across the skin, but because smaller molecules were preferentially transported, the anticoagulant activity of molecules transported was lower than that in the donor compartment. Fluxes achieved by low-voltage iontophoresis having the same time-averaged current were an order of magnitude lower than those achieved with electroporation (28). Delivery of heparin by microporation is discussed in Section 3.5.4.

5.7.3   DOXEPIN

Doxepin complexed by hydroxyl propyl-β-cyclodextrin was delivered by electroporation (120 V; 10 ms pulses) across porcine epidermis. A sustained drug release effect was observed during *in vitro* studies, and results were confirmed *in vivo* in the rat model using pinprick scores to show prolonged analgesic activity. This approach could potentially be used for treatment of chronic pain in a condition like postherpetic neuralgia, a common complication following an incident of herpes zoster with pain lasting as long as several months (102).

REFERENCES

1. M. R. Prausnitz, V. G. Bose, R. Langer, and J. C. Weaver. Electroporation of mammalian skin—A mechanism to enhance transdermal drug delivery, *Proc. Natl. Acad. Sci.*, 90:10504–10508 (1993).
2. D. A. Treco and R. F. Selden. Non-viral gene therapy, *Mol. Med. Today*, 1:314–321 (1995).
3. J. C. Weaver. Electroporation theory—Concepts and mechanisms. In J. A. Nickoloff (ed), *Electroporation protocols for microorganisms*, Humana Press, Totowa, NJ, 1995, pp. 1–26.
4. T. C. Tomov. Quantitative dependence of electroporation on the pulse parameters, *Bioelectrochem. Bioenerg.*, 37:101–107 (1995).
5. J. C. Weaver. Electroporation: A general phenomenon for manipulating cells and tissues, *J. Cellular Biochem.*, 51:426–435 (1993).
6. T. Y. Tsong. Electroporation of cell membranes, *Biophys. J.*, 60:297–306 (1991).
7. M. R. Prausnitz. A practical assessment of transdermal drug delivery by skin electroporation, *Adv. Drug Deliv. Rev.*, 35:61–76 (1999).
8. M. J. Jaroszeski, R. Gilbert, C. Nicolau, and R. Heller. *In vivo* gene delivery by electroporation, *Adv. Drug Deliv. Rev.*, 35:131–137 (1999).
9. L. C. Smith and J. L. Nordstrom. Advances in plasmid gene delivery and expression in skeletal muscle, *Curr. Opin. Mol. Ther.*, 2:150–154 (2000).
10. T. Tamura and T. Sakata. Application of *in vivo* electroporation to cancer gene therapy, *Curr. Gene Ther.*, 3:59–64 (2003).
11. J. J. Escobar-Chavez, D. Bonilla-Martinez, M. A. Villegas-Gonzalez, and A. L. Revilla-Vazquez. Electroporation as an efficient physical enhancer for skin drug delivery, *J. Clin. Pharmacol.*, 49:1262–1283 (2009).
12. A. M. Bodles-Brakhop, R. Heller, and R. Draghia-Akli. Electroporation for the delivery of DNA-based vaccines and immunotherapeutics: Current clinical developments, *Mol. Ther.*, 17:585–592 (2009).
13. H. Inada, A. H. Ghanem, and W. I. Higuchi. Studies on the effects of applied voltage and duration on human epidermal membrane alteration/recovery and the resultant effects upon iontophoresis, *Pharm. Res.*, 11:687–697 (1994).

14. J. C. Weaver and Y. Chizmadzhev. Electroporation. In C. Polte and E. Postow (eds), *Biological effects of electromagnetic fields*, CRC Press, Boca Raton, FL, 1996, pp. 247–274.
15. N. A. Charoo, Z. Rahman, M. A. Repka, and S. N. Murthy. Electroporation: An avenue for transdermal drug delivery, *Curr. Drug Deliv.*, 7:125–136 (2010).
16. T. Chen, R. Langer, and J. C. Weaver. Skin electroporation causes molecular transport across the stratum corneum through localized transport regions, *J. Investig. Dermatol. Symp. Proc.*, 3:159–165 (1998).
17. U. Pliquett and J. C. Weaver. Electroporation of human skin: Simultaneous measurement of changes in the transport of two fluorescent molecules and in the passive electrical properties, *Bioelectrochem. Bioenerg.*, 39:1–12 (1996).
18. M. R. Prausnitz, J. A. Gimm, R. H. Guy, R. Langer, J. C. Weaver, and C. Cullander. Imaging regions of transport across human stratum corneum during high-voltage and low-voltage exposures, *J. Pharm. Sci.*, 85:1363–1370 (1996).
19. D. A. Edwards, M. R. Prausnitz, R. Langer, and J. C. Weaver. Analysis of enhanced transdermal transport by skin electroporation, *J. Control. Release*, 34:211–221 (1995).
20. M. R. Prausnitz. Do high-voltage pulses cause changes in skin structure? *J. Control. Release*, 40:321–326 (1996).
21. D. B. Bommannan, J. Tamada, L. Leung, and R. O. Potts. Effect of electroporation on transdermal iontophoretic delivery of luteinizing hormone releasing hormone (LHRH) *in vitro*, *Pharm. Res.*, 11:1809–1814 (1994).
22. M. R. Prausnitz, U. Pliquett, R. Langer, and J. C. Weaver. Rapid temporal control of transdermal drug delivery by electroporation, *Pharm. Res.*, 11:1834–1837 (1994).
23. M. R. Prausnitz. The effects of electric current applied to skin: A review for transdermal drug delivery, *Advan. Drug Delivery. Rev.*, 18:395–425 (1996).
24. M. R. Prausnitz. The effect of pulsed electrical protocols on skin damage, sensation and pain. *Proc. Int. Symp. Control. Rel. Bioact. Mater.*, 24:25–26 (1997).
25. U. Pliquett, R. Langer, and J. C. Weaver. Changes in the passive electrical properties of human stratum corneum due to electroporation, *Biochim. Biophys. Acta*, 1239:111–121 (1995).
26. S. A. Gallo, A. R. Oseroff, P. G. Johnson, and S. W. Hui. Characterization of electric-pulse-induced permeabilization of porcine skin using surface electrodes, *Biophys. J.*, 72:2805–2811 (1997).
27. U. Pliquett, M. R. Prausnitz, Y. A. Chizmadzhev, and J. C. Weaver. Measurement of rapid release kinetics for drug delivery, *Pharm. Res.*, 12:549–555 (1995).
28. M. R. Prausnitz, E. R. Edelman, J. A. Gimm, R. Langer, and J. C. Weaver. Transdermal delivery of heparin by skin electroporation, *Biotechnology*, 13:1205–1209 (1995).
29. J. C. Weaver, R. Vanbever, T. E. Vaughan, and M. R. Prausnitz. Heparin alters transdermal transport associated with electroporation, *Biochem. Biophys. Res. Commun.*, 234:637–640 (1997).
30. R. Vanbever, M. R. Prausnitz, and V. Preat. Macromolecules as novel transdermal transport enhancers for skin electroporation, *Pharm. Res.*, 14:638–644 (1997).
31. T. E. Zewert, U. F. Pliquett, R. Vanbever, R. Langer, and J. C. Weaver. Creation of transdermal pathways for macromolecule transport by skin electroporation and a low toxicity, pathway-enlarging molecule, *Bioelectrochem. Bioenerg.*, 49:11–20 (1999).
32. R. Vanbever, E. LeBoulenge, and V. Preat. Transdermal delivery of fentanyl by electroporation. 1. Influence of electrical factors, *Pharm. Res.*, 13:559–565 (1996).
33. A. S. Khan, L. C. Smith, R. V. Abruzzese, K. K. Cummings, M. A. Pope, P. A. Brown, and R. Draghia-Akli. Optimization of electroporation parameters for the intramuscular delivery of plasmids in pigs, *DNA Cell Biol.*, 22:807–814 (2003).
34. A. S. Khan, M. A. Pope, and R. Draghia-Akli. Highly efficient constant-current electroporation increases *in vivo* plasmid expression, *DNA Cell Biol.*, 24:810–818 (2005).

35. H. Hebel, H. Attra, A. Khan, and R. Draghia-Akli. Successful parallel development and integration of a plasmid-based biologic, container/closure system and electrokinetic delivery device, *Vaccine*, 24:4607–4614 (2006).

36. R. Draghia-Akli, A. S. Khan, M. A. Pope, and P. A. Brown. Innovative electroporation for therapeutic and vaccination applications, *Gene Ther. Mol. Biol.*, 9:329–338 (2005).

37. R. Draghia-Akli, A. S. Khan, P. A. Brown, M. A. Pope, L. Wu, L. Hirao, and D. B. Weiner. Parameters for DNA vaccination using adaptive constant-current electroporation in mouse and pig models, *Vaccine*, 26:5230–5237 (2008).

38. C. Curcio, A. S. Khan, A. Amici, M. Spadaro, E. Quaglino, F. Cavallo, G. Forni, and R. Draghia-Akli. DNA immunization using constant-current electroporation affords long-term protection from autochthonous mammary carcinomas in cancer-prone transgenic mice, *Cancer Gene Ther.*, 15:108–114 (2008).

39. M. Kotulska, J. Basalyga, M. B. Derylo, and P. Sadowski. Metastable pores at the onset of constant-current electroporation, *J. Membr. Biol.*, 236:37–41 (2010).

40. M. Rosati, A. Valentin, R. Jalah, V. Patel, A. von Gegerfelt, C. Bergamaschi, C. Alicea, D. Weiss, J. Treece, R. Pal, P. D. Markham, E. T. Marques, J. T. August, A. Khan, R. Draghia-Akli, B. K. Felber, and G. N. Pavlakis. Increased immune responses in rhesus macaques by DNA vaccination combined with electroporation, *Vaccine*, 26:5223–5229 (2008).

41. Inovio, http://www.inovio.com (accessed August 6, 2010).

42. T. E. Tjelle, R. Salte, I. Mathiesen, and R. Kjeken. A novel electroporation device for gene delivery in large animals and humans, *Vaccine*, 24:4667–4670 (2006).

43. M. P. Fons. Next generation vaccines: Antigen-encoding DNA delivered via electroporation, *Drug Del. Technol.*, 9:32–35 (2009).

44. K. E. Dolter, C. F. Evans, and D. Hannaman. *In vivo* delivery of nucleic acid-based agents with electroporation, *Drug Del. Technol.*, 10:37–41 (2010).

45. G. Widera and D. Rabussay. Electroporation mediated drug and DNA delivery in oncology and gene therapy, *Drug Del. Technol.*, 2:34–40 (2002).

46. P. J. Canatella and M. R. Prausnitz. Prediction and optimization of gene transfection and drug delivery by electroporation, *Gene Ther.*, 8:1464–1469 (2001).

47. B. Rubinsky. Irreversible electroporation in medicine, *Technol. Cancer Res. Treat.*, 6:255–260 (2007).

48. U. Pliquett, E. A. Gift, and J. C. Weaver. Determination of the electric field and anomalous heating caused by exponential pulses with aluminum electrodes in electroporation experiments, *Bioelectrochem. Bioenerg.*, 39:39–53 (1996).

49. S. B. Dev and G. A. Hofmann. Electrochemotherapy—A novel method of cancer treatment, *Cancer Treatment Rev.*, 20:105–115 (1994).

50. R. Vanbever, N. Lecouturier, and V. Preat. Transdermal delivery of metoprolol by electroporation, *Pharm. Res.*, 11:1657–1662 (1994).

51. U. Pliquett and J. C. Weaver. Transport of a charged molecule across the human epidermis due to electroporation, *J. Control. Release*, 38:1–10 (1996).

52. T. E. Zewert, U. F. Pliquett, R. Langer, and J. C. Weaver. Transdermal transport of DNA antisense oligonucleotides by electroporation, *Biochem. Biophys. Res. Commun.*, 212:286–292 (1995).

53. R. Heller, M. J. Jaroszeski, L. F. Glass, J. L. Messina, D. P. Rapaport, R. C. DeConti, N. A. Fenske, R. A. Gilbert, L. M. Mir, and D. S. Reintgen. Phase I/II trial for the treatment of cutaneous and subcutaneous tumors using electrochemotherapy, *Cancer*, 77:964–971 (1996).

54. M. R. Prausnitz, C. S. Lee, C. H. Liu, J. C. Pang, T. P. Singh, R. Langer, and J. C. Weaver. Transdermal transport efficiency during skin electroporation and iontophoresis, *J. Control. Release*, 38:205–217 (1996).

55. M. R. Prausnitz, D. S. Seddick, A. A. Kon, V. G. Bose, S. Frankenburg, S. N. Klaus, R. Langer, and J. C. Weaver. Methods for *in vivo* tissue electroporation using surface electrodes, *Drug Delivery*, 1:125–131 (1993).

56. A. Jadoul, V. Regnier, and V. Preat. Influence of ethanol and propylene glycol addition on the transdermal delivery by iontophoresis and electroporation, *Pharm. Res.*, 14:S-308–S-309 (1997).

57. L. Heller and M. L. Lucas. Delivery of plasmid DNA by *in vivo* electroporation, *Gene Ther. Mol. Biol.*, 5:55–60 (2000).

58. D. Miklavcic, L. M. Mir, and V. P. Thomas. Electroporation-based technologies and treatments, *J. Membr. Biol.*, 236:1–2 (2010).

59. K. Mori, T. Hasegawa, S. Sato, and K. Sugibayashi. Effect of electric field on the enhanced skin permeation of drugs by electroporation, *J. Control. Release*, 90:171–179 (2003).

60. E. Gronevik, F. V. von Steyern, J. M. Kalhovde, T. E. Tjelle, and I. Mathiesen. Gene expression and immune response kinetics using electroporation-mediated DNA delivery to muscle, *J. Gene Med.*, 7:218–227 (2005).

61. T. W. Wong, C. H. Chen, C. C. Huang, C. D. Lin, and S. W. Hui. Painless electroporation with a new needle-free microelectrode array to enhance transdermal drug delivery, *J. Control. Release*, 110:557–565 (2006).

62. S. O. Choi, Y. C. Kim, J. H. Park, J. Hutcheson, H. S. Gill, Y. K. Yoon, M. R. Prausnitz, and M. G. Allen. An electrically active microneedle array for electroporation, *Biomed. Microdevices*, 12:263–273 (2010).

63. D. Carroll, M. Tramer, H. McQuay, B. Nye, and A. Moore. Transcutaneous electrical nerve stimulation in labour pain: A systematic review, *Br. J. Obstet. Gynaecol.*, 104:169–175 (1997).

64. J. A. Craig, M. B. Cunningham, D. M. Walsh, G. D. Baxter, and J. M. Allen. Lack of effect of transcutaneous electrical nerve stimulation upon experimentally induced delayed onset muscle soreness in humans, *Pain*, 67:285–289 (1996).

65. L. Pizzamiglio, G. Vallar, and L. Magnotti. Transcutaneous electrical stimulation of the neck muscles and hemineglect rehabilitation, *Restor. Neurol. Neurosci.*, 10:197–203 (1996).

66. M. Torry, A. Wilcock, B. G. Cooper, and A. E. Tattersfield. The effect of chest wall transcutaneous electrical nerve stimulation on dyspnoea, *Resp. Physiol.*, 104:23–28 (1996).

67. M. F. Stevens, U. Linstedt, B. Neruda, P. Lipfert, and H. Wulf. Effect of transcutaneous electrical nerve stimulation on onset of axillary plexus block, *Anaesthesia*, 51:916–919 (1996).

68. R. K. Garnsworthy, R. L. Gully, P. Kenins, and R. A. Westerman. Transcutaneous electrical stimulation and the sensation of prickle, *J. Neurophys.*, 59:1116–1127 (1988).

69. R. Vanbever and V. Preat. *In vivo* efficacy and safety of skin electroporation, *Adv. Drug Del. Rev.*, 35:77–88 (1999).

70. L. Zhang and D. P. Rabussay. Clinical evaluation of safety and human tolerance of electrical sensation induced by electric fields with non-invasive electrodes, *Bioelectrochemistry*, 56:233–236 (2002).

71. P. Quaglino, C. Mortera, S. Osella-Abate, M. Barberis, M. Illengo, M. Rissone, P. Savoia, and M. G. Bernengo. Electrochemotherapy with intravenous bleomycin in the local treatment of skin melanoma metastases, *Ann. Surg. Oncol.*, 15:2215–2222 (2008).

72. V. Sharma, K. Stebe, J. C. Murphy, and L. Tung. Poloxamer 188 decreases susceptibility of artificial lipid membranes to electroporation, *Biophys. J.*, 71:3229–3241 (1996).

73. Y. Gauthier. Stress and skin: Experimental approaches, *Pathol. Biol.*, 44:882–887 (1996).

74. J. E. Riviere, N. A. Monteiroriviere, R. A. Rogers, D. Bommannan, J. A. Tamada, and R. O. Potts. Pulsatile transdermal delivery of LHRH using electroporation: Drug delivery and skin toxicology, *J. Control. Release*, 36:229–233 (1995).

75. S. B. Dev. Killing cancer cells with a combination of pulsed electric fields and chemotherapeutic agents, *Cancer Watch*, 3:12–14 (1994).

76. D. P. Rabussay, G. S. Nanda, and P. M. Goldfarb. Enhancing the effectiveness of drug-based cancer therapy by electroporation (electropermeabilization), *Technol. Cancer Res. Treat.*, 1:71–82 (2002).

77. A. Gothelf, L. M. Mir, and J. Gehl. Electrochemotherapy: Results of cancer treatment using enhanced delivery of bleomycin by electroporation, *Cancer Treat. Rev.*, 29:371–387 (2003).

78. M. Maty, G. Sersa, J. R. Garbay, J. Gehl, C. G. Collins, M. Snoj, V. Bilard, P. F. Geertsen, J. O. Larkin, D. Miklavcic, I. Pavlovic, S. M. Paulin-Kosir, M. Cemazar, N. Morsli, D. M. Soden, Z. Rudolf, C. Robert, G. C. O'Sullivan, and L. M. Mir. Electrochemotherapy—An easy, highly effective and safe treatment of cutaneous and subcutaneous metastases: Results of ESOPE (European standard operating procedures of electrochemotherapy) study, *EJC Supplements*, 4:3–13 (2006).

79. G. Sersa. The state-of-the-art of electrochemotherapy before the ESOPE study; advantages and clinical uses, *EJC Supplements*, 4:52–59 (2006).

80. J. O. Larkin, C. G. Collins, S. Aarons, M. Tangney, M. Whelan, S. O'Reily, O. Breathnach, D. M. Soden, and G. C. O'Sullivan. Electrochemotherapy: Aspects of preclinical development and early clinical experience, *Ann. Surg.*, 245:469–479 (2007).

81. S. B. Dev and G. A. Hofmann. Clinical applications of electroporation. In P. T. Lynch and M. R. Davey (eds), *Electrical manipulation of cells*, Chapman & Hall, London, 1996, pp. 185–199.

82. R. Heller, M. Jaroszeski, R. Perrott, J. Messina, and R. Gilbert. Effective treatment of B16 melanoma by direct delivery of bleomycin using electrochemotherapy, *Melanoma Res.*, 7:10–18 (1997).

83. M. J. Jaroszeski, R. Gilbert, R. Perrott, and R. Heller. Enhanced effects of multiple treatment electrochemotherapy, *Melanoma Res.*, 6:427–433 (1996).

84. R. A. Gilbert, M. J. Jaroszeski, and R. Heller. Novel electrode designs for electrochemotherapy, *Biochim. Biophys. Acta*, 1334:9–14 (1997).

85. A. K. Banga and M. R. Prausnitz. Assessing the potential of skin electroporation for the delivery of protein- and gene-based drugs, *Trends in Biotechnology*, 16:408–412 (1998).

86. Y. Tokudome and K. Sugibayashi. The synergic effects of various electrolytes and electroporation on the *in vitro* skin permeation of calcein, *J. Control. Release*, 92:93–101 (2003).

87. S. Bose, W. R. Ravis, Y. J. Lin, L. Zhang, G. A. Hofmann, and A. K. Banga. Electrically-assisted transdermal delivery of buprenorphine, *J. Control. Release*, 73:197–203 (2001).

88. S. Chang, G. A. Hofmann, L. Zhang, L. J. Deftos, and A. K. Banga. The effect of electroporation on iontophoretic transdermal delivery of calcium regulating hormones. *J. Control. Release*, 66, 127–133 (2000).

89. R. Conjeevaram, A. K. Banga, and L. Zhang. Electrically modulated transdermal delivery of fentanyl, *Pharm. Res.*, 19:440–444 (2002).

90. A. C. Hirsch, R. S. Upasani, and A. K. Banga. Factorial design approach to evaluate interactions between electrically assisted enhancement and skin stripping for delivery of tacrine, *J. Control. Release*, 103:113–121 (2005).

91. A. R. Denet and V. Preat. Transdermal delivery of timolol by electroporation through human skin, *J. Control. Release*, 88:253–262 (2003).

92. M. P. Cruz, S. L. Eeckhoudt, R. K. Verbeeck, and V. Preat. Transdermal delivery of flurbiprofen in the rat by iontophoresis and electroporation, *Pharm. Res.*, 14:S-309 (1997).

93. S. Wang, M. Kara, and T. R. Krishnan. Topical delivery of cyclosporin A coevaporate using electroporation technique, *Drug Dev. Ind. Pharm.*, 23:657–663 (1997).

94. L. Zhang, E. Nolan, S. Kreitschitz, and D. P. Rabussay. Enhanced delivery of naked DNA to the skin by non-invasive *in vivo* electroporation, *Biochim. Biophys. Acta*, 1572:1–9 (2002).

95. G. A. Hofmann, W. V. Rustrum, and K. S. Suder. Electro-incorporation of microcarriers as a method for the transdermal delivery of large molecules, *Bioelectrochem. Bioenerg.*, 38:209–222 (1995).

96. L. Zhang, L. N. Li, Z. L. An, R. M. Hoffman, and G. A. Hofmann. *In vivo* transdermal delivery of large molecules by pressure-mediated electroincorporation and electroporation: A novel method for drug and gene delivery, *Bioelectrochem. Bioenerg.*, 42:283–292 (1997).

97. L. Zhang and G. A. Hofmann. Electric pulse mediated transdermal drug delivery, *Proc. Int. Symp. Control. Rel. Bioact. Mater.*, 24:27–28 (1997).

98. S. N. Murthy, Y. L. Zhao, S. W. Hui, and A. Sen. Electroporation and transcutaneous extraction (ETE) for pharmacokinetic studies of drugs, *J. Control. Release*, 105:132–141 (2005).

99. K. Cheng, P. P. Tarjan, Y. C. Thio, and P. M. Mertz. *In vivo* 3-D distributions of electric fields in pig skin with rectangular pulse electrical current stimulation (RPECS), *Bioelectromagnetics*, 17:253–262 (1996).

100. J. Kost, U. Pliquett, S. Mitragotri, A. Yamamoto, R. Langer, and J. Weaver. Synergistic effect of electric field and ultrasound on transdermal transport, *Pharm. Res.*, 13:633–638 (1996).

101. G. L. Xiong, D. Y. Quan, and H. I. Maibach. Effects of penetration enhancers on *in vitro* percutaneous absorption of low molecular weight heparin through human skin, *J. Control. Release*, 42:289–296 (1996).

102. S. M. Sammeta, S. R. Vaka, and S. N. Murthy. Transcutaneous electroporation mediated delivery of doxepin-HPCD complex: A sustained release approach for treatment of postherpetic neuralgia, *J. Control. Release,* 142:361–367 (2010).

6 Sonophoresis for Intradermal and Transdermal Drug Delivery

6.1 INTRODUCTION

Healthy young adults typically can hear sounds between the frequencies of 40 Hz to 20,000 Hz (20 kHz). Ultrasound consists of cyclic sound pressure waves produced above this frequency (i.e., >20 kHz). Ultrasound is typically classified based on frequency as (1) high-frequency or diagnostic ultrasound (above 3 MHz), (2) medium-frequency or therapeutic ultrasound (1 to 3 MHz), and (3) low-frequency or power ultrasound (20 to 100 kHz). High-frequency ultrasound has well-known applications for clinical imaging such as in sonography for fetal imaging. Use of high-frequency ultrasound has also been reported for drug delivery applications (1). Medium-frequency or therapeutic ultrasound has been used for drug delivery in physical therapy clinics. At these frequencies, ultrasonic energy is more readily absorbed by tissue; therefore, a local heating effect often accompanies treatment, which can possibly offer some relief to patients with muscle soreness and related conditions even in the absence of any drug molecule. The temperature can rise by a few degrees Celsius and may also be directly responsible for the increased percutaneous absorption. Therapeutic ultrasound has been reported to have its primary effect on the diffusion coefficient of the drug rather than on its partition coefficient (2). Low-frequency ultrasound (LFU) uses frequencies in the range of 20 to 100 kHz and is now being recognized as being the most efficient for transdermal delivery applications (3–5). This is believed to be due to the cavitation that occurs at these frequencies. Sonophoresis or phonophoresis is the use of ultrasound to increase the permeation of drugs into or through the skin. Frequencies in the range of 20 kHz to 16 MHz have been investigated for this purpose. The drug may be delivered simultaneously while ultrasound is being applied, or the skin may be pretreated with ultrasound followed by application of the drug (5–8).

6.2 FACTORS AFFECTING SONOPHORESIS

Ultrasonic pressure waves are oscillatory and longitudinal in nature in the sense that the direction of propagation is the same as the direction of oscillation. These waves are characterized mainly by their frequency and amplitude. Cavitation, which plays a major role in enhancing skin permeability by LFU (see Section 6.3), varies inversely with frequency (9). Amplitude is often measured as the peak-wave pressure in units of Pascals or as intensity in units of W/cm^2. If ultrasound is applied in a pulse mode rather than continuously, then an additional parameter, the duty cycle, comes into play which represents the fraction of application time. Intensity (I) of application is dependent on the acoustic energy (E) emitted and the speed of sound (c) in the applied medium: $I = cE$ (10). The total energy delivered from the transducer then depends on the intensity and the net exposure time (4). Therefore, the main factors that can be controlled are the frequency, intensity, duty cycle, and application time.

In order to transfer ultrasound energy to the body, a contact or coupling medium is needed, because ultrasound is completely reflected by air. The coupling media used can be a gel or water, mineral oil, water-miscible creams, or emulsions. Emulsions can disperse the ultrasound waves at the oil–water interfaces, which may result in a reduction of intensity or localized heating (11). Many clinical studies have used cream-based preparations in the past, but a gel would be superior with respect to transmissivity of ultrasound (12). The application of gel on top of an occlusive dressing covering dexamethasone cream applied on the skin of healthy subjects has also been reported for sonophoresis treatment (13). The transmissivity is indicated by the acoustic impedance, and a coupling media is best for sonophoresis when its acoustic impedance is similar to that of skin (1.6×10^6 $kg/m^2/s$) (e.g., water or a gel having water). As air has very low (0.0004 $\times 10^6$ $kg/m^2/s$) acoustic impedance, ultrasound is reflected by air; therefore, a coupling media is necessary, and it should not have air bubbles. The ultrasonic propagation properties of excised human skin have been investigated by measuring acoustical speed, attenuation, and backscatter as a function of frequency. The speed of sound had a mean value of 1645 m s^{-1} in the epidermis and 1595 m s^{-1} in the dermis (14).

In addition to frequency and energy, other factors that affect sonophoretic enhancement include transducer geometry and gas concentration in the coupling medium (4). The distance between the transducer and the skin is another parameter that affects sonophoresis. The ultrasound intensity passes through a series of complex maxima and minima near the transducer but then decreases as a function of distance. In a study that measured skin conductivity, it was found to be inversely proportional to the distance from the horn (15). It was reported that ultrasound and sodium lauryl sulfate (SLS) exhibit a synergistic effect on transdermal transport (16–18). A 90-minute application of ultrasound from a 1% SLS solution induced a 200-fold increase in transport of mannitol across pig skin. In contrast, SLS alone enhanced permeation only threefold, and ultrasound alone enhanced permeation only eightfold. The synergistic effect was attributed to enhanced delivery of SLS into the skin and enhanced dispersion of SLS within the skin (16). In addition to these mechanisms, it has been shown that the simultaneous application of ultrasound and SLS also leads to a modification of the pH profile of the stratum corneum (17).

6.3   MECHANISMS OF ULTRASOUND-MEDIATED TRANSDERMAL DELIVERY

LFU has been shown to be most effective for transdermal delivery due to acoustic cavitation, which is the formation and collapse of gaseous bubbles caused by the local variations in acoustic pressure induced by the ultrasound waves. Using a high-pressure cell that was constructed to completely suppress cavitation during LFU, it was confirmed that cavitation is the primary mechanism by which LFU permeabilizes the skin (19). LFU is known to create microbubbles in water and tissue, and this acoustic cavitation is generally considered to create water channels in lipid bilayers. The resulting disorder in stratum corneum then allows the transport of hydrophilic drugs (9,20,21). It was reported that a significant fraction (~30%) of the intercellular lipids of the stratum corneum are removed during LFU (22), and pathways with a wide range of pore sizes are created though the size of the existing pores does not change (23,24). At the frequencies (typically 20 to 100 kHz) and pressure amplitudes (typically 1 to 2.4 bar) used in LFU, the radius of the cavitation bubbles is estimated to be about 10 to 100 μm, and these large bubbles are unlikely to generate in the 15 μm or so thick stratum corneum. Therefore, cavitation mostly occurs in the coupling media and creates disturbances that alter skin permeability (4,19). However, this may only be true for LFU. An early study with medium-frequency ultrasound using confocal microscopy suggested that cavitation occurs in the keratinocytes of the stratum corneum. Oscillations of the cavitation bubbles possibly enhance transdermal transport by inducing disorder in the stratum corneum lipids (25). Another report described the creation of 20-μm-sized defects in the stratum corneum by continuous application of 168 kHz ultrasound (26). Similarly, high-frequency treatment of skin at 16 MHz for 20 minutes was reported to damage the skin, presumably due to cavitational effects (27). In another study, a frequency of 3.5 MHz applied at high intensity was reported to damage follicular structures (28). Cavitation may be stable or inertial. Stable cavitation involves periodic growth and oscillations of bubbles which occur at low-pressure amplitudes and when the size of the bubbles is not at the resonant size for the ultrasound frequency being used. It can generate some convective transport with local swirling fluid due to the vibration, but the resulting microstreaming is generally not sufficient to enhance skin permeability though it can disrupt weak fluid lipid bilayers such as that of red blood cells. In contrast, inertial cavitation can induce major changes in skin permeability and involves what is described in the literature as a "violent event" that occurs at relatively high-pressure amplitudes when bubbles are near the resonant size for the frequency being used (29). When the inertial cavitation bubbles collapse near skin, they generate a shock wave, vigorous microstreaming and microjets that can permeabilize the skin reversibly (30). Amplitude of the shock wave decreases rapidly with distance. During LFU treatment of skin at a particular frequency, there is a threshold intensity before which skin permeabilization is not seen. After the threshold intensity is reached, enhancement increases with intensity until another threshold, the decoupling intensity, is reached. The threshold energy is highly dependent on the frequency and increases by about 130-fold as the frequency is increased from about 20 to 93 kHz (4,31). Within the range (20 to 100 kHz) of LFU, lower frequencies produce more enhancement, but it has been recognized in recent years that this enhancement may be localized to certain areas termed *localized transport pathways* (LTRs) (5,32). At higher frequencies, transport is more homogeneous, but as discussed, it will also need more energy. An optimum frequency seems to be around 60 kHz where a reasonable energy dose can produce significant enhancement in skin permeability (4). This is perhaps why the marketed SonoPrep® device (see Section 6.5) uses a 55-kHz frequency. Formation of LTRs at low frequencies became evident when recent studies used colored or fluorescent permeants or quantum dots (18) and noted the presence of millimeter-size discrete domains or regions that were highly permeable and may occupy about 5% of the total skin area exposed to ultrasound. Most (but not all) of the studies that reported the formation of LTRs were done at a frequency of 20 kHz applied with a 5s:5s on:off duty cycle at intensities ranging from 2.5 to 7.5 W/cm^2, and these were reviewed by Kushner et al. (32). Diffusion masked experiments where hydrophobic silicone high vacuum grease is applied to LTRs or non-LTR regions have been used along with measurements of skin electrical resistivity to study the relative permeability of different regions of the skin. LTRs can be up to 80-fold more permeable than the non-LTR region of skin treated with LFU, and two to three orders of magnitude more permeable than untreated skin. Also, the electrical resistivity of skin treated with LFU may decrease about 170-fold in the non-LTR region and over 5000-fold in the LTR region (5,32,33).

Another noncavitational mechanism involved in ultrasound-mediated delivery is acoustic streaming (different from microstreaming), which is a large-scale convective motion, but this is generally believed not to be a major factor in enhancing skin permeability. However, it was reported to be a contributing factor to sonophoretic transport enhancement (34). Using mannitol as a model hydrophilic permeant, it was shown that convection can play an important role during low LFU conditions when using human heat-stripped skin. In contrast, convective flow was not an important contributor to transport when using full-thickness pig skin (35). Thermal effects that accompany sonophoresis may also have a role in enhancing permeation (36), especially under conditions other than LFU. However, even for delivery of mannitol under LFU conditions, it was reported that thermal effects accounted for 25% of the enhancement with the rest attributed to cavitation (37). Several of these mechanisms may act in synergy during sonophoresis (20).

6.4 DELIVERY OF LIPOPHILIC DRUGS

Sonophoresis seems to work better for delivery of hydrophilic drugs (38). In a study by Morimoto et al., the authors used lipophilic rhodamine B dye, and it was seen that unlike the hydrophilic dextrans, the dye did not transport into the skin by sonophoresis, even in areas of skin directly under the ultrasound horn (39). In an earlier study that utilized nine drug molecules with varying octanol/water partition coefficients, flux of lipophilic drugs across freshly excised hairless rat skin after sonication was the same as that before sonication. In contrast, the flux of hydrophilic drugs was increased about sevenfold after sonication performed at 150 kHz and 2 W/cm^2 (40). A lack of understanding of the different behavior of hydrophilic versus lipophilic drugs under sonophoresis has resulted in a lot of literature, especially in the early years, where the authors interpreted these differences as variations in the efficacy of sonophoresis and even questioned its usefulness. Mitragotri et al. (41) proposed an equation to predict the relative sonophoretic enhancement, e ([sonophoretic permeability/passive permeability] −1) by an equation based on the drug's octanol–water partition coefficient ($K_{o/w}$) and passive permeability (P) in units of cm/h:

$$e = \frac{K_{o/w}^{0.75}}{(4 \times 10^4)\ P}$$

Predicted e values of 13 drugs were calculated and found to be in agreement with published literature (41). In an early study by Machet et al., *in vitro* transport of digoxin across human skin was not enhanced by sonophoresis performed at 3.3 MHz. Transport across mouse skin was also not enhanced at an intensity of 1 W/cm^2 but was enhanced at 3 W/cm^2 (42). This was perhaps because digoxin is a lipophilic drug molecule, though frequency used is also an important factor. Another study found that ultrasound at 150 kHz increased the permeability of the polar compound, antipyrine, across excised hairless rat skin but for isosorbide dinitrate, only a slight increase was seen which was attributed to thermal effects of ultrasound (43).

6.5 SONOPHORESIS DEVICES

Ultrasound is generated by a sonicator device that generates an electrical AC signal at the desired frequency and amplitude and applies the signal across a piezoelectric crystal (transducer) to generate ultrasound waves. The piezoelectric principle describes the behavior of certain ceramics like lead zirconate, titanate, or crystallized quartz which develop electrical charges on the surface when a pressure is applied to them or expand or contract when a voltage is applied across them. These dimensional changes in the transducer convert electrical energy to mechanical energy in the form of oscillations that then generate acoustic waves. The thickness of the crystal is chosen in a way that it will resonate at the operating frequency of the sonicator device. Some examples of published reports include devices that operate at frequencies of 20 kHz (3,19,21,44), 36 kHz (45), 41 kHz (39), 48 kHz (46,47), and around 55 kHz (24), including the SonoPrep device at 55 kHz (48,49). Higher frequencies have also been used in published reports, including 150 kHz (40,43), 300 kHz (50), around 1 MHz (51–55), around 3 MHz (13,42,52,56), and 16 MHz (27).

SonoPrep is a low-frequency (55 kHz) ultrasound device that was marketed by the former Sontra Medical Corporation in 2004, and a second-generation device was later marketed in 2006. Sontra Medical Corporation is now Echo Therapeutics (Franklin, Massachusetts) and is developing an even more compact device for delivery of lidocaine. At the tip, the SonoPrep device has a cylindrical ultrasonic horn inside a housing that positions the horn 7.5 mm above the skin. The ultrasonic horn contained in the hand piece vibrates at 55,000 times per second (55 kHz). The housing gets filled with a coupling buffer consisting of phosphate buffered saline (PBS) and 1% sodium lauryl sulfate (SLS), and the cavitation field in the buffer creates random 25- to 125-μm

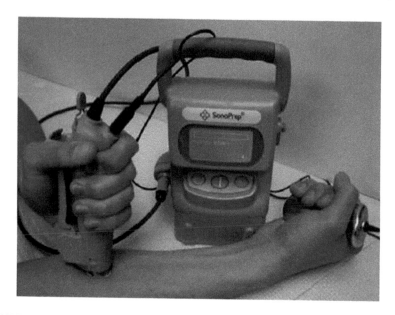

FIGURE 6.1   Sonication of the subject's volar forearm by the SonoPrep® device. The hand piece interacts with the main console of the device to determine the duration of sonication by measuring the drop in skin resistance. (Reprinted from *Ultrasound Med. Biol.*, 35:1405–1408, J. Gupta and M. R. Prausnitz, Recovery of Skin Barrier Properties after Sonication in Human Subjects [49]. Copyright 2009, with permission from Elsevier.) **(See color insert.)**

crevices and pores in the stratum corneum. A closed loop feedback measurement of the drop in skin impedance (see Figure 6.1) controls the duration of the ultrasound application (typically, 5 to 30 seconds). The SonoPrep is still used today even though new units are no longer commercially available. A commercially available medium-frequency device from the Chattanooga Group (Vista, California) operates at 1 and 3.3 MHz frequencies and operates in continuous or pulsed mode at 10%, 20%, 50%, or 100% (57). A Patch-Cap sonophoresis device based on U-strip technology was being developed by the former Dermisonics, Inc. (58), but it seems that this product is no longer being developed. Ultrasound devices are also available from several other vendors, including Enraf-Nonius (Rotterdam, the Netherlands), Labthermics Technologies, Inc. (Champaign, Illinois), and RichMar (Chattanooga, Tennessee). In addition, several investigators have used custom-built devices. A lightweight cymbal transducer design was described in the literature. It consists of two metal caps epoxied onto a lead zirconate-titanate ceramic, creating a shallow cavity beneath the caps and the ceramic. The flexing of the end caps by the radial motion of the ceramic is the main mode of vibration, and four cymbal transducers are connected in an array (21,59,60).

6.6   SKIN IRRITATION AND ELECTRICAL PROPERTIES

Ultrasonic irradiation of the skin will change its electrochemical properties. It was reported that constant current iontophoresis (0.1 mA/cm^2) of excised hairless rat skin after pretreatment by ultrasound significantly increased the flux of benzoate anion

(BA) through skin compared to that without pretreatment. Flux during iontophoresis was dependent on the duration of ultrasonic pretreatment. The difference in electric potential across the skin during iontophoresis was lower if the skin was pretreated with ultrasound, suggesting that ultrasound causes structural disorder in the stratum corneum lipids, leading to increased aqueous region in the stratum corneum and increased diffusivity of benzoate ion in the skin (61). Human skin samples exposed to ultrasound intensity of 5.2 W/cm^2 exhibited epidermal detachment and edema, but no changes were observed when the intensity was 2.5 W/cm^2, applied up to 10 minutes of continuous mode or 1 hour of pulsed operation at 10% duty cycle (0.1s on:0.9 s off) (5,32). Similarly, a study in dogs reported that low-frequency sonophoresis was safe, unless applied at a high intensity when it can cause second-degree burns, most likely due to localized heating (62). Ultrasound-induced damage to skin may also be species dependent. Exposure of ultrasound at 48 kHz (0.5 W/cm^2) for 5 minutes did not damage human skin but caused damage to hairless mice skin (47). High-frequency treatment of skin at 16 MHz for 20 minutes has been reported to alter the cellular morphology of the skin, but no damage was reported at 5 minutes (27). Application of ultrasound in pulsed mode can reduce thermal effects on the skin, but it may also reduce cavitational effects or may require a higher intensity to achieve the same cavitational effects. The higher intensity, in turn, will increase the thermal effects again.

In a human study with the SonoPrep device (Figure 6.1), the recovery of the skin barrier following sonophoresis treatment was followed by monitoring electrical impedance which, in turn, was shown to correlate with drug flux. Impedance of skin dropped from >200 kΩ-cm^2 to less than 10 kΩ-cm^2. Skin was observed to gain back some impedance within 7 hours. However, if the skin was occluded, this recovery was delayed for up to 42 hours. Upon removal of occlusion at 42 hours, the impedance increased rapidly (49). Interestingly, we observed a similar effect of occlusion on pore closure following skin microporation, and this was discussed in Chapter 3. Another human study used a 36-kHz frequency to delivery histamine into the anterior side of the subject's forearms. Echographic measurements of the skin (see Section 2.5) showed successful delivery of histamine into the arm, as indicated by an increase in dermal edema (Figure 6.2). In addition to proof of concept of delivery, a potential use of delivery of histamine by sonophoresis could be its use as a positive control in allergy testing as an alternative to prick tests. Self-reported pain was also evaluated in this study on a 0 (no pain) to 100 (maximal pain) VAS score, and large variation (5 to 40) was reported by the subjects in their perception of pain. It was suggested that this may be due to differences in hair follicle density of the subjects, as pain may be resulting from gaseous microbubbles entrapped at the root of the hair follicles (45).

6.7   TRANSDERMAL DELIVERY APPLICATION OF SONOPHORESIS

Due to the multiplicity of interactions of ultrasound with cells and tissues, it has found uses in a diverse set of medical applications including bone-fracture healing, lipoplasty, tumor ablation, dissolving blood clots, gene therapy, ocular delivery, nail delivery, chemopotentiation, and use of stabilized microbubbles injected into systemic circulation as second-generation ultrasound contrast agents (29,38,63,64).

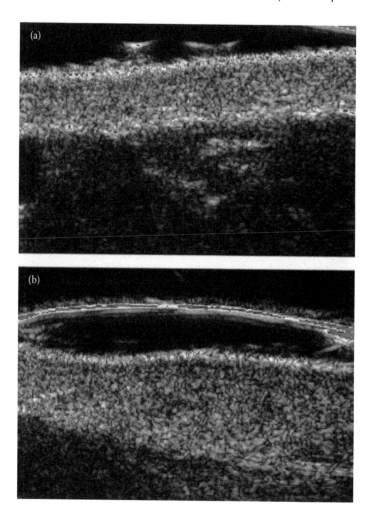

FIGURE 6.2 High-resolution ultrasound imaging of skin thickness (a) before treatment, and (b) after 30 minutes of histamine application by sonophoresis. (Reprinted from *Int. J. Pharm.*, 385:37–41, A. Maruani et al., Efficiency of Low-Frequency Ultrasound Sonophoresis in Skin Penetration of Histamine: A Randomized Study in Humans [45]. Copyright 2010, with permission from Elsevier.) **(See color insert.)**

Cerevast Therapeutics (Redmond, Washington) is developing sonolysis products to dissolve blood clots, aided by 1- to 2-micron-sized gaseous microspheres that can penetrate the fibrin matrix of blood clots and undergo cavitation when exposed to the ultrasound field (65). However, the focus of this chapter is on the applications of ultrasound in drug delivery into or through the skin.

6.7.1 Delivery of Small Molecules

Some of the drug classes investigated for delivery by sonophoresis include analgesics, anti-inflammatory agents, anesthetics, antibiotics, anticancer agents, corticosteroids, vasodilators, and hormones, and these were recently reviewed (66). Other reviews compiled a list of drugs delivered by ultrasound (8,10,38). Some specific examples of drugs investigated for delivery by sonophoresis include dexamethasone (13), 5-fluorouracil (55), hydrocortisone (52), ibuprofen (53), lidocaine (46), and prednisolone (67). Echo Therapeutics (Franklin, Massachusetts) completed clinical trials involving ultrasound treatment followed by application of 4% lidocaine cream for quick onset of dermal anesthesia and reported successful results. The company now plans to submit a 510(k) premarket notification to the U.S. Food and Drug Administration (FDA) in collaboration with Ferndale Pharma Group, Inc. (Ferndale, Michigan) (68). The appearance of prednisolone following the bioconversion of prednisolone 21-acetate in skin was reported to decrease with increasing the product of the intensity of ultrasound and the duration for which it is applied, suggesting possible deactivation of skin enzymes by ultrasound (67). In a study that delivered mannitol as a model hydrophilic molecule, the authors compared three *in vitro* skin models (full-thickness and dermatomed pig skin as well as heat separated human skin) to the *in vivo* pig model. Using low-frequency sonophoresis, the authors reported that living skin *in vivo* is more sensitive to ultrasound treatment and good *in vitro–in vivo* correlation was observed only when *in vitro* skin was treated with different ultrasound energies to reach the same skin electrical resistance value as that of *in vivo* pig skin (69). Another study delivered mannitol and inulin and reported that permeation was increased by 5- to 20-fold within 1 to 2 hours following ultrasound application for 3 to 5 minutes (51). A low-frequency massage device has been investigated for delivery of sodium benzoate, ketoprofen, and diclofenac sodium (70).

6.7.2 Delivery of Insulin

Several studies reported the successful delivery of insulin by sonophoresis. In some early studies, Tachibana et al. delivered insulin to hairless mice at 48 kHz and to diabetic rabbits at 105 kHz and reported a rapid drop in blood glucose levels (71,72). Using a lightweight cymbal transducer (see Section 6.5) at 20 kHz, Smith and coworkers reported the delivery of insulin *in vitro* across human skin (73) and *in vivo* to rats (74,75), rabbits (60), or pigs (21). When ultrasound was applied to rats, a 20-minute exposure had the same effect as a 60-minute exposure (74), and a subsequent study demonstrated that an exposure time of just 5 to 10 minutes is sufficient to deliver a clinically significant dose of insulin to reduce blood glucose levels (75). When ultrasound was applied to large pigs for 1 hour for delivery of human insulin, blood glucose levels were found to drop over a period of 90 minutes (Figure 6.3), which was statistically significant as compared to the control group (21).

6.7.3 Delivery of Other Macromolecules

The ability of ultrasound to enable protein delivery into or through skin has now been recognized for some time (76), and an early study demonstrated the delivery of the macromolecules insulin, interferon, and erythropoietin across human skin *in vitro* by low-frequency (20 kHz) sonophoresis applied at high intensities (77). Delivery of luteinizing hormone-releasing hormone, inulin, and dextran by low-frequency

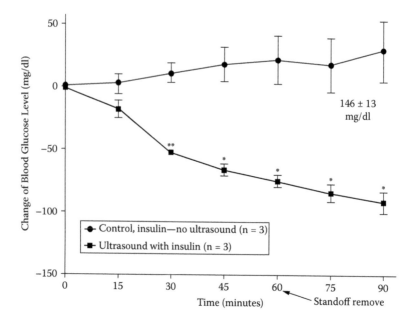

FIGURE 6.3 Blood glucose levels in pigs over time in control (circles) versus treatment (rectangles) group. Treatment group was treated with sonophoresis at 20 kHz for 60 minutes. (With kind permission from Springer Netherlands: *Pharm. Res.*, Ultrasound-Mediated Transdermal Insulin Delivery in Pigs Using a Lightweight Transducer, 24:1396–1401, 2007, E. J. Park, J. Werner, and N. B. Smith [21]. Copyright 2007.)

sonophoresis was also reported (24). Morimoto et al. reported on the delivery of FITC-labeled dextrans up to 38 kDa across hairless rat skin by using low-frequency (41 kHz) sonophoresis. By taking confocal images of skin at different depths, intense fluorescence was observed under the ultrasound horn without any specific localization pattern (Figure 6.4). Fluorescence outside the horn was similar to the passive control. In boundary areas, some fluorescence was seen along the crack-like structures. Sonophoresis was also shown to increase transport of tritiated water in this study, and it was suggested that sonophoresis increases the transport of hydrophilic drugs across skin by inducing convective flow. The authors hypothesized that swelling of corneocytes and formation of water domains within the corneocytes is involved (39). High-frequency ultrasound at 16 MHz for just 5 minutes was reported to deliver a hydrophilic colloidal tracer, lanthanum hydroxide through the epidermis to the upper dermis in the hairless guinea pig model (27). Penetration of particles, quantum dots (20 nm) well beyond the stratum corneum was also reported by using low-frequency sonophoresis (18). Another study (78) reported the diffusion of 51 kDa poly-*l*-lysine-fluorescein isothiocyanate and even particles up to 25 μm into skin. This study used ultrasound at a frequency of 20 kHz and at a high intensity of 19 W/cm^2, probably resulting in the formation of LTRs (see Section 6.3). Low-frequency ultrasound has also found application for transcutaneous immunization (79). It was reported to enhance the immune response induced by topical application of tetanus toxoid, which was partly attributed to enhanced delivery of the antigen into the skin

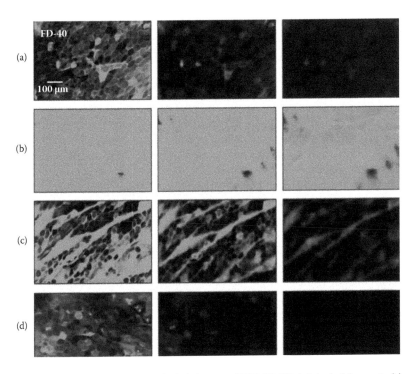

FIGURE 6.4 Delivery of FITC-labeled dextran (MW 38 kDa) into hairless rat skin as imaged by confocal microscopy up to a depth of 20 μm by (a) passive diffusion over 12 hours or by sonication at 41 kHz (b) directly under the horn (c) in boundary area; or (d) in area outside the horn. (Reprinted from *J. Control. Release*, 103:587–597, Y. Morimoto et al., Elucidation of the Transport Pathway in Hairless Rat Skin Enhanced by Low-Frequency Sonophoresis Based on the Solute-Water Transport Relationship and Confocal Microscopy [39]. Copyright 2005.) **(See color insert.)**

but also partly attributed to activation of immune cells by low-frequency ultrasound (80). A detailed discussion of transdermal delivery of vaccines by physical enhancement techniques can be found in Chapter 9.

6.7.4 CLINICAL STUDIES

Sonophoresis has been used by physiotherapists for delivery of drugs like corticosteroids for over 40 years, though some of the early studies were conducted on a subjective and nonquantitative basis (81). These studies have typically used medium-frequency ultrasound in patients at an intensity less than 3 W cm^{-2} for a typical treatment period of 10 minutes or less. In a study with 26 patients scheduled for surgery for knee disorders, sonophoresis was performed with a commercially available 2.5% ketoprofen (log P 0.97) gel in continuous (1 MHz, 1.5 W/cm^2 for 5 minutes) or pulsed mode (100 Hz, 20% duty cycle) just before surgery. During surgery, a biopsy of adipose tissue and synovial tissue was taken. It was found that ketoprofen did not accumulate in the fat tissue but was found in the synovial tissue for both the continuous and pulsed treatment groups. A third control group that received sham sonophoresis did not have the drug in the synovial tissue. Blood samples were also taken, but the concentration of ketoprofen in plasma was negligible in all three groups. Because synovial tissue is a major site of inflammation, the presence of the drug here is of clinical importance (12). Another study in 19 patients delivered ketoprofen by sonophoresis to the elbow joint and reported that it is useful for the treatment of epicondylitis (82). Similarly, another study with 80 patients having knee osteoarthritis found that both continuous and pulsed sonophoresis of diclofenac gel was more effective for pain relief and functional status of the patients as compared to just topical application of the gel (83).

In a proof of concept–type study in 10 healthy volunteers, delivery of nicotinate esters by sonophoresis was demonstrated by vasodilatation as a pharmacodynamic parameter, monitored by laser Doppler velocimetry (see Section 2.5). When sonophoresis was applied at 3 MHz for 5 minutes at an intensity of 1 W/cm^2 using a gel base, an enhanced vasodilation response was observed. It was suggested that the mechanism of delivery involved disordering of the structured lipids in the stratum corneum (56,84). Ultrasound has also been investigated for extraction of interstitial fluid as a predictor of blood glucose levels in both *in vitro* and human studies. Glucose levels in the extracted interstitial fluid correlated well with blood glucose concentrations over the range of 50 to 250 mg/dl (85,86). Similar studies have also been done by reverse iontophoresis and are discussed in detail in Section 4.8. Echo Therapeutics (Franklin, Massachusetts) is developing an ultrasound-based glucose monitoring system for hospital patients (68).

6.8   COMBINATION APPROACHES

Sonophoresis has been used in combination with other physical enhancement techniques, such as iontophoresis, electroporation, chemical enhancers, and microneedles. A synergy between iontophoresis and ultrasound is expected, because each technique enhances transport through different mechanisms (87). Investigators reported on the synergistic effect of iontophoresis and sonophoresis using model drug molecules such as calcein, heparin, and sulfordamine (88,89). A recent study (50) investigated the combination of iontophoresis and sonophoresis using seven model chemicals and reported that synergistic effects are seen only for uncharged molecules, suggesting a possible role of convective flow in the enhancement. Sonophoresis has also been used in combination with chemical enhancers (44,52,54). A synergistic effect was reported when chemical enhancers and sonophoresis were used together for delivery of tizanidine hydrochloride, a centrally acting muscle relaxant with low oral bioavailability (44). In a study that combined a microneedle roller with sonophoresis, it was reported that the combination increased the delivery of glycerol 2.3-fold as compared to microneedling alone. Glycerol can reduce tissue turbidity by dehydration and partial replacement of interstitial fluid to allow better imaging for light diagnosis and therapy (90). Studies investigating other combination approaches were also reported, such as sonophoresis with hollow microneedles (91), with microdermabrasion (92), and with liposomes (93).

REFERENCES

1. D. Bommannan, H. Okuyama, P. Stauffer, and R. H. Guy. Sonophoresis. I. The use of high-frequency ultrasound to enhance transdermal drug delivery, *Pharm. Res.*, 9:559–564 (1992).
2. S. Mitragotri. Effect of therapeutic ultrasound on partition and diffusion coefficients in human stratum corneum, *J. Control. Release*, 71:23–29 (2001).
3. S. Mitragotri, D. Blankschtein, and R. Langer. Transdermal drug delivery using low-frequency sonophoresis, *Pharm. Res.*, 13:411–420 (1996).
4. S. Mitragotri and J. Kost. Low-frequency sonophoresis: A review, *Adv. Drug Deliv. Rev.*, 56:589–601 (2004).
5. M. Ogura, S. Paliwal, and S. Mitragotri. Low-frequency sonophoresis: Current status and future prospects, *Adv. Drug Deliv. Rev.*, 60:1218–1223 (2008).
6. P. Tyle and P. Agrawala. Drug delivery by phonophoresis, *Pharm. Res.*, 6:355–361 (1989).
7. I. Lavon and J. Kost. Ultrasound and transdermal drug delivery, *Drug Discov. Today*, 9:670–676 (2004).
8. N. B. Smith. Perspectives on transdermal ultrasound mediated drug delivery, *Int. J. Nanomedicine*, 2:585–594 (2007).
9. H. Ueda, M. Mutoh, T. Seki, D. Kobayashi, and Y. Morimoto. Acoustic cavitation as an enhancing mechanism of low-frequency sonophoresis for transdermal drug delivery, *Biol. Pharm. Bull.*, 32:916–920 (2009).
10. L. Machet and A. Boucaud. Phonophoresis: Efficiency, mechanisms and skin tolerance, *Int. J. Pharm.*, 243:1–15 (2002).
11. E. Vranic. Sonophoresis—Mechanisms and application, *Bosn. J. Basic Med. Sci.*, 4:25–32 (2004).
12. B. Cagnie, E. Vinck, S. Rimbaut, and G. Vanderstraeten. Phonophoresis versus topical application of ketoprofen: Comparison between tissue and plasma levels, *Phys. Ther.*, 83:707–712 (2003).
13. S. Saliba, D. J. Mistry, D. H. Perrin, J. Gieck, and A. Weltman. Phonophoresis and the absorption of dexamethasone in the presence of an occlusive dressing, *J. Athl. Train.*, 42:349–354 (2007).
14. C. M. Moran, N. L. Bush, and J. C. Bamber. Ultrasonic propagation properties of excised human skin, *Ultrasound Med. Biol.*, 21:1177–1190 (1995).
15. T. Terahara, S. Mitragotri, J. Kost, and R. Langer. Dependence of low-frequency sonophoresis on ultrasound parameters; distance of the horn and intensity, *Int. J. Pharm.*, 235:35–42 (2002).
16. S. Mitragotri, D. Ray, J. Farrell, H. Tang, B. Yu, J. Kost, D. Blankschtein, and R. Langer. Synergistic effect of low-frequency ultrasound and sodium lauryl sulfate on transdermal transport, *J. Pharm. Sci.*, 89:892–900 (2000).
17. I. Lavon, N. Grossman, and J. Kost. The nature of ultrasound–SLS synergism during enhanced transdermal transport, *J. Control. Release*, 107:484–494 (2005).
18. S. Paliwal, G. K. Menon, and S. Mitragotri. Low-frequency sonophoresis: Ultrastructural basis for stratum corneum permeability assessed using quantum dots, *J. Invest. Dermatol.*, 126:1095–1101 (2006).
19. H. Tang, C. C. Wang, D. Blankschtein, and R. Langer. An investigation of the role of cavitation in low-frequency ultrasound-mediated transdermal drug transport, *Pharm. Res.*, 19:1160–1169 (2002).
20. I. Lavon, N. Grossman, J. Kost, E. Kimmel, and G. Enden. Bubble growth within the skin by rectified diffusion might play a significant role in sonophoresis, *J. Control. Release*, 117:246–255 (2007).

21. E. J. Park, J. Werner, and N. B. Smith. Ultrasound mediated transdermal insulin delivery in pigs using a lightweight transducer, *Pharm. Res.*, 24:1396–1401 (2007).

22. R. Alvarez-Roman, G. Merino, Y. N. Kalia, A. Naik, and R. H. Guy. Skin permeability enhancement by low frequency sonophoresis: Lipid extraction and transport pathways, *J. Pharm. Sci.*, 92:1138–1146 (2003).

23. A. Tezel, A. Sens, and S. Mitragotri. A theoretical analysis of low-frequency sonophoresis: Dependence of transdermal transport pathways on frequency and energy density, *Pharm. Res.*, 19:1841–1846 (2002).

24. A. Tezel, A. Sens, and S. Mitragotri. Description of transdermal transport of hydrophilic solutes during low-frequency sonophoresis based on a modified porous pathway model, *J. Pharm. Sci.*, 92:381–393 (2003).

25. S. Mitragotri, D. A. Edwards, D. Blankschtein, and R. Langer. A mechanistic study of ultrasonically-enhanced transdermal drug delivery, *J. Pharm. Sci.*, 84:697–706 (1995).

26. J. Wu, J. Chappelow, J. Yang, and L. Weimann. Defects generated in human stratum corneum specimens by ultrasound, *Ultrasound Med. Biol.*, 24:705–710 (1998).

27. D. Bommannan, G. K. Menon, H. Okuyama, P. M. Elias, and R. H. Guy. Sonophoresis. II. Examination of the mechanism(s) of ultrasound-enhanced transdermal drug delivery, *Pharm. Res.*, 9:1043–1047 (1992).

28. V. M. Meidan, A. D. Docker, A. D. Walmsley, and W. J. Irwin. Low intensity ultrasound as a probe to elucidate the relative follicular contribution to total transdermal absorption, *Pharm. Res.*, 15:85–92 (1998).

29. W. G. Pitt. Defining the role of ultrasound in drug delivery, *Am. J. Drug Del.*, 1:27–42 (2003).

30. A. Tezel and S. Mitragotri. Interactions of inertial cavitation bubbles with stratum corneum lipid bilayers during low-frequency sonophoresis, *Biophys. J.*, 85:3502–3512 (2003).

31. S. Mitragotri, J. Farrell, H. Tang, T. Terahara, J. Kost, and R. Langer. Determination of threshold energy dose for ultrasound-induced transdermal drug transport, *J. Control. Release*, 63:41–52 (2000).

32. J. Kushner, D. Blankschtein, and R. Langer. Heterogeneity in skin treated with low-frequency ultrasound, *J. Pharm. Sci.*, 97:4119–4128 (2008).

33. J. Kushner, D. Blankschtein, and R. Langer. Experimental demonstration of the existence of highly permeable localized transport regions in low-frequency sonophoresis, *J. Pharm. Sci.*, 93:2733–2745 (2004).

34. M. Mutoh, H. Ueda, Y. Nakamura, K. Hirayama, M. Atobe, D. Kobayashi, and Y. Morimoto. Characterization of transdermal solute transport induced by low-frequency ultrasound in the hairless rat skin, *J. Control. Release*, 92:137–146 (2003).

35. H. Tang, S. Mitragotri, D. Blankschtein, and R. Langer. Theoretical description of transdermal transport of hydrophilic permeants: Application to low-frequency sonophoresis, *J. Pharm. Sci.*, 90:545–568 (2001).

36. J. P. Simonin. On the mechanisms of *in vitro* and *in vivo* phonophoresis, *J. Control. Release*, 33:125–141 (1995).

37. G. Merino, Y. N. Kalia, M. B. Delgado-Charro, R. O. Potts, and R. H. Guy. Frequency and thermal effects on the enhancement of transdermal transport by sonophoresis, *J. Control. Release*, 88:85–94 (2003).

38. R. Rao and S. Nanda. Sonophoresis: Recent advancements and future trends, *J. Pharm. Pharmacol.*, 61:689–705 (2009).

39. Y. Morimoto, M. Mutoh, H. Ueda, L. Fang, K. Hirayama, M. Atobe, and D. Kobayashi. Elucidation of the transport pathway in hairless rat skin enhanced by low-frequency sonophoresis based on the solute-water transport relationship and confocal microscopy, *J. Control. Release*, 103:587–597 (2005).

40. H. Ueda, K. Sugibayashi, and Y. Morimoto. Skin penetration-enhancing effect of drugs by phonophoresis, *J. Control. Release*, 37:291–297 (1995).
41. S. Mitragotri, D. Blankschtein, and R. Langer. An explanation for the variation of the sonophoretic transdermal transport enhancement from drug to drug, *J. Pharm. Sci.*, 86:1190–1192 (1997).
42. L. Machet, J. Pinton, F. Patat, B. Arbeille, L. Pourcelot, and L. Vaillant. *In vitro* phonophoresis of digoxin across hairless mice and human skin: Thermal effect of ultrasound, *Int. J. Pharm.*, 133:39–45 (1996).
43. H. Ueda, M. Ogihara, K. Sugibayashi, and Y. Morimoto. Difference in the enhancing effects of ultrasound on the skin permeation of polar and non-polar drugs, *Chem. Pharm. Bull. Tokyo*, 44:1973–1976 (1996).
44. S. Mutalik, H. S. Parekh, N. M. Davies, and N. Udupa. A combined approach of chemical enhancers and sonophoresis for the transdermal delivery of tizanidine hydrochloride, *Drug Deliv.*, 16:82–91 (2009).
45. A. Maruani, E. Vierron, L. Machet, B. Giraudeau, and A. Boucaud. Efficiency of low-frequency ultrasound sonophoresis in skin penetration of histamine: A randomized study in humans, *Int. J. Pharm.*, 385:37–41 (2010).
46. K. Tachibana and S. Tachibana. Use of ultrasound to enhance the local anesthetic effect of topically applied aqueous lidocaine, *Anesthesiology*, 78:1091–1096 (1993).
47. N. Yamashita, K. Tachibana, K. Ogawa, N. Tsujita, and A. Tomita. Scanning electron microscopic evaluation of the skin surface after ultrasound exposure, *Anat. Rec.*, 247:455–461 (1997).
48. A. Farinha, S. Kellogg, K. Dickinson, and T. Davison. Skin impedance reduction for electrophysiology measurements using ultrasonic skin permeation: Initial report and comparison to current methods, *Biomed. Instrum. Technol.*, 40:72–77 (2006).
49. J. Gupta and M. R. Prausnitz. Recovery of skin barrier properties after sonication in human subjects, *Ultrasound Med. Biol.*, 35:1405–1408 (2009).
50. T. Hikima, S. Ohsumi, K. Shirouzu, and K. Tojo. Mechanisms of synergistic skin penetration by sonophoresis and iontophoresis, *Biol. Pharm. Bull.*, 32:905–909 (2009).
51. D. Levy, J. Kost, Y. Meshulam, and R. Langer. Effect of ultrasound on transdermal drug delivery to rats and guinea pigs, *J. Clin. Invest.*, 83:2074–2078 (1989).
52. V. M. Meidan, M. F. Docker, A. D. Walmsley, and W. J. Irwin. Phonophoresis of hydrocortisone with enhancers: An acoustically defined model, *Int. J. Pharm.*, 170:157–168 (1998).
53. R. Brucks, M. Nanavaty, D. Jung, and F. Siegel. The effect of ultrasound on the *in vitro* penetration of ibuprofen through human epidermis, *Pharm. Res.*, 6:697–701 (1989).
54. M. E. Johnson, S. Mitragotri, A. Patel, D. Blankschtein, and R. Langer. Synergistic effects of chemical enhancers and therapeutic ultrasound on transdermal drug delivery, *J. Pharm. Sci.*, 85:670–679 (1996).
55. V. M. Meidan, A. D. Walmsley, M. F. Docker, and W. J. Irwin. Ultrasound-enhanced diffusion into coupling gel during phonophoresis of 5-fluorouracil, *Int. J. Pharm.*, 185:205–213 (1999).
56. J. C. McElnay, H. A. Benson, R. Harland, and J. Hadgraft. Phonophoresis of methyl nicotinate: A preliminary study to elucidate the mechanism of action, *Pharm. Res.*, 10:1726–1731 (1993).
57. http://www.chattgroup.com (accessed September 26, 2010).
58. B. K. Redding. Comparison of a new two-part transdermal drug delivery system, the Patch-Cap, to conventional passive transdermal patches, *Drug Delivery Report*, Autumn/Winter:55–57 (2005).
59. E. Maione, K. K. Shung, R. J. Meyer, Jr., J. W. Hughes, R. E. Newnham, and N. B. Smith. Transducer design for a portable ultrasound enhanced transdermal drug-delivery system, *IEEE Trans. Ultrason. Ferroelectr. Freq. Control*, 49:1430–1436 (2002).

60. S. Lee, B. Snyder, R. E. Newnham, and N. B. Smith. Noninvasive ultrasonic transdermal insulin delivery in rabbits using the lightweight cymbal array, *Diabetes Technol. Ther.*, 6:808–815 (2004).

61. H. Ueda, M. Ogihara, K. Sugibayashi, and Y. Morimoto. Change in the electrochemical properties of skin and the lipid packing in stratum corneum by ultrasonic irradiation, *Int. J. Pharm.*, 137:217–224 (1996).

62. A. J. Singer, C. S. Homan, A. L. Church, and S. A. McClain. Low-frequency sonophoresis: Pathologic and thermal effects in dogs, *Acad. Emerg. Med.*, 5:35–40 (1998).

63. S. Mitragotri. Healing sound: The use of ultrasound in drug delivery and other therapeutic applications, *Nat. Rev. Drug Discov.*, 4:255–260 (2005).

64. S. Paliwal and S. Mitragotri. Therapeutic opportunities in biological responses of ultrasound, *Ultrasonics*, 48:271–278 (2008).

65. http://www.cerevast.com (accessed September 26, 2010).

66. J. J. Escobar-Chavez, D. Bonilla-Martinez, M. A. Villegas-Gonzalez, I. M. Rodriguez-Cruz, and C. L. Dominguez-Delgado. The use of sonophoresis in the administration of drugs throughout the skin, *J. Pharm. Pharm. Sci.*, 12:88–115 (2009).

67. T. Hikima, Y. Hirai, and K. Tojo. Effect of ultrasound application on skin metabolism of prednisolone 21-acetate, *Pharm. Res.*, 15:1680–1683 (1998).

68. http://www.echotx.com (accessed September 26, 2010).

69. H. Tang, D. Blankschtein, and R. Langer. Effects of low-frequency ultrasound on the transdermal permeation of mannitol: Comparative studies with *in vivo* and *in vitro* skin, *J. Pharm. Sci.*, 91:1776–1794 (2002).

70. H. Sakurai, Y. Takahashi, and Y. Machida. Influence of low-frequency massage device on transdermal absorption of ionic materials, *Int. J. Pharm.*, 305:112–121 (2005).

71. K. Tachibana and S. Tachibana. Transdermal delivery of insulin by ultrasonic vibration, *J. Pharm. Pharmacol.*, 43:270–271 (1991).

72. K. Tachibana. Transdermal delivery of insulin to alloxan-diabetic rabbits by ultrasound exposure, *Pharm. Res.*, 9:952–954 (1992).

73. N. B. Smith, S. Lee, E. Maione, R. B. Roy, S. McElligott, and K. K. Shung. Ultrasound-mediated transdermal transport of insulin *in vitro* through human skin using novel transducer designs, *Ultrasound Med. Biol.*, 29:311–317 (2003).

74. N. B. Smith, S. Lee, and K. K. Shung. Ultrasound-mediated transdermal *in vivo* transport of insulin with low-profile cymbal arrays, *Ultrasound Med. Biol.*, 29:1205–1210 (2003).

75. S. Lee, R. E. Newnham, and N. B. Smith. Short ultrasound exposure times for noninvasive insulin delivery in rats using the lightweight cymbal array, *IEEE Trans. Ultrason. Ferroelectr. Freq. Control*, 51:176–180 (2004).

76. J. Kost. Ultrasound induced delivery of peptides, *J. Cont. Rel.*, 24:247–255 (1993).

77. S. Mitragotri, D. Blankschtein, and R. Langer. Ultrasound-mediated transdermal protein delivery, *Science*, 269:850–853 (1995).

78. L. J. Weimann and J. Wu. Transdermal delivery of poly-L-lysine by sonomacroporation, *Ultrasound Med. Biol.*, 28:1173–1180 (2002).

79. A. Dahlan, H. O. Alpar, P. Stickings, D. Sesardic, and S. Murdan. Transcutaneous immunisation assisted by low-frequency ultrasound, *Int. J. Pharm.*, 368:123–128 (2009).

80. A. Tezel, S. Paliwal, Z. Shen, and S. Mitragotri. Low-frequency ultrasound as a transcutaneous immunization adjuvant, *Vaccine*, 23:3800–3807 (2005).

81. V. M. Meidan, A. D. Walmsley, and W. J. Irwin. Phonophoresis—Is it a reality? *Int. J. Pharm.*, 118:129–149 (1995).

82. A. Cabak, M. Maczewska, M. Lyp, J. Dobosz, and U. Gasiorowska. The effectiveness of phonophoresis with ketoprofen in the treatment of epicondylopathy, *Ortop. Traumatol. Rehabil.*, 7:660–665 (2005).

83. S. Deniz, O. Topuz, N. S. Atalay, A. Sarsan, N. Yildiz, G. Findikoglu, O. Karaca, and F. Ardic. Comparison of the effectiveness of pulsed and continuous diclofenac phonophoresis in treatment of knee osteoarthritis, *J. Phys. Ther. Sci.*, 21:331–336 (2009).

84. H. A. E. Benson, J. C. McElnay, R. Harland, and J. Hadgraft. Influence of ultrasound on the percutaneous absorption of nicotinate esters, *Pharm. Res.*, 8:204–209 (1991).

85. S. Mitragotri, M. Coleman, J. Kost, and R. Langer. Analysis of ultrasonically extracted interstitial fluid as a predictor of blood glucose levels, *J. Appl. Physiol.*, 89:961–966 (2000).

86. J. Kost, S. Mitragotri, R. A. Gabbay, M. Pishko, and R. Langer. Transdermal monitoring of glucose and other analytes using ultrasound, *Nat. Med.*, 6:347–350 (2000).

87. S. Mitragotri. Synergistic effect of enhancers for transdermal drug delivery, *Pharm. Res.*, 17:1354–1359 (2000).

88. J. Kost, U. Pliquett, S. Mitragotri, A. Yamamoto, R. Langer, and J. Weaver. Synergistic effect of electric field and ultrasound on transdermal transport, *Pharm. Res.*, 13:633–638 (1996).

89. L. Le, J. Kost, and S. Mitragotri. Combined effect of low-frequency ultrasound and iontophoresis: Applications for transdermal heparin delivery, *Pharm. Res.*, 17:1151–1154 (2000).

90. J. Yoon, D. Park, T. Son, J. Seo, J. S. Nelson, and B. Jung. A physical method to enhance transdermal delivery of a tissue optical clearing agent: Combination of microneedling and sonophoresis, *Lasers Surg. Med.*, 42:412–417 (2010).

91. B. Chen, J. Wei, and C. Iliescu. Sonophoretic enhanced microneedles array (SEMA)— Improving the efficiency of transdermal drug delivery, *Sensors and Actuators B: Chemical*, 145:54–60 (2010).

92. J. Dudelzak, M. Hussain, R. G. Phelps, G. J. Gottlieb, and D. J. Goldberg. Evaluation of histologic and electron microscopic changes after novel treatment using combined microdermabrasion and ultrasound-induced phonophoresis of human skin, *J. Cosmet. Laser Ther.*, 10:187–192 (2008).

93. A. Dahlan, H. O. Alpar, and S. Murdan. An investigation into the combination of low frequency ultrasound and liposomes on skin permeability, *Int. J. Pharm.*, 379:139–142 (2009).

7 How to Go about Selecting the Optimal Enhancement Method for Transdermal Delivery of a Specific Drug Molecule
Case Studies

7.1 INTRODUCTION

The gold standard for bringing a product to market is perhaps still an oral dosage form such as a tablet or a capsule. However, in many cases, this is not possible due to a variety of reasons (e.g., in the case of a large protein, the enzymatic and absorption barriers of the gastrointestinal (GI) tract will not allow any significant blood levels to be achieved following oral administration. In such cases, a company will often bring the product to market in an injectable dosage form. These parenteral dosage forms have their own disadvantages related to costs associated with administering the injection in a supervised clinical setting as well as compliance issues related to fear of needles as well as the need for multiple injections over short periods of time. Companies are therefore always exploring noninvasive alternate routes of administration. Transdermal delivery is one such promising alternative, especially because the patch is now a household word. The credit for the latter goes to over 15 different drugs that have been successfully commercialized, including the well-known nicotine patches. Sometimes, a company may also look at transdermal and other delivery routes for marketing advantages when a patent is about to expire and generic competition is forthcoming.

Once a company decides that a patch is the way to go, the company has more decisions to make on the type of patch to be developed. For potent, moderately lipophilic drugs with short half-lives, the choice is relatively clear that a passive patch is needed. Then the company will start looking at related considerations such as the choice of the pressure-sensitive adhesive based on drug loading levels to reach near saturation, time of application, and so on. However, for other drug molecules that do not normally diffuse through the skin, an active patch needs to be developed using one of the technologies discussed in this book, such as iontophoresis, microneedles, sonophoresis, and so on. Big pharma will generally turn to small companies that specialize in these technologies. However, it is often difficult for the technology evaluation teams or similar groups formed in the companies to evaluate such technologies side by side for selection of the best technology. It is often not realized that the best technology depends on the specific drug molecule under consideration based on several factors including the physicochemical properties of the drug molecule, as well as clinical, cost, and marketing considerations. Specialty companies will often make good sales presentations to executives of big pharma, and then it becomes even more difficult to pick the best technology to sign a business deal. Commercial and technical factors such as the ability of the partner to scale up the technology and protection of intellectual property by having several issued patents also become very important. In fact, the choice between two competing companies who have equally effective technologies may come down to these commercial factors, and it will be important to consider which company is most likely to get FDA approval for the drug/technology combination being considered. Some existing clinical data will increase the chance of success and accelerate timelines, though it may also make the technology more expensive to acquire or license. If big pharma is considering acquiring the company, they should consider if the company has several products in the pipeline or just one. Commercial development of such active patches is discussed in Chapter 10, and for these patches, design considerations completely change from loading saturated drug levels in a pressure-sensitive adhesive to developing a patch or a patch/device configuration that is unique to the enhancement technology being utilized.

7.2   TECHNOLOGY COMPARISON AND LIMITATIONS

While selecting enhancement technologies for the delivery of hydrophilic drugs, small conventional molecules, or macromolecules, both benefits and limitations should be taken into consideration. For example, iontophoretic delivery is generally believed to have an upper cutoff of around 10 to 12 kDa for the size of the molecule that can be delivered. Also, the drug needs to have some aqueous solubility and ideally should be charged, though neutral drugs can also be delivered via electroosmosis. Iontophoresis also allows for modulated delivery by controlling the current. An ideal candidate for iontophoresis would have a high isoelectric point so that the drug can be delivered under anode, to take advantage of electroosmosis and can maintain its positive charge as it passes into skin where the pH is around 3 to 4 (see Chapter 4). Finally, a therapeutic rationale needs to exist to develop a transdermal iontophoretic (or any other) patch. Unlike iontophoresis, there are no size limits with skin microporation technologies, provided the drug is dissolved in some vehicle and not present as particulates. The dimensions of pores created in skin by microneedles or other skin microporation technologies are typically around 50 to 200 μm. In contrast, the hydrodynamic radius of even the largest of molecules, such as monoclonal antibodies, is in the Å range. In fact, even small particulates can be delivered via these microchannels. It should also be noted that particles as large as around 5 μm can enter via hair follicles (see Section 3.2.6). A perfect example of selection of an appropriate candidate for delivery using microporation technology is "vaccines." Coated microneedles may be particularly attractive (see Chapters 3 and 10) because they will allow for quick mass immunization programs in developing countries using low-cost and easy-to-use disposable microneedle arrays coated with stabilized dry antigens.

Sonophoresis is another technology that works well for the delivery of hydrophilic macromolecules (see Chapter 6). Unlike iontophoresis and microporation, commercial development of sonophoresis lags behind due to fewer players in the field. Perhaps this may be due to issues relating to hurdles in miniaturization of sonophoresis devices, unlike an iontophoresis patch with self-contained button batteries or a simple microneedle array. However, other technologies like thermal or radio-frequency ablation (see Chapter 3) have used handheld devices to pretreat skin prior to placement of the patch, and a similar design is possible for sonophoresis as well. Perhaps it is just a matter of time before more companies will pursue this technology. There are several publications that describe two or more of the enhancement techniques being combined, but such combinations may make the patch/device more complex and regulatory approval more unlikely.

7.3 PASSIVE VERSUS ACTIVE PATCHES FOR TRANSDERMAL AND INTRADERMAL DELIVERY

For passive delivery, moderately lipophilic drugs (typically, log P in the range of 1 to 3) are desired. Most marketed patches at this time are for moderately lipophilic potent drug molecules. As discussed in Chapter 1, marketed passive patches include those for buprenorphine, capsaicin, clonidine, estradiol, fentanyl, granisetron, methylphenidate, rivastigmine, rotigotine, selegiline, nicotine, nitroglycerin, oxybutynin, scopolamine, and testosterone, with several brands available for many of these drugs. Patches with estrogen/progestin combination for birth control, and lidocaine patches for topical use are also available. Passive delivery generally works best for drugs with a molecular weight (MW) <500; therefore, the exact molecular weight in the relatively narrow range of 100 to 500 becomes less important (1). However, as we start to use physical enhancement techniques to deliver macromolecules all the way up to monoclonal antibodies (around 150 kDa), the MW will become more important depending on which specific technology is being utilized. For example, as discussed earlier, iontophoretic delivery is generally limited to molecules with a MW of less than around 12 kDa. For example, we found that iontophoresis can produce an excellent, therapeutically relevant, modulated delivery profile for calcitonin (3.6 kDa), but no delivery was seen when iontophoresis was used for human growth hormone (22 kDa) or the IgG (150 kDa) molecule. However, it is not realized by many that typically it is not the MW that necessitates the need for active delivery systems but rather it is the hydrophilicity of the molecule. Most organic molecules that are moderately lipophilic have low MW, so it is not often that one comes across a need to deliver a lipophilic macromolecule. In most cases, when one realizes that passive delivery will not work, the molecule happens to be hydrophilic, and it may be a small molecule or a macromolecule. In both cases, passive delivery will generally not work, and an active patch technology needs to be explored.

There are several additional considerations when intradermal delivery is being considered. Delivery of drug directly into skin for local effects is likely to have significantly reduced dosage requirements as compared to other conventional dosage forms like oral or parenteral delivery, as the drug is directly being administered to the site of action. Furthermore, the side effects are likely to be reduced dramatically. However, the required dosage to be delivered topically is often not known. For example, psoriasis is a common skin disorder characterized by hyperproliferation of epidermal cells, leading to plaques and erythematous papules. Methotrexate is one drug used for treatment which acts by inhibiting mitotic activity though it is administered systemically and has the associated risk of hepatotoxicity along with other side effects. Delivery of methotrexate directly into the skin is being investigated (see Section 7.4.3); however, the dose for topical administration is not known. When considering racemic drugs, the enantiomeric form with the lowest melting point (highest solubility) will have higher skin flux (1). Some other aspects to be considered include skin irritation and effect of heat on drug delivery (discussed in Chapter 10). Other general considerations are not being discussed in this chapter.

7.4 CASE STUDIES

In this section, a few examples of specific drug molecules will be discussed which have different physicochemical properties and can be delivered by different enhancement technologies. In addition to a brief literature review on the efforts to deliver the molecule, some comments will be made about which technology may be suitable for the particular drug molecule being discussed. However, it is hard to make such generalizations, and exceptions will exist depending upon several factors. Drugs that have been delivered primarily by one technology have been discussed in the relevant chapter (e.g., studies to deliver apomorphine by iontophoresis, all the way from *in vitro* investigations to a study in patients with advanced Parkinson's, are discussed in Chapter 4).

7.4.1 MODERATELY LIPOPHILIC

Moderately lipophilic (log P: 1 to 3) drugs can partition into the stratum corneum and can then further partition out into the more hydrophilic dermis and blood circulation. Therefore, these drugs can be formulated into passive patches. Patches are generally used for systemic delivery, but it is also possible to deliver a drug systemically using a topical formulation such as a gel. For example, testosterone, which is on the upper end of being moderately lipophilic (log P 3.3), is commercially available in two gel formulations as well as two transdermal patches (2,3). A microemulsion formulation has also been recently investigated (3). Use of enhancers such as isopropyl myristate (4), and sunscreen agents that were found to be safe penetration enhancers, such as octisalate or padimate O, are also being investigated (5). Additionally, a commercial testosterone spray product is under clinical development. Therefore, it can be seen that several different types of formulations can coexist in the marketplace. For moderately lipophilic drugs that have significant passive permeation to reach therapeutic levels, physical enhancement techniques such as iontophoresis, sonophoresis, or microporation are generally not needed. Many of these techniques may also not be feasible (e.g., iontophoresis needs the drug to have a reasonably high solubility in water). In general, most of these enhancement techniques are meant to enable delivery of hydrophilic drug molecules. Chemical enhancers are sometimes used if therapeutic levels are falling a little short of the desired levels. However, only very few enhancers can possibly receive regulatory approval due to irritation and other safety issues, so the choice of the enhancer should be made carefully (see Section 1.4.1). In some cases, drugs that are already on the market may possibly benefit from skin enhancement techniques for improved delivery profiles, such as reduced lag time, modulated delivery, or other unique advantages specific to a particular drug molecule.

7.4.1.1 Lidocaine

Lidocaine (log P: 2.3) is one of the very few examples of a locally acting drug available in a patch form for relief of pain. However, considerable efforts are still directed toward achieving faster onset of action or deeper penetration into skin. The successful use of iontophoresis for local anesthesia, using lidocaine as the drug of choice, has been reported in the dental; ear, nose, and throat (ENT); ophthalmic; and physical medicine literature (6). Iontophoresis of lidocaine is useful for painless injections and painless venipuncture (7–9). Lidocaine exists as a base (MW 234.3) or hydrochloride salt (MW 270.8). It is very stable in solution and has an acid dissociation constant of 7.84 (determined potentiometrically) at 25°C (10). At physiological pH, lidocaine hydrochloride is a positive drug due to protonation of the amine nitrogen (11). A basic study on the factors affecting iontophoretic delivery of lidocaine across human stratum corneum has been reported (12). Iontophoretic delivery of lidocaine is often performed for local anesthesia or in combination with dexamethasone for the treatment of acute musculoskeletal inflammation. Lidocaine will be about 95% ionized at a pH of about 6 and thus can be delivered under the positive electrode. Iontophoretic treatment has been reported to result in high concentrations of lidocaine in the underlying tissues as compared to passive application to rat skin (13). Lidocaine anesthesia induced by iontophoresis was compared with the effects produced by an injection or simple topical application in 27 subjects. It was found that the anesthesia induced by iontophoresis has a shorter duration (about 5 minutes) and lower clinical efficacy than tissue infiltration via injection, but iontophoresis effects were significantly better than topical application of lidocaine (14). Iontophoresis of lidocaine has also been compared to a commercially available (EMLA®, APP Pharmaceuticals, Schaumburg, Illinois) topical anesthetic cream. EMLA cream (lidocaine 2.5% and prilocaine 2.5%) is an emulsion in which the oil phase is a eutectic mixture of lidocaine and prilocaine in a ratio of 1:1 by weight. A greater degree of anesthesia was reported by iontophoretic treatment for 30 minutes as compared to EMLA left on the skin for 30 or 60 minutes (15). Absorption of lidocaine by the cutaneous microvasculature may further be controlled by coiontophoresis of vasoactive compounds. For example, coiontophoresis with the vasodilator tolazoline has been found to increase *in vivo* systemic absorption of lidocaine, while the vasoconstrictor norepinephrine decreased the iontophoretic absorption of lidocaine (16). Conversely, lower systemic absorption can lead to higher cutaneous depot. Thus, coiontophoresis of norepinephrine resulted in increased concentration of lidocaine in skin up to a depth of 3 mm, while tolazoline decreased tissue concentration of lidocaine (17). Potential use of iontophoretic delivery of lidocaine for acute wound healing has also been investigated (18), and a sutured skin incision model has been reported for such studies (19). Currently, a combination of 2% lidocaine HCl and 1:100,000 epinephrine is approved for dermal anesthesia, and another prefilled device with 10% lidocaine and 0.1% epinephrine is also approved for anesthesia prior to superficial dermatological procedures (see Section 10.6). A commercial device for delivery of lidocaine by sonophoresis is also available and is discussed in Chapter 6.

7.4.2 Highly Lipophilic

Highly lipophilic drugs will readily enter the skin but then may not partition out from the stratum corneum into the more hydrophilic dermis and blood circulation. In some cases, skin deposition is desired, and this may not be a problem. For example, a highly lipophilic drug such as adapalene (synthetic derivative of vitamin A [retinol], log P 8.04) may not permeate across the skin but is used for acne. Thus, permeation just into the superficial layers of skin may be all that is needed, and there is a product on the market for this drug (20). In some cases, using skin enhancement techniques may be justified to enable the systemic delivery of drugs with high lipophilicity. For example, the drug could be formulated as nanoparticles which could then be delivered via microchannels created in skin by the various microporation technologies (see Chapter 3). However, such approaches tend to be more academic in nature at this time.

7.4.2.1 Buprenorphine

Drugs that need to be delivered systemically may still be developed as passive patches if the lipophilicity is not very high. For example, a passive patch for buprenorphine (log P: 3.9) has been on the European market for several years, and recently it was approved by the U.S. Food and Drug Administration (FDA) as well. Buprenorphine is a synthetic opiate analgesic that is marketed for relief of moderate to severe pain. It also has use for opiate addiction therapy as it has lower abuse potential, less respirational effects, less physical dependence, and milder withdrawal symptoms compared to methadone. Oral dosing is impractical due to low absorption and large first-pass metabolism. Studies with transdermal delivery of buprenorphine had shown that transdermal delivery can provide therapeutic plasma levels for sustained analgesic effect (21,22). Iontophoretic delivery could provide advantages to achieve higher plasma levels, better reproducibility, and rapid onset time. However, buprenorphine is a lipophilic drug with low water solubility, posing challenges to delivery by iontophoresis. It carries a positive charge below pH 9 with an aqueous solubility of its hydrochloride salt being 15 mg/ml at pH 4 and less than 0.1 mg/ml at pH 7. Nevertheless, in an *in vivo* study on weanling Yorkshire swine, buprenorphine was successfully delivered to achieve therapeutic doses. When delivered from the anode (0.2 mA/cm^2 for 24 hours) with a concentration of 10 mg/ml, plasma concentrations were much higher than those delivered by passive control, and the dose delivered after 24 hours was 0.75 ± 0.05 mg (23).

7.4.2.2 Fentanyl

Fentanyl is a potent (80 times more potent than morphine) synthetic opioid widely used as an analgesic and as a narcotic analgesic supplement in general or regional anesthesia due to its rapid onset, short duration of action, and high potency. Due to its extensive first-pass metabolism, it is only available via parenteral and transdermal routes. Bolus injections are not desirable as plasma concentrations will reach toxic levels and then decline rapidly due to its short half-life. Controlled release is thus desirable, and transdermal delivery of fentanyl has been widely investigated (24–27) and was approved for marketing in the United States in 1991. Currently, it is the only opioid commercially available in a transdermal form and is approved for the management of chronic pain, especially cancer pain. The patch can release fentanyl for 3 days, upon application, providing effective pain relief during this period. Fentanyl appears in blood within a few hours of applying the patch, with serum levels leveling off after 12 to 24 hours and then staying steady over the remaining 2 days (28). Because this patch has a slow onset and a long duration of action, efforts are underway to develop a patch with a faster onset and shorter duration of action for the control of postoperative pain by passive, iontophoresis, and electroporation approaches (29–32). Titration of dose for individual patients is desirable (33) and may be accomplished by iontophoretic delivery (34–38). Electrical and physico-chemical factors affecting the iontophoretic delivery of fentanyl across hairless rat skin have been investigated. Iontophoresis was more effective from an acidic pH and the flux was found to increase with increasing current density, increasing duration of current application, and increasing drug concentration in the donor compartment (35). Clinically significant doses of fentanyl citrate can be administered to humans. Analgesic doses of fentanyl were administered by iontophoresis for delivery periods of 2 hours. Mean times for detectable plasma concentration were 33 and 19 minutes after treatment at 1 and 2 mA, respectively, with corresponding maximum concentrations being 0.76 and 1.59 ng/ml after 122 and 119 minutes, respectively (39). A wearable iontophoretic patch for delivery of fentanyl was recently marketed but then withdrawn (see Section 10.6). It is perhaps possible to bring this or another similar iontophoretic patch back to market, but the risks have to be weighed against the benefits considering that fentanyl is a controlled substance with significant risk of abuse. Also, recent deaths related to overdose from fentanyl patches due to leakage or applied heat (see Section 10.3) will most likely bring very close scrutiny for regulatory approval of any active energy fentanyl device, even if it is designed for hospital use. This would normally be a hindrance to bringing such an active patch to market, but fentanyl has a strong market share, with the passive Duragesic® patch having sales exceeding $1 billion. Therefore, there is still incentive for such efforts in spite of the drawbacks. Further, use of active technologies may be beneficial in overcoming some of the problems encountered in the past. For example, delivery of fentanyl citrate salt by microporation has been investigated to avoid the formation of a depot in the skin, thereby eliminating the possibility of dose dumping in cases when skin temperature rises, which would otherwise be fatal (see Section 10.5).

7.4.3 Hydrophilic Small Molecules

7.4.3.1 Acyclovir

An example of the importance of commercial factors on product development of delivery systems is provided by the development of an iontophoretic delivery system for acyclovir (log P: 1.6) which was being developed by the former Transport Pharmaceuticals (Framingham, Massachusetts). The device was intended for cold sores (oral herpes) therapy using a handheld fountain pen–sized device that was applied to a herpetic lesion under the lips for a single 10-minute treatment (40). Phase 2a clinical trials demonstrated that the product is significantly more effective than the products currently approved to treat oral herpes. However, intellectual property rights were then purchased by NITRICbio, Inc. (Bristol, Pennsylvania). This technology falls outside the strategic focus of the new holder of the rights, but they will seek a partner to continue this development (41). Acyclovir (MW 225), a synthetic analog of 2′-deoxyguanosine, is an ampholyte drug with two ionizable groups: pK_a of 2.4 for the basic group and 9.2 for the acidic group. It is predominantly positively charged below pH 2, predominantly negatively charged above pH 9.5, and neutral in the pH 3 to 8.5 range. Currently, the commercially marketed dermatological formulations of acyclovir have poor skin penetration. In the clinical study referred to earlier, involving the handheld iontophoretic device, a cream formulation was used, but it was later reported that only the acyclovir (0.3%) dissolved in the water phase of the 5% cream was available for iontophoretic delivery. A new formulation with a gel, at pH 11, was then designed for cathodal iontophoresis which provided higher acyclovir concentrations in the dermis (42). Some of the early literature reported passive delivery to be low (43) and used iontophoresis to increase delivery. It was reported that at pH 3, acyclovir was delivered primarily by electrorepulsion as 20% of the drug was in protonated form. When anodal iontophoresis was applied at pH 7.4, it was primarily delivered by electroosmosis, because the drug was mostly unionized at this pH (44). In a study with rabbits where iontophoretic delivery of acyclovir was monitored by microdialysis, the concentrations of acyclovir in the dialysate increased with increasing current density, but no drug was detected in the plasma, suggesting that systemic exposure was negligible (45). A combination of enhancers and iontophoresis has also been found to be promising to enhance the delivery of acyclovir into skin (46–48). Delivery of other antiviral agents is discussed in Section 4.12.6.

7.4.3.2 Methotrexate

Other than its use as an antineoplastic agent in high doses, methotrexate has been used in low doses as an immunosuppressant in the treatment of psoriasis and rheumatoid arthritis. Psoriasis is a common skin disease affecting 2% of the Caucasian population and is characterized by the presence of plaques resulting from the increased mitotic activity of keratinocytes. It is typically administered via oral and parenteral routes that are associated with hepatotoxicity and other adverse effects after prolonged therapy. Topical delivery would be preferred, but methotrexate is a hydrophilic molecule (log P: 1.85) that is ionized at physiological pH and therefore is not expected to have any significant passive permeation. Therefore, enhancement methods such as iontophoresis, electroporation, and microneedles have been investigated for intradermal delivery of methotrexate. As methotrexate (pK_a values: 5.6, 4.8, and 3.8) is negatively charged at physiological pH and has low solubility in acidic pH, iontophoretic delivery has been investigated under the cathode (49). We also delivered methotrexate by cathodal iontophoresis in the hairless rat model, *in vivo*, and monitored delivery by a microdialysis probe inserted 0.54 mm deep into

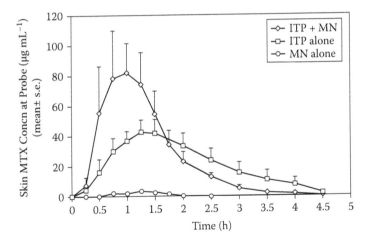

FIGURE 7.1 Concentration versus time profile of methotrexate delivered into skin, *in vivo*, by iontophoresis (ITP), microneedles (MN), and ITP and MN combined; measured by an intradermal microdialysis probe. (V. Vemulapalli et al., Synergistic Effect of Iontophoresis and Soluble Microneedles for Transdermal Delivery of Methotrexate, *J. Pharm. Pharmacol.*, 60:27–33, 2008 [50]. Copyright Wiley-VCH Verlag GmbH & Co. Reproduced with permission.)

the skin, as imaged by ultrasound using a 20-MHz probe. Delivery was increased 14-fold with iontophoresis alone and 25-fold with a combination of iontophoresis and microneedles, as compared to the concentration achieved by microneedles alone ($p < 0.05$) (Figure 7.1) (50). In a study by Stagni and Shukla in the rabbit model, they reported that systemic exposure to methotrexate increased linearly with current density, but the amount deposited in skin did not increase with increasing current density (51). For drugs that form a depot in the skin, the opposite effect may be observed in that increasing current may deposit more drug in the skin, though corresponding increasing amounts are not observed in the receptor compartment (in case of *in vitro* studies) or in blood (in case of *in vivo* studies). Insulin is one such molecule (see Section 7.4). Electroporation-assisted delivery of methotrexate through pig skin was reported (52). A microemulsion formulation for successful delivery of methotrexate was also reported (53). As mentioned earlier, the topical dosage for methotrexate is not known. However, we have shown that even if we look at oral dosage, the desired blood levels of 337 ng ml^{-1} can be achieved by a 32.5 cm^2 iontophoretic patch, though in reality, the dosage required for topical delivery is likely to be much lower (50). Also, a case report published in the literature found that iontophoretic delivery of methotrexate was effective in a patient for the management of a localized recalcitrant form of psoriasis (54).

7.4.4 HYDROPHILIC MACROMOLECULES

Delivery of hydrophilic macromolecules by different enhancement technologies has been discussed throughout this book. Clinical development of parathyroid hormone for osteoporosis using microporation technologies is discussed in Chapter 10. Investigations to deliver calcitonin, another protein for osteoporosis, are summarized in Chapter 8. In this section, the focus will be on efforts directed toward transdermal delivery of insulin.

7.4.4.1 Insulin

Currently, there are about 280 million diabetics worldwide; therefore, insulin delivery systems have attracted much attention over the years. Many start-up companies do some initial work with insulin to establish a proof of concept for their technology. Insulin is used because of its market potential and as a model protein molecule to demonstrate the technological abilities of the company. In reality, insulin is anything but a model protein due to several unique problems relating to its aggregation behavior and isoelectric point. However, the choice of insulin may be dictated by the size of the insulin market; not so much for the market share, but more so because venture capital firms can relate to insulin delivery to provide start-up funds. However, various insulin analogs and devices have already been marketed, so the insulin market is very competitive, and insurance may not reimburse for new delivery devices unless it can be demonstrated that the device provides a clearly unmet need. The marketing and withdrawal of an insulin inhalation product illustrates this point and had ripple effects on other companies that were in advanced clinical trials with inhalable insulin but then decided to abandon their programs. Insulin, being a protein, cannot be administered orally. It is currently administered by subcutaneous injection, but repeated injections are required due to its short half-life. Even with long-acting preparations, the inconvenience of making an injection still exists. Alternate noninvasive delivery systems are therefore desirable (55). Transdermal delivery is one possibility, but because insulin is a hydrophilic macromolecule, enhancement techniques are needed. Insulin associated with transfersomes (also see Section 1.4) has also been reported to be effective for permeation across skin (56,57). Enhancers such as dimethylsulfoxide (DMSO), urea, and bile salts were also able to enhance the flux (58). Efforts to deliver insulin by sonophoresis are discussed in Section 6.7. Other efforts, especially those related to iontophoretic approaches and microporation, are discussed here.

7.4.4.1.1 Iontophoretic Delivery of Insulin: Fact or Fiction?

Research on the transdermal iontophoretic delivery of insulin for systemic effects started in the 1980s, with the first publication appearing in 1984. Since then, several reports have been published. However, these studies have not conclusively established the feasibility of delivering intact insulin across human skin in therapeutically meaningful amounts. This is because different investigators have used different methods, and there are several variables to be considered before data can be interpreted. First, the molecular weight of the polypeptide being administered is not certain. At a molecular weight of about 6000, insulin would be expected to have sufficient permeability under iontophoresis to make the technology feasible. However, most commercially available insulin products actually exist in hexameric form, so we may really be trying to deliver a protein with a molecular weight of about 36 kDa, which is most likely too high to be within the scope of iontophoretic delivery. Furthermore, the pI of insulin (5.3) falls in the region of skin pI (4 to 6). This poses major hurdles to its delivery as discussed in general for peptides (see Chapter 8). Although there is no evidence for precipitation of insulin in the skin, the formation of a depot upon iontophoresis has been suggested by several investigators. However, the depot effect could also result simply due to accumulation of insulin in the less accessible regions of skin and its subsequent slow leaching from those regions. The self-association behavior of insulin may also be implicated in its depot effect. While insulin circulates in blood in low concentrations (10^{-8} to 10^{-11} M) and brings about its biological effects as a monomer, it dimerizes at higher concentrations, and in the presence of zinc ions, further assembles into hexamers (59). Thus, insulin exists as a hexamer, as commonly used, but its absorption may require a monomeric form. Absorption of insulin following subcutaneous injection or infusion has been shown to depend on the volume and concentration of the injected or infused insulin and has been modeled taking into account its aggregation status and binding in tissue (60). Slow release of monomeric or dimeric insulin from accumulated hexameric insulin in skin can also account for the depot effect. If monomeric analogs of insulin or formulations in which insulin exists as monomer are used, it could lead to better absorption from the subcutaneous site. In the first reported study (61) on transdermal iontophoretic delivery of insulin for systemic effects, it was attempted to deliver regular soluble insulin to human volunteers. Iontophoresis of commercially available insulin was done on eight volunteers, but negative results were obtained even after repeating the study on three occasions. In 1986, Kari tried to deliver regular, soluble insulin by iontophoresis to New Zealand white rabbits made diabetic by injection of alloxan. Hair was removed from the back of animals using an electric clipper, a process that most likely disrupted the stratum corneum, as acknowledged by the author. Using currents of <1 mA on a surface area of 6.2 cm^2, regular pork-zinc insulin was delivered. The blood glucose levels decreased and serum insulin concentrations increased within 1 hour of turning the current on, but these changes continued even after the current was turned off, suggesting accumulation of insulin in skin and subcutaneous tissues (62). Feasibility of iontophoretic delivery of insulin in diabetic rabbits following disruption of stratum corneum was also demonstrated (63). In another study involving 26 diabetic albino rabbits, a therapeutic dose of human insulin was reported to be successfully delivered across intact stratum corneum. Active patches contained insulin, and an electrical circuit was established when the patch was applied on the skin, delivering a direct current of 0.4 mA over a period of 14 hours. Control patches, which had an equal amount of insulin but without the corresponding electrical circuit, were also used. After placement of patches, animals with active patches were found to have significant elevation in serum insulin levels ($p < 0.05$) with a corresponding reduction in blood glucose levels ($p < 0.01$), while no changes were observed in controls. High-performance liquid chromatography (HPLC) analysis of insulin in the patch, prior to application, immediately after application, and 9 and 18 hours after application indicated that insulin was chemically stable in the patch over the course of the study (64). Several other investigators demonstrated the *in vivo* iontophoretic delivery of insulin in rats (65–70). In one study, regular and hairless rats were made diabetic by injection of streptozotocin, and purified pork insulin was iontophoretically delivered using a commercially available iontophoresis device. There was some evidence for the formation of a reservoir in the skin, as the blood glucose levels continued to decrease, even 2 to 3 hours after the iontophoresis treatment was stopped (65). In a study that used freshly excised rabbit inner pinna skin, the addition of skin to a control buffer increased insulin degradation from 9% to 40% during the first 8 hours. When current was applied, the degradation was 70% during the first 8 hours. All insulin was completely degraded within 24 hours, irrespective of the conditions used. Also, almost all insulin was degraded as it passed through the skin during iontophoresis (71). The observation that no permeation was seen across untreated human skin in one reported study (72) conflicts with several published reports of successful iontophoretic delivery of insulin across animal skin. This may be due to the fact that human skin is known to be less permeable as compared to animal skin. As discussed, monomeric analogs of insulin may be better candidates for its iontophoretic delivery. It was shown that clinically relevant amounts of insulin could be iontophoretically delivered, after wiping the skin with absolute alcohol. In this study, a series of monomeric analogs of human insulin could be iontophoretically delivered across hairless mouse skin, while only low and insignificant flux was observed with regular hexameric human insulin (73). A more important modification may be the development of insulin analogs with modified pI. In a study with several insulin analogs, all analogs with pI inside the pH range of skin resulted in undetectable flux. However, sulfated insulin (pI: 2), an analog with pI outside the skin pH range, showed measurable flux in a porcine skin flap model. When cathodal iontophoresis was conducted for 4 hours at 0.4 mA, using a 2-cm^2 reservoir, the total recovery was about one quarter of a unit of insulin. Under these conditions, the biological response in streptozotocin-induced diabetic rabbits resulted in significant glucose decline of 200 to 300 mg/dl, proving that at least some biological activity was retained in the sulfated insulin delivered iontophoretically. Calibration against a subcutaneous dose and comparison with skin flap data suggested that a large fraction of the dose was biologically active (74). Another technique known as reverse iontophoresis (see Section 4.8) can be exploited for noninvasive glucose measurement by its iontophoretic extraction from the subcutaneous tissue (75). Although this step has tremendous commercial value for noninvasive glucose monitoring, the next logical step would be to link this glucose measurement to a closed-loop biofeedback system for iontophoretic delivery of insulin.

As is evident from the above discussion, extensive work needs to be performed to make transdermal iontophoretic delivery of insulin a viable technique for the treatment of diabetics. *In vivo* studies are complicated by biological variations because blood glucose and plasma insulin levels can fluctuate due to several factors including, but not limited to, circadian rhythms, anesthesia, and stress. Also, these studies were conducted in laboratory animals, and results may not be extrapolated to human skin. For example, rabbit skin could contain a very high hair follicle density, a crucial difference because iontophoretic transport takes place through these shunt pathways (76). On the other hand, poor experimental design and the lack of sensitive analytical techniques used for most *in vitro* studies failed to assess the stability of insulin during transport. A significant portion of radioactivity could very well be insulin degradation products instead of the intact labeled insulin. The depot effect observed by several investigators could be a disadvantage for modulation of delivery or development of a biofeedback system. However, iontophoresis would still be useful to load skin tissues with insulin in a noninvasive manner, for a relatively prolonged release similar to the long-acting preparations on market. It was suggested that porcine insulin, following dissociation to a dimer in a 0.1 M glycine-HCl buffer (pH 4), was absorbed *in vivo* through Wistar rat skin, as estimated by blood glucose levels. The best absorption was obtained with a liposomal formulation containing two penetration enhancers (77). The combination of iontophoresis with penetration enhancers or other pretreatment of skin (78,79) may also be feasible, though it is likely to result in a complex system that is difficult to characterize. In contrast to iontophoresis, delivery of insulin by skin microporation will not suffer from most of these limitations.

7.4.4.1.2   *Delivery of Insulin by Microporation*

As discussed, efforts for iontophoretic delivery of insulin suffer from the drawback that the physicochemical properties of the molecule, especially its pI and hexameric nature, are not favorable for iontophoretic transport. However, insulin delivery via microchannels will avoid these drawbacks. A microneedle-based insulin delivery system for diabetes can potentially provide a biofeedback system where microneedles can be used to monitor glucose levels by extraction of interstitial fluid and then deliver the desired insulin dose to maintain blood glucose levels. Glucose sensors typically require about 0.5 to 2 μL of interstitial fluid, which is achievable (80). Several studies have been reported on microneedle-mediated transdermal delivery of insulin (81–86). Prausnitz and his group used both solid (87) and hollow (88) microneedles for delivery of insulin to diabetic rats and reported a drop in blood glucose levels by as much as 50% to 80%. Self-dissolving microneedles have also been investigated for insulin delivery. Ito et al. prepared insulin-loaded dextrin microneedles and demonstrated lowering of plasma glucose levels in mice (89). In a study with diabetic rats, microneedle rollers were used to deliver insulin, and it was reported that blood glucose levels decreased continuously throughout the 3-hour insulin delivery period (86). In a study with human subjects, it was reported that intradermal insulin delivery with microneedles led to rapid absorption by the rich capillary network of papillary dermis, leading quickly to high peak insulin concentrations (90).

7.4.5   THERAPEUTIC CATEGORIES

As a final note on case studies, two therapeutic categories, nonsteroidal anti-inflammatory drugs (NSAIDs) and antihypertensive/cardiovascular agents will be discussed. This is to bring out the similarities and differences in compounds within a series which may be structurally or functionally similar and the possible approaches that may be used to deliver them efficiently.

7.4.5.1   Nonsteroidal Anti-Inflammatory Drugs (NSAIDs)

Oral administration of NSAIDs may not be preferable, as they cause gastric mucosal damage that may result in ulceration or bleeding. For some NSAIDs such as ketoprofen, these adverse GI effects occur in about 10% to 30% of patients receiving the drug and may be severe enough to require discontinuance of drug therapy in about 5% to 15% of patients. These adverse side effects could be avoided by topical administration for intradermal delivery, where a much lower dosage will be needed. In situations where systemic circulation is desired, transdermal delivery will avoid the GI side effects seen with oral administration. This prompted investigations on the percutaneous absorption of nonsteroidal anti-inflammatory agents. Studies with ketorolac acid (91–93), indomethacin (94), diclofenac (95–97), ibuprofen and naproxen (98), piroxicam (99), and flurbiprofen (100) were performed, and the results are promising. Based on *in vitro* passive permeation studies with human skin, predictions suggest that ketorolac will provide plasma concentrations that at steady state would be nearest to the therapeutic concentration (101). Though bioavailability from a topical site is expected to be much lower than oral absorption, topical application could result in a high local drug concentration at the diseased site. Furthermore, because blood supply to diseased sites may be reduced, the dosage requirements for topical delivery could be less than those for oral delivery. Some potent NSAIDs may have sufficient permeability to be useful in topical formulations (102), but others may require enhancement aids for percutaneous absorption. This relatively low percutaneous penetration of certain anti-inflammatory agents may be increased by iontophoresis or sonophoresis. In a study on the delivery of indomethacin to pigs, iontophoresis (0.1 mA/cm^2) increased the C_{max} levels in blood from 32 (controls) to 82 ng/ml and increased urinary excretion over 5 hours from 29.4 to 181.1 ng/cm^2. On the other hand, sonophoresis did not improve indomethacin absorption. This study also looked at delivery of indomethacin to seven human subjects. Using skin of the back (1380 cm^2), plasma levels increased from 43 ng/ml (control) to 221 ng/ml after 1 hour of iontophoresis. Urinary excretion over 5 hours increased from 18.1 (controls) to 97.6 ng/cm^2 (103). Clinical studies for iontophoretic delivery of NSAIDs are based on relief of clinical symptoms (96,97,104,105), and there are very few studies that actually measure the level of drug in tissue or blood following iontophoresis. In a collaborative study done by the author on iontophoretic delivery of ketoprofen to human subjects (106,107), iontophoresis was conducted using 14.2-cm^2 electrodes, at a current density of 0.28 mA/cm^2 (4 mA/14 cm^2). Ketoprofen (300 mg/ml) in phosphate buffer (pH 7.4) with 20% ethanol was sterilized before use and delivered at the volar surface of the wrist under cathode. Treatment parameters were set at a maximum current of 4 milliamperes (mA) for 40 minutes resulting in a total of 160 milliampere*minutes. Forty minutes of cathodal iontophoresis was required to obtain measurable ketoprofen levels in all subjects. Ketoprofen was detected in the arm ipsilateral to the iontophoresis. Total ketoprofen concentration in the arm contralateral to the iontophoresis was below detectability at 30 minutes following termination of the iontophoresis. Accumulation of conjugated and unconjugated ketoprofen in the urine was also examined, with 75% of the drug being excreted 8 hours post-iontophoresis. In this investigation, following iontophoresis, no adverse effects were observed in any of the subjects due to the applied current. A mild tingling sensation was felt by the subjects as the current was increased. A slight redness of skin was observed under the electrodes, and this was resolved within a few hours (107). When the same electronic dose (160 mA*min) was applied at 2 mA current applied for 80 minutes (instead of 4 mA for 40 minutes), the resulting serum levels were found to be reduced 10-fold, while the total ketoprofen and conjugates excreted in the urine were reduced to one half. The reasons for these differences are not clear but could be related to changes in skin resistance and its recovery, local hyperemia, or formation of skin depot as a function of magnitude and duration of current application. This study suggests that the use of total exposure (mA*min) to represent drug dosage may be inaccurate, contrary to what is commonly used in physical therapy clinics (106). The use of mA*min to represent dose may also result in problems due to variation in delivery from electrodes of different manufacturers, differences in formulation, and so on. In the ketoprofen studies done by us, there was no significant difference between the R and S ketoprofen enantiomer concentrations in either the serum or the urine (108). Iontophoretic delivery of ketoprofen was also reported in the rat (109) and pig (110) models, and the use of a chemical enhancer approach (111) was investigated, with all studies reporting positive results.

7.4.5.2 Antihypertensive/Cardiovascular Agents

Transdermal delivery of antihypertensive and other cardiovascular drugs offers several advantages. For nitroglycerin, the transdermal route avoids the unreliability of drug absorption from the gastrointestinal tract, precludes first-pass metabolism of the drug through the liver, and provides continuous, prolonged delivery with a 24-hour patch, even though the drug has a half-life of just a few minutes. For clonidine, a 7-day patch has the additional advantage of minimizing the dose-dependent side effects that occur with the oral form, yet providing adequate antihypertensive effect (112). Both nitroglycerin and clonidine are on the market as passive delivery patches. Transdermal delivery of other cardiovascular agents has also been investigated (113). For cardiovascular drugs, particularly for those indicated for hypertension, extended-release formulations can have a significant impact on reducing patient noncompliance through less frequent dosing. Because of large differences in first-pass effects, plasma concentration of propranolol can vary as much as 10- to 20-fold between individuals following oral administration. More importantly, the extended-release tablet on the market cannot be conveniently used to individualize dosage for a patient or readily adjust dosage for a different indication. Iontophoretic delivery might be useful to overcome these drawbacks and to possibly develop a self-regulated system where delivery of propranolol is linked to blood pressure. A link to blood pressure rather than plasma concentration is sought, as there is no simple correlation between dose or plasma level and therapeutic effect. This long-term goal seems realistic, as the feasibility of a noninvasive measurement of blood flow in response to propranolol treatment has been established by dual-beam pulsed-wave Doppler technique (114,115). There have been several investigations on the passive transdermal absorption of propranolol and other β-blockers (116–118). These studies have attempted to use penetration enhancers such as fatty acids (119) and alkane or alkanols (120). Collagen membranes have also been investigated to fabricate transdermal delivery systems for propranolol (121), and artificial membranes have been used to optimize iontophoretic delivery of timolol (122). However, passive permeation of propranolol hydrochloride is unlikely to be successful, as propranolol hydrochloride is a "hydrophilic" permeant, and skin is more permeable to "lipophilic" permeants. If propranolol permeates the skin as a free base, the partition coefficient for the free base is still in the order of 100-fold less than other drugs such as diazepam or indomethacin (120). The use of prodrugs of propranolol has shown some promise to increase the amounts delivered across skin (123). Alternatively, the technique of iontophoresis can overcome these disadvantages. Furthermore, it may allow individualization of dosage by adjusting the current parameters. Propranolol hydrochloride has been iontophoretically delivered *in vitro* across rat abdominal skin, and delivery was found to increase with increasing drug concentration applied, current density, duration of application, and duty cycle of pulsed DC applied. Iontophoretic delivery was significantly higher than passive diffusion (124). We investigated the comparative pharmacokinetics of five β-blockers and found that therapeutically relevant levels of propranolol were delivered *in vivo* in hairless rats within 1 hour of iontophoresis (125). A potential barrier to commercialization of iontophoretic delivery systems for β-blockers could be the potential for skin irritation. A high correlation has been found between the cumulative amounts of β-blockers permeating through the stratum corneum of guinea pig skin and the degree of erythema, a skin irritation reaction quantitated by a chromameter (126). We reported that a liposomal formulation of propranolol base reduced or eliminated the skin irritation of the base in a solution formulation relative to the nonliposomal solution formulation of the base (125).

Factors affecting the iontophoretic delivery and reversibility of skin permeability of another antihypertensive drug, verapamil, were investigated. Following iontophoresis with pulsed DC for just 10 minutes, the time required for 50% of the drug to be released from the skin depot was as high as 8 to 10 hours when low drug concentrations were used and about 20 hours for high drug concentration (127,128). The potential of iontophoresis was also investigated for other delivery of cardiovascular drugs such as sotalol (129) and metoprolol (130–132). Following iontophoretic delivery of metoprolol to spontaneously hypertensive rabbits, systolic pressure was reduced from a pretreatment pressure of 126 ± 9 mm Hg to a posttreatment pressure of 86 ± 11 mm Hg ($P < 0.05$) within 2 hours (133). Electroporation-mediated enhancement of metoprolol delivery through skin was also investigated. A study by Denet et al. compared the effect of electroporation on iontophoretic delivery of a lipophilic (timolol) and a hydrophilic (atenolol) β-blocker and found that the lipophilic cation, timolol, accumulated in the skin to decrease electroosmosis (134). The effect of drug accumulation in skin on electroosmosis is discussed in Section 8.5. Another hypertensive agent, captopril, is an orally effective angiotensin I–converting enzyme inhibitor, but it has a relatively short elimination half-life in plasma and has been investigated for transdermal delivery (135). A monophasic iontophoretic device was investigated for delivery of captopril in rabbits. Blood pressure reduction was evident within 10 minutes, and the effects were pronounced within 40 minutes. Feedback circuitry was used to alter the current to change the antihypertensive drug flux in order to attempt to maintain blood pressure at a target level. This attempted autoregulation failed possibly due to the loading of the drug in a skin reservoir that provided a source of drug that continued to act long after the iontophoretic patch was removed from the skin surface. Thus, autoregulation using simple feedback techniques may not be feasible, but predictive tools may be required to attempt autoregulation (136). In another study, iontophoretic delivery of an inotropic catecholamine to dogs has been compared to intravenous infusion. Iontophoresis was found to deliver the drug, achieving the same degree of cardiac contractility and steady-state plasma concentrations as via an intravenous infusion (137). Microneedles have also been used for delivery of antihypertensive agents. We investigated delivery of nicardipine HCl, a calcium channel blocker used in the treatment of hypertension which undergoes extensive hepatic first-pass metabolism and therefore has a rationale for transdermal delivery. Pretreatment of skin with maltose microneedles increased both *in vitro* and *in vivo* flux of nicardipine across hairless rat skin (138).

REFERENCES

1. A. C. Williams. *Transdermal and topical drug delivery*, Pharmaceutical Press, London, 2003.
2. R. G. Harrison and J. E. Stover. Transdermal testosterone for treatment of female sexual dysfunction, *Drug Del. Technol.*, 5:63–65 (2005).
3. R. M. Hathout, T. J. Woodman, S. Mansour, N. D. Mortada, A. S. Geneidi, and R. H. Guy. Microemulsion formulations for the transdermal delivery of testosterone, *Eur. J. Pharm Sci.*, 40:188–196 (2010).
4. M. L. Leichtnam, H. Rolland, P. Wuthrich, and R. H. Guy. Identification of penetration enhancers for testosterone transdermal delivery from spray formulations, *J. Control. Release*, 113:57–62 (2006).
5. J. A. Nicolazzo, T. M. Morgan, B. L. Reed, and B. C. Finnin. Synergistic enhancement of testosterone transdermal delivery, *J. Control. Release*, 103:577–585 (2005).
6. E. J. Henley. Transcutaneous drug delivery: Iontophoresis, phonophoresis, *Crit. Rev. Ther. Drug Carr. Syst.*, 2:139–151 (1991).
7. T. Petelenz, I. Axenti, T. J. Petelenz, J. Iwinski, and S. Dubel. Mini set for iontophoresis for topical analgesia before injection, *J. Clin. Pharmacol. Ther. Toxicol.*, 22:152–155 (1984).
8. S. B. Arvidsson, R. H. Ekroth, M. C. Hansby, A. H. Lindholm, and G. William-Olsson. Painless venipuncture. A clinical trial of iontophoresis of lidocaine for venipuncture in blood donors, *Acta Anaesthesiol. Scand.*, 28:209–210 (1984).
9. L. Zeltzer, M. Regalado, L. S. Nichter, D. Barton, S. Jennings, and L. Pitt. Iontophoresis versus subcutaneous injection: A comparison of two methods of local anesthesia delivery in children, *Pain*, 44:73–78 (1991).
10. K. Groningsson, J. E. Lindgren, E. Lundberg, R. Sandberg, and A. Wahlen. Lidocaine base and hydrochloride, *Anal. Prof. Drug Subst.*, 14:207–243 (1985).
11. S. S. Kamath and L. P. Gangarosa. Electrophoretic evaluation of the mobility of drugs suitable for iontophoresis, *Meth. Find. Exp. Clin. Pharmacol.*, 17:227–232 (1995).
12. O. Siddiqui, M. S. Roberts, and A. E. Polack. The effect of iontophoresis and vehicle pH on the *in vitro* permeation of lignocaine through human stratum corneum, *J. Pharm. Pharmacol.*, 37:732–735 (1985).
13. P. Singh and M. S. Roberts. Iontophoretic transdermal delivery of salicyclic acid and lidocaine to local subcutaneous structures, *J. Pharm. Sci.*, 82:127–131 (1993).
14. J. Russo, A. G. Lipman, T. J. Comstock, B. C. Page, and R. L. Stephen. Lidocaine anesthesia: Comparison of iontophoresis, injection and swabbing, *Am. J. Hosp. Pharm.*, 37:843–847 (1980).
15. S. S. Greenbaum and E. F. Bernstein. Comparison of iontophoresis of lidocaine with a eutectic mixture of lidocaine and prilocaine (EMLA) for topically administered local anesthesia, *J. Dermatol. Surg. Oncol.*, 20:579–583 (1994).
16. J. E. Riviere, B. Sage, and P. L. Williams. Effects of vasoactive drugs on transdermal lidocaine iontophoresis, *J. Pharm. Sci.*, 80:615–620 (1991).
17. J. E. Riviere, N. A. Monteiro-Riviere, and A. O. Inman. Determination of lidocaine concentrations in skin after transdermal iontophoresis: Effects of vasoactive drugs, *Pharm. Res.*, 9:211–214 (1992).

18. A. A. Ernst, J. Pomerantz, T. G. Nick, J. Limbaugh, and M. Landry. Lidocaine via ionto-phoresis in laceration repair: A preliminary safety study, *Am. J. Emerg. Med.*, 13:17–20 (1995).

19. H. J. Holovics, C. R. Anderson, B. S. Levine, H. W. Hui, and C. E. Lunte. Investigation of drug delivery by iontophoresis in a surgical wound utilizing microdialysis, *Pharm. Res.*, 25:1762–1770 (2008).

20. L. Trichard, M. B. Delgado-Charro, R. H. Guy, E. Fattal, and A. Bochot. Novel beads made of alpha-cyclodextrin and oil for topical delivery of a lipophilic drug, *Pharm Res.*, 25:435–440 (2008).

21. S. D. Roy, E. Roos, and K. Sharma. Transdermal delivery of buprenorphine through cadaver skin, *J. Pharm. Sci.*, 83(2):126–130 (1994).

22. I. R. Wilding, S. S. Davis, G. H. Rimoy, P. Rubin, T. Kuriharabergstrom, V. Tipnis, B. Berner, and J. Nightingale. Pharmacokinetic evaluation of transdermal buprenorphine in man, *Int. J. Pharm.*, 132:81–87 (1996).

23. J. Denuzzio, K. Boericke, D. Sutter, A. Mcfarland, D. Dey, R. Cesarini, E. Monty, D. Colville, R. Bock, M. O'Connell, and B. Sage. Iontophoretic delivery of buprenorphine, *Proc. Int. Symp. Control. Rel. Bioact. Mater.*, 23:285–286 (1996).

24. S. D. Roy and G. L. Flynn. Transdermal delivery of narcotic analgesics: pH, anatomical, and subject influences on cutaneous permeability of fentanyl and sufentanil, *Pharm. Res.*, 7:842–847 (1990).

25. M. A. Simmonds. Transdermal fentanyl: Clinical development in the United States, *Anti-Cancer. Drug*, 6:35–38 (1995).

26. R. Payne, S. Chandler, and M. Einhaus. Guidelines for the clinical use of transdermal fentanyl, *Anti-Cancer. Drug*, 6:50–53 (1995).

27. B. Donner and M. Zenz. Transdermal fentanyl: A new step on the therapeutic ladder, *Anti-Cancer. Drug*, 6:39–43 (1995).

28. M. A. Southam. Transdermal fentanyl therapy: System design, pharmacokinetics and efficacy, *Anti-Cancer. Drug*, 6:29–34 (1995).

29. R. Miguel, J. M. Kreitzer, D. Reinhart, P. S. Sebel, J. Bowie, G. Freedman, and J. B. Eisenkraft. Postoperative pain control with a new transdermal fentanyl delivery system: A multicenter trial, *Anesthesiology*, 83:470–477 (1995).

30. P. Fiset, C. Cohane, S. Browne, S. C. Brand, and S. L. Shafer. Biopharmaceutics of a new transdermal fentanyl device, *Anesthesiology*, 83:459–469 (1995).

31. S. D. Roy, M. Gutierrez, G. L. Flynn, and G. W. Cleary. Controlled transdermal delivery of fentanyl: Characterizations of pressure-sensitive adhesives for matrix patch design, *J. Pharm. Sci.*, 85:491–495 (1996).

32. R. Conjeevaram, A. K. Banga, and L. Zhang. Electrically modulated transdermal deliv-ery of fentanyl, *Pharm. Res.*, 19:440–444 (2002).

33. W. Korte, N. deStoutz, and R. Morant. Day-to-day titration to initiate transdermal fenta-nyl in patients with cancer pain: Short- and long-term experiences in a prospective study of 39 patients, *J. Pain Symptom. Mgmt.*, 11:139–146 (1996).

34. S. Thysman and V. Preat. *In vivo* iontophoresis of fentanyl and sufentanil in rats: Pharmacokinetics and acute antinociceptive effects, *Anesth. Analg.*, 77:61–66 (1993).

35. S. Thysman, C. Tasset, and V. Preat. Transdermal iontophoresis of fentanyl—Delivery and mechanistic analysis, *Int. J. Pharm.*, 101:105–113 (1994).

36. J. E. Chelly. An iontophoretic, fentanyl HCl patient-controlled transdermal system for acute postoperative pain management, *Expert Opin. Pharmacother.*, 6:1205–1214 (2005).

37. I. Power. Fentanyl HCl iontophoretic transdermal system (ITS): Clinical application of iontophoretic technology in the management of acute postoperative pain, *Br. J. Anaesth.*, 98:4–11 (2007).

38. H. S. Minkowitz, J. Yarmush, M. T. Donnell, P. H. Tonner, C. V. Damaraju, and R. J. Skowronski. Safety and tolerability of fentanyl iontophoretic transdermal system: Findings from a pooled data analysis of four clinical trials, *J. Opioid Manag.*, 6:203–210 (2010).

39. M. A. Ashburn, J. Streisand, J. Zhang, G. Love, M. Rowin, S. Niu, J. K. Kievit, J. R. Kroep, and M. J. Mertens. The iontophoresis of fentanyl citrate in humans, *Anesthesiology*, 82:1146–1153 (1995).

40. E. M. Morrel, S. L. Spruance, and D. I. Goldberg. Topical iontophoretic administration of acyclovir for the episodic treatment of herpes labialis: A randomized, double-blind, placebo-controlled, clinic-initiated trial, *Clin. Infect. Dis.*, 43:460–467 (2006).

41. NITRICbio, http://www.nitricbio.com (accessed October 16, 2010).

42. C. Shukla, P. Friden, R. Juluru, and G. Stagni. *In vivo* quantification of acyclovir exposure in the dermis following iontophoresis of semisolid formulations, *J. Pharm. Sci.*, 98:917–925 (2009).

43. F. Yamashita, Y. Koyama, H. Sezaki, and M. Hashida. Estimation of a concentration profile of acyclovir in the skin after topical administration, *Int. J. Pharm.*, 89:199–206 (1993).

44. N. M. Volpato, P. Santi, and P. Colombo. Iontophoresis enhances the transport of acyclovir through nude mouse skin by electrorepulsion and electroosmosis, *Pharm. Res.*, 12:1623–1627 (1995).

45. G. Stagni, M. E. Ali, and D. Weng. Pharmacokinetics of acyclovir in rabbit skin after i.v.-bolus, ointment, and iontophoretic administrations, *Int. J. Pharm.*, 274:201–211 (2004).

46. U. T. Lashmar and J. Manger. Investigation into the potential for iontophoresis facilitated transdermal delivery of acyclovir, *Int. J. Pharm.*, 111:73–82 (1994).

47. S. S. Vaghani, M. Gurjar, S. Singh, S. Sureja, S. Koradia, N. P. Jivani, and M. M. Patel. Effect of iontophoresis and permeation enhancers on the permeation of an acyclovir gel, *Curr. Drug Deliv.*, 7:329–333 (2010).

48. S. Nicoli, M. Eeman, M. Deleu, E. Bresciani, C. Padula, and P. Santi. Effect of lipopeptides and iontophoresis on aciclovir skin delivery, *J. Pharm. Pharmacol.*, 62:702–708 (2010).

49. M. J. Alvarez-Figueroa, M. B. Delgado-Charro, and J. Blanco-Mendez. Passive and iontophoretic transdermal penetration of methotrexate, *Int. J. Pharm.*, 212:101–107 (2001).

50. V. Vemulapalli, Y. Yang, P. M. Friden, and A. K. Banga. Synergistic effect of iontophoresis and soluble microneedles for transdermal delivery of methotrexate, *J. Pharm. Pharmacol.*, 60:27–33 (2008).

51. G. Stagni and C. Shukla. Pharmacokinetics of methotrexate in rabbit skin and plasma after iv-bolus and iontophoretic administrations, *J. Control. Release*, 93:283–292 (2003).

52. T. W. Wong, Y. L. Zhao, A. Sen, and S. W. Hui. Pilot study of topical delivery of methotrexate by electroporation, *Br. J. Dermatol.*, 152:524–530 (2005).

53. M. J. Alvarez-Figueroa and J. Blanco-Mendez. Transdermal delivery of methotrexate: Iontophoretic delivery from hydrogels and passive delivery from microemulsions, *Int. J. Pharm.*, 215:57–65 (2001).

54. S. B. Tiwari, B. C. Kumar, N. Udupa, and C. Balachandran. Topical methotrexate delivered by iontophoresis in the treatment of recalcitrant psoriais—A case report, *Int. J. Dermatol.*, 42:157–159 (2003).

55. Y. W. Chien and A. K. Banga. Potential developments in systemic delivery of insulin, *Drug Develop. Ind. Pharm.*, 15:1601–1634 (1989).

56. G. Cevc, A. Schatzlein, and G. Blume. Transdermal drug carriers: Basic properties, optimization and transfer efficiency in the case of epicutaneously applied peptides, *J. Control. Release*, 36:3–16 (1995).

57. G. Cevc. Transdermal drug delivery of insulin with ultradeformable carriers, *Clin. Pharmacokinet.*, 42:461–474 (2003).
58. V. U. Rao and A. N. Misra. Enhancement of iontophoretic permeation of insulin across human cadaver skin, *Pharmazie*, 49:538–539 (1994).
59. R. C. Bi, Z. Dauter, E. Dodson, G. Dodson, F. Giordano, and C. Reynolds. Insulin's structure as a modified and monomeric molecule, *Biopolymers*, 23:391–395 (1984).
60. E. Mosekilde, K. S. Jensen, C. Binder, S. Pramming, and B. Thorsteinsson. Modeling absorption kinetics of subcutaneous injected soluble insulin, *J. Pharmacokinet. Biopharm.*, 17:67–87 (1989).
61. R. L. Stephen, T. J. Petelenz, and S. C. Jacobsen. Potential novel methods for insulin administration: I. Iontophoresis, *Biomed. Biochim. Acta*, 43:553–558 (1984).
62. B. Kari. Control of blood glucose levels in alloxan-diabetic rabbits by iontophoresis of insulin, *Diabetes*, 35:217–221 (1986).
63. B. C. Shin and H. B. Lee. Iontophoretic delivery of insulin: Effects of skin treatments, *Proc. Intern. Symp. Control. Rel. Bioact. Mater.*, 21:373–374 (1994).
64. B. R. Meyer, H. L. Katzeff, J. C. Eschbach, J. Trimmer, S. B. Zacharias, S. Rosen, and D. Sibalis. Transdermal delivery of human insulin to albino rabbits using electrical current, *Am. J. Med. Sci.*, 297:321–325 (1989).
65. O. Siddiqui, Y. Sun, J. C. Liu, and Y. W. Chien. Facilitated transdermal transport of insulin, *J. Pharm. Sci.*, 76:341–345 (1987).
66. Y. W. Chien, P. Lelawongs, O. Siddiqui, Y. Sun, and W. M. Shi. Facilitated transdermal delivery of therapeutic peptides and proteins by iontophoretic delivery devices, *J. Control. Release*, 13:263–278 (1990).
67. J. C. Liu, Y. Sun, O. Siddiqui, and Y. W. Chien. Blood glucose control in diabetic rats by transdermal iontophoretic delivery of insulin, *Int. J. Pharm.*, 44:197–204 (1988).
68. Y. Sun and H. Xue. Transdermal delivery of insulin by iontophoresis, *Ann. N. Y. Acad. Sci.*, 618:596–598 (1991).
69. T. Sophie and P. Veronique. Iontophoretic mobility of insulin across hairless rat skin, *Proc. Intern. Symp. Control. Rel. Bioact. Mater.*, 18:609–610 (1991).
70. O. Pillai and R. Panchagnula. Transdermal delivery of insulin from poloxamer gel: *Ex vivo* and *in vivo* skin permeation studies in rat using iontophoresis and chemical enhancers, *J. Control. Release*, 89:127–140 (2003).
71. Y. Y. Huang and S. M. Wu. Stability of peptides during iontophoretic transdermal delivery, *Int. J. Pharm.*, 131:19–23 (1996).
72. V. Srinivasan, W. I. Higuchi, S. M. Sims, A. H. Ghanem, and C. R. Behl. Transdermal iontophoretic drug delivery: Mechanistic analysis and application to polypeptide delivery, *J. Pharm. Sci.*, 78:370–375 (1989).
73. L. Langkjaer, J. Brange, G. M. Grodsky, and R. H. Guy. Iontophoresis of monomeric insulin analogues *in vitro*: Effects of insulin charge and skin pretreatment, *J. Control. Release*, 51:47–56 (1998).
74. B. H. Sage, C. R. Bock, J. D. Denuzzio, and R. A. Hoke. Technological and developmental issues of iontophoretic transport of peptide and protein drugs. In V. H. L. Lee, M. Hashida, and Y. Mizushima (eds), *Trends and future perspectives in peptide and protein drug delivery*, Harwood Academic, Chur, Switzerland, 1995, pp. 111–134.
75. G. Rao, P. Glikfeld, and R. H. Guy. Reverse iontophoresis: Development of a noninvasive approach for glucose monitoring, *Pharm. Res.*, 10:1751–1755 (1993).
76. A. K. Banga and Y. W. Chien. Iontophoretic delivery of drugs: Fundamentals, developments and biomedical applications, *J. Control. Release*, 7:1–24 (1988).
77. T. Ogiso, S. Nishioka, and M. Iwaki. Dissociation of insulin oligomers and enhancement of percutaneous absorption of insulin, *Biol. Pharm. Bull.*, 19:1049–1054 (1996).
78. J. Hao, D. Li, S. Li, and J. Zheng. The effects of some penetration enhancers on the transdermal iontophoretic delivery of insulin *in vitro*, *J. Chin. Pharm. Sci.*, 5:88–92 (1996).

79. S. K. Rastogi and J. Singh. Passive and iontophoretic transport enhancement of insulin through porcine epidermis by depilatories: Permeability and Fourier transform infrared spectroscopy studies, *AAPS. PharmSciTech.*, 4:E29 (2003).
80. P. Khanna, J. A. Strom, J. I. Malone, and S. Bhansali. Microneedle-based automated therapy for diabetes mellitus, *J. Diabetes Sci. Technol.*, 2:1122–1129 (2008).
81. H. J. G. E. Gardeniers, R. Luttge, E. J. W. Berenschot, M. J. de Boer, S. Y. Yeshurun, M. Hefetz, R. Oever, and A. Berg. Silicon micromachined hollow microneedles for trans-dermal liquid transport, *J. Microelectromechanical Syst.*, 12:855–862 (2003).
82. L. Nordquist, N. Roxhed, P. Griss, and G. Stemme. Novel microneedle patches for active insulin delivery are efficient in maintaining glycaemic control: An initial comparison with subcutaneous administration, *Pharm. Res.*, 24:1381–1388 (2007).
83. N. Roxhed, B. Samel, L. Nordquist, P. Griss, and G. Stemme. Painless drug delivery through microneedle-based transdermal patches featuring active infusion, *IEEE Trans. Biomed. Eng*, 55:1063–1071 (2008).
84. A. Davidson, B. Al-Qallaf, and D. B. Das. Transdermal drug delivery by coated microneedles: Geometry effects on effective skin thickness and drug permeability, *Chem. Engg. Res. Design*, 86:1196–1206 (2008).
85. H. Chen, H. Zhu, J. Zheng, D. Mou, J. Wan, J. Zhang, T. Shi, Y. Zhao, H. Xu, and X. Yang. Iontophoresis-driven penetration of nanovesicles through microneedle-induced skin microchannels for enhancing transdermal delivery of insulin, *J. Control. Release*, 139:63–72 (2009).
86. C. P. Zhou, Y. L. Liu, H. L. Wang, P. X. Zhang, and J. L. Zhang. Transdermal delivery of insulin using microneedle rollers *in vivo*, *Int. J. Pharm.*, 392:127–133 (2010).
87. W. Martanto, S. P. Davis, N. R. Holiday, J. Wang, H. S. Gill, and M. R. Prausnitz. Transdermal delivery of insulin using microneedles *in vivo*, *Pharm. Res.*, 21:947–952 (2004).
88. S. P. Davis, W. Martanto, M. G. Allen, and M. R. Prausnitz. Hollow metal microneedles for insulin delivery to diabetic rats, *IEEE Trans. Biomed. Eng.*, 52:909–915 (2005).
89. Y. Ito, E. Hagiwara, A. Saeki, N. Sugioka, and K. Takada. Feasibility of microneedles for percutaneous absorption of insulin, *Eur. J. Pharm. Sci.*, 29:82–88 (2006).
90. J. Gupta. Microneedles for transdermal drug delivery in human subjects, *Ph.D. Thesis, Georgia Institute of Technology*, (2009).
91. S. D. Roy and E. Manoukian. Permeability of ketorolac acid and its ester analogs (pro-drug) through human cadaver skin, *J. Pharm. Sci.*, 83:1548–1553 (1994).
92. S. D. Roy, D. J. Chatterjee, E. Manoukian, and A. Divor. Permeability of pure enantiom-ers of ketorolac through human cadaver skin, *J. Pharm. Sci.*, 84:987–990 (1995).
93. S. D. Roy, E. Manoukian, and D. Combs. Absorption of transdermally delivered ketoro-lac acid in humans, *J. Pharm. Sci.*, 84:49–52 (1995).
94. C. H. Liu, H. O. Ho, M. C. Hsieh, T. D. Sokoloski, and M. T. Sheu. Studies on the in-vitro percutaneous penetration of indomethacin from gel systems in hairless mice, *J. Pharm. Pharmacol.*, 47:365–372 (1995).
95. Y. Obata, K. Takayama, Y. Maitani, Y. Machida, and T. Nagai. Effect of ethanol on skin permeation of nonionized and ionized diclofenac, *Int. J. Pharm.*, 89:191–198 (1993).
96. J. P. Famaey, G. Broux, D. Cleppe, A. Deroulez, F. Duckerts, F. Evrard, L. Goethals, J. Vanhecke, L. Verbruggen, and J. M. Tyberghein. Ionisation with voltaren: A multi-centre trial, *J. Belge Med. Phys. Rehabil.*, 5:55–60 (1982).
97. L. Vecchini and E. Grossi. Ionization with diclofenac sodium in rheumatic disorders: A double-blind placebo-controlled trial, *J. Int. Med. Res.*, 12:346–350 (1984).
98. W. J. Irwin, F. D. Sanderson, and A. L. W. Po. Percutaneous absorption of ibuprofen and naproxen: Effect of amide enhancers on transport through rat skin, *Int. J. Pharm.*, 66:243–252 (1990).

99. S. Santoyo, A. Arellano, P. Ygartua, and C. Martin. Penetration enhancer effects on the *in vitro* percutaneous absorption of piroxicam through rat skin, *Int. J. Pharm.*, 117:219–224 (1995).

100. S. Uchida, T. Morishita, Y. Ikeda, and T. Akashi. Anti-inflammatory effect of flurbiprofen tape applied percutaneously to rats with adjuvant-induced arthritis, *Jpn. J. Pharmacol.*, 69:37–41 (1995).

101. J. A. Cordero, L. Alarcon, E. Escribano, R. Obach, and J. Domenech. A comparative study of the transdermal penetration of a series of nonsteroidal antiinflammatory drugs, *J. Pharm. Sci.*, 86:503–508 (1997).

102. L. Tessari, L. Ceciliani, A. Belluati, G. Letizia, U. Martorana, L. Pagliara, A. Pognani, G. Thovez, A. Siclari, G. Torri, L. Solimeno, and E. Montull. Aceclofenac cream versus piroxicam cream in the treatment of patients with minor traumas and phlogistic affections of soft tissues: A double-blind study, *Curr. Ther. Res.*, 56:702–712 (1995).

103. H. Pratzel, P. Dittrich, and W. Kukovetz. Spontaneous and forced cutaneous absorption of indomethacin in pigs and humans, *J. Rheumatol.*, 13:1122–1125 (1986).

104. U. Garagiola, U. Dacatra, F. Braconaro, E. Porretti, A. Pisetti, and V. Azzolini. Iontophoretic administration of pirprofen or lysine soluble aspirin in the treatment of rheumatic diseases, *Clinical Ther.*, 10:553–558 (1988).

105. R. Saggini, M. Zoppi, F. Vecchiet, L. Gatteschi, G. Obletter, and M. Giamberardino. Comparison of electromotive drug administration with ketorolac or with placebo in patients with pain from rheumatic disease: A double-masked study, *Clinical Ther.*, 18:1169–1174 (1996).

106. P. C. Panus, S. B. Kulkarni, J. Campbell, W. R. Ravis, and A. K. Banga. Effect of iontophoretic current and application time on transdermal delivery of ketoprofen in man, *Pharm. Sci.*, 2:467–469 (1996).

107. P. C. Panus, J. Campbell, S. B. Kulkarni, R. T. Herrick, W. R. Ravis, and A. K. Banga. Transdermal iontophoretic delivery of ketoprofen through human cadaver skin and in humans, *J. Control. Release*, 44:113–121 (1997).

108. C. M. Heard and K. R. Brain. Does solute stereochemistry influence percutaneous penetration? *Chirality*, 7:305–309 (1995).

109. Y. Tashiro, Y. Kato, E. Hayakawa, and K. Ito. Iontophoretic transdermal delivery of ketoprofen: Effect of iontophoresis on drug transfer from skin to cutaneous blood, *Biol. Pharm. Bull.*, 23:1486–1490 (2000).

110. P. C. Panus, K. E. Ferslew, B. Tober-Meyer, and R. L. Kao. Ketoprofen tissue permeation in swine following cathodic iontophoresis. *Phys. Ther.*, 79:40–49 (1999).

111. M. T. Garcia, C. H. da Silva, D. C. de Oliveira, E. C. Braga, J. A. Thomazini, and M. V. Bentley. Transdermal delivery of ketoprofen: The influence of drug-dioleylphosphatidylcholine interactions, *Pharm. Res.*, 23:1776–1785 (2006).

112. J. E. Shaw. Pharmacokinetics of nitroglycerin and clonidine delivered by the transdermal route, *Am. Heart J.*, 108:217–223 (1984).

113. D. Kobayashi, T. Matsuzawa, K. Sugibayashi, Y. Morimoto, M. Kobayashi, and M. Kimura. Feasibility of use of several cardiovascular agents in transdermal therapeutic systems with l-menthol-ethanol system on hairless rat and human skin, *Biol. Pharm. Bull.*, 16:254–258 (1993).

114. A. Luca, J. C. Garciapagan, F. Feu, J. C. Lopeztalavera, M. Fernandez, C. Bru, J. Bosch, and J. Rodes. Noninvasive measurement of femoral blood flow and portal pressure response to propranolol in patients with cirrhosis, *Hepatology*, 21:83–88 (1995).

115. S. Bornmyr, H. Svensson, B. Lilja, and G. Sundkvist. Skin temperature changes and changes in skin blood flow monitored with laser Doppler flowmetry and imaging: A methodological study in normal humans, *Clin. Physiol.*, 17:71–81 (1997).

116. P. G. Green, J. Hadgraft, and G. Ridout. Enhanced *in vitro* skin permeation of cationic drugs, *Pharm. Res.*, 6:628–632 (1989).

117. Y. Maitani, A. Coutelegros, Y. Obata, and T. Nagai. Prediction of skin permeabilities of diclofenac and propranolol from theoretical partition coefficients determined from cohesion parameters, *J. Pharm. Sci.*, 82:416–420 (1993).

118. R. Krishna and J. K. Pandit. Carboxymethylcellulose-sodium based transdermal drug delivery system for propranolol, *J. Pharm. Pharmacol.*, 48:367–370 (1996).

119. T. Ogiso and M. Shintani. Mechanism for the enhancement effect of fatty acids on the percutaneous absorption of propranolol, *J. Pharm. Sci.*, 79:1065–1071 (1990).

120. M. Hori, H. I. Maibach, and R. H. Guy. Enhancement of propranolol hydrochloride and diazepam skin absorption *in vitro*. II. Drug, vehicle, and enhancer penetration kinetics, *J. Pharm. Sci.*, 81:330–333 (1992).

121. D. Thacharodi and K. P. Rao. Collagen membrane controlled transdermal delivery of propranolol hydrochloride, *Int. J. Pharm.*, 131:97–99 (1996).

122. D. F. Stamatialis, H. H. Rolevink, and G. H. Koops. Controlled transport of timolol maleate through artificial membranes under passive and iontophoretic conditions, *J. Control. Release*, 81:335–345 (2002).

123. S. Ahmed, T. Imai, and M. Otagiri. Stereoselective hydrolysis and penetration of propranolol prodrugs: *In vitro* evaluation using hairless mouse skin, *J. Pharm. Sci.*, 84:877–883 (1995).

124. A. Nanda and R. K. Khar. Enhancement of percutaneous absorption of propranolol hydrochloride by iontophoresis, *Drug Develop. Ind. Pharm.*, 20:3033–3044 (1994).

125. R. Conjeevaram, A. Chaturvedula, G. V. Betageri, G. Sunkara, and A. K. Banga. Iontophoretic *in vivo* transdermal delivery of beta-blockers in hairless rats and reduced skin irritation by liposomal formulation, *Pharm. Res.*, 20:1496–1501 (2003).

126. I. Kobayashi, K. Hosaka, T. Ueno, H. Maruo, M. Kamiyama, C. Konno, and M. Gemba. Relationship between amount of beta-blockers permeating through the stratum corneum and skin irritation after application of beta-blocker adhesive patches to guinea pig skin, *Biol. Pharm. Bull.*, 20:421–427 (1997).

127. L. Wearley, J. C. Liu, and Y. W. Chien. Iontophoresis facilitated transdermal delivery of verapamil. I. *In vitro* evaluation and mechanistic studies, *J. Control. Release*, 8:237–250 (1989).

128. L. Wearley, J. C. Liu, and Y. W. Chien. Iontophoresis facilitated transdermal delivery of verapamil. II. Factors affecting the reversibility of skin permeability, *J. Control. Release*, 9:231–242 (1989).

129. V. Labhasetwar, T. Underwood, S. P. Schwendeman, and R. J. Levy. Iontophoresis for modulation of cardiac drug delivery in dogs, *Proc. Natl. Acad. Sci. USA*, 92:2612–2616 (1995).

130. K. Okabe, H. Yamaguchi, and Y. Kawai. New iontophoretic transdermal administration of the beta blocker metoprolol, *J. Control. Release*, 4:79–85 (1986).

131. S. Thysman, V. Preat, and M. Roland. Factors affecting iontophoretic mobility of metoprolol, *J. Pharm. Sci.*, 81:670–675 (1992).

132. S. Ganga, P. Ramarao, and J. Singh. Effect of Azone on the iontophoretic transdermal delivery of metoprolol tartrate through human epidermis *in vitro*, *J. Control. Release*, 42:57–64 (1996).

133. C. A. Zakzewski and J. K. Li. Pulsed mode constant current iontophoretic transdermal metoprolol tartrate delivery in established acute hypertensive rabbits, *J. Control. Release*, 17:157–162 (1991).

134. A. R. Denet, B. Ucakar, and V. Preat. Transdermal delivery of timolol and atenolol using electroporation and iontophoresis in combination: A mechanistic approach, *Pharm. Res.*, 20:1946–1951 (2003).

135. P. C. Wu, Y. B. Huang, H. H. Lin, and Y. H. Tsai. Percutaneous absorption of captopril from hydrophilic cellulose gel through excised rabbit skin and human skin, *Int. J. Pharm.*, 145:215–220 (1996).

136. C. A. Zakzewski, J. K. J. Li, D. W. Amory, J. C. Jensen, and E. Kalatzis-Manolakis. Design and implementation of a constant-current pulsed iontophoretic stimulation device, *Med. Biol. Eng. Comput.*, 34:484–488 (1996).
137. J. E. Sanderson, R. W. Caldwell, J. Hsiao, R. Dixon, and R. R. Tuttle. Noninvasive delivery of a novel inotropic catecholamine iontophoretic versus intravenous infusion in dogs, *J. Pharm. Sci.*, 76:215–218 (1987).
138. C. S. Kolli and A. K. Banga. Characterization of solid maltose microneedles and their use for transdermal delivery, *Pharm. Res.*, 25:104–113 (2008).

8 Transdermal Delivery of Peptides and Proteins

8.1 INTRODUCTION

Over the last several years, therapeutic peptides and proteins have gained increasing importance as a result of rapid strides in the biotechnology industry. Therapeutic application and market introduction of this new generation of therapeutic agents requires parallel development of efficient delivery systems by the pharmaceutical industry. Because of their polypeptide nature, peptide and protein drugs are destroyed in the gastrointestinal (GI) tract and must be administered parenterally. These are invasive routes that involve the inconvenience of injections and add to the cost of the health-care system if administered under medical supervision. Thus, noninvasive methods of administration would be preferred (1). Skin, with its accessibility, enormous surface area, and possibility for site targeting, offers a potential means for noninvasive delivery. Several drugs are now available on the market as transdermal patches. However, none of these is a peptide or protein drug. This is because skin is ordinarily permeable only to small lipophilic molecules, a criterion readily fulfilled by drugs, such as nitroglycerin, scopolamine, clonidine, and nicotine. Peptide and protein drugs, being hydrophilic and macromolecular in nature, do not readily permeate the skin. The transdermal route offers some distinct advantages for the delivery of peptide drugs, in addition to the general advantages of transdermal delivery discussed in Chapter 1. Because these drugs have short half-lives, the greatest benefit would be the fact that the transdermal route provides a continuous mode of administration, somewhat similar to that provided by an intravenous infusion. Transdermal delivery of peptides may become feasible if assisted by enhancement means such as iontophoresis, microporation, electroporation, jet injectors, phonophoresis, or chemical enhancers (2,3). Each of these techniques has advantages and disadvantages. Iontophoresis is generally believed to work best for polypeptides with a molecular weight less than ~10 kDa, but recent literature has reported delivery of proteins up to 14 kDa (4,5). On the other hand, microchannels created by microporation technologies are several microns in dimensions and, therefore, will enhance delivery, as there are no limitations with respect to molecular size.

Iontophoresis has been widely used for topical delivery of conventional drugs (see Chapter 4). Similarly, it can be used for topical delivery of peptides if any local indications exist. For instance, delivery of anti-infective peptides for local effect may be feasible. However, perhaps the most common use of therapeutic proteins for local effect on skin would be the area of wound healing. The potential use of iontophoresis in this field remains to be seen, because iontophoresis is normally used on intact skin, not broken skin. Nevertheless, iontophoresis has shown promise for the delivery of lidocaine for wound healing. The most promising agent for topical dermatological use is the epidermal growth factor (EGF), a polypeptide of 53 amino acid residues, which has shown promise in healing open and burn wounds (6). It has been found that the presence of protease inhibitors is required in the formulation to stabilize EGF at the wound site (7,8). Formulations of fibroblast growth factor were also reported to accelerate wound healing in a diabetic mouse model (9,10). The topical application of interferon for controlling eruptions of herpes has also been reported (11,12). However, this chapter will mostly discuss only the systemic delivery of therapeutic peptides. Iontophoretic enhancement provides further benefit for peptide delivery, because, theoretically speaking, the rate of peptide delivery can be initiated, terminated, or accurately controlled or modulated merely by switching the current on and off or adjusting the current application parameters, respectively (13–16). This would be especially useful because a pulsatile delivery may be required for some peptides, as opposed to constant delivery. Also, skin is relatively low in proteolytic activity, as compared to other mucosal routes, thereby reducing degradation at the site of administration. A multitude of factors need to be carefully considered for a proper design of a transdermal drug delivery system for delivery of peptides. These include considerations of the charges on the peptide molecule and the skin in relation to the environmental pH. The presence of proteolytic enzymes in the skin and the binding of peptides in skin to form a reservoir (or depot) are some of the other important considerations. When employing coated microneedles for peptide/protein delivery, the amount of peptide that can be coated onto a microneedle array is limited, and this aspect has to be considered (see Chapter 3).

Preformulation data should be generated for formulation development of a dosage form or the design of a drug delivery system to achieve optimum stability and maximum bioavailability. Peptide/protein drugs are known to have a strong tendency to get adsorbed to a variety of surfaces, and this must be carefully evaluated as it could lead to misleading interpretations of data. Certain additives can be used to minimize such surface adsorption. Another potential problem is self-aggregation of peptide/protein molecules. A careful choice of electrode, pH, buffer, and ionic strength is required when iontophoresis is employed. The possibility of a reversible electrode such as silver-silver chloride reacting with peptides to cause precipitation or discoloration should be considered. The charge on the peptide molecule can be controlled by adjusting the solution pH; thus, delivery can be manipulated for either cathodal or anodal iontophoresis. The use of hydrogels could provide a pragmatic choice for a delivery system (see Section 10.7). It is important to have a sensitive, stability indicating, validated assay to characterize the permeation profiles of the drug. Because proteins are very potent drugs, only small quantities are used, and the amount of peptide permeating through the skin into receptor in an *in vitro* setup is generally too low to be within the sensitivity of HPLC assay with regular ultraviolet (UV) detector. Radiolabeled peptides are thus commonly used, but the use of radiotracers can lead to erroneous results, as the assay may be measuring proteolytic fragments rather than the intact protein. The use of a radiochemical detector on an HPLC may help to solve this problem. Similarly, once distributed in the volume of distribution, the plasma concentrations during *in vivo* studies may be below the assay limits for some analytical methods. Analytical techniques such as radioimmunoassay and

enzyme-linked immunosorbent assay may be used to characterize the plasma levels
of peptides following transdermal delivery.

8.2 STRUCTURAL CONSIDERATIONS OF POLYPEPTIDES
RELEVANT TO TRANSDERMAL DELIVERY

A basic understanding of peptide structure is required to fully appreciate the complexities involved in their transport into or across skin, especially for iontophoretic
transport. The 20 different naturally occurring amino acids are the building blocks
of peptides and proteins. All amino acids (except glycine) have a chiral carbon and
are zwitterions, both in solution and in solid state. The ionizable sites in the peptide,
such as the imidazolyl nitrogen of histidine, phenol hydroxyl of tyrosine, or guanidine nitrogen of arginine, should be known. Each functional group has a characteristic dissociation constant, K, whose negative logarithm is called the pK_a. Since pK_a
values were measured at fixed ionic strength, they are sometimes called, "apparent dissociation constants, pK_a'." The constant pK_1' usually refers to the most acidic
group. For example, the dissociation constants for glycine, pK_1' (COOH) and pK_2'
(NH_3^+) are 2.34 and 9.60, respectively. Aspartic and glutamic acid have a negative
charge, while lysine and arginine have a positive charge at the physiological pH
of 7.4. The imidazole group in histidine carries a partial positive charge at pH 7.4.
Serine and threonine have side chains that carry no charge at any pH but are polar in
nature. In contrast, tryptophan, phenylalanine, and isoleucine have side chains that
are more hydrocarbon-like in character. The pH at which all the molecules are in the
zwitterionic form so that there is no charge on the molecule is called the isoelectric
point (pI) of the amino acid. Of the 20 amino acids, 15 have an isoelectric point (pI)
near 6. The three basic amino acids have higher pI values, and the two acidic ones
have lower pI values (16). Amino acids join with each other by peptide bonds to
form polymers referred to as peptides or proteins. The distinction between peptides
and proteins is somewhat arbitrary. Typically, peptides contain fewer than 20 amino
acids and have a molecular weight less than 5000 Da, while proteins contain 50 or
more amino acids. Between these two categories are polypeptides that contain about
20 to 50 amino acids. Therapeutic proteins are generally referred to as globular proteins, as they are of nearly spherical shape in solution. A protein has several levels
of structure, generally referred to as the primary, secondary, tertiary, and quaternary
structures. The sequence of covalently bonded amino acids in the polypeptide chain
is known as the primary structure and is determined genetically by the sequence
of nucleotides in DNA. Once a polypeptide chain is formed, there is only one residue with a free amino group (N-terminal) and only one with a free carboxyl group
(C-terminal). Peptide and protein chains are always written with the N-terminal residue on the left. The turns and loops of the polypeptide chain constitute the secondary structure. Alpha helices and beta sheets are the common conformations of the
secondary structure, with the rest being tight turns, small loops, and random coils.
In the α-helix, a single protein chain twists like a coiled spring. Each turn contains
3.6 amino acid residues, with the side chains pointing out from the helix. In the
β-sheet conformation, the protein backbone is nearly fully extended, and β-sheets
can be either parallel or antiparallel. In the parallel β-sheet, two or more peptide

chains align their backbones in the same general direction. The antiparallel β-sheet has adjacent chains having parallel but opposite orientations. A fully extended polypeptide chain of about 60 residues may be about 200 Å long, but the folded globular protein may be just about 30 Å in diameter. Tertiary structure of a protein is the overall packing-in space of the various elements of secondary structure. Protein folds in a specific fashion, but it is believed that the specific folding is determined by the amino acid sequence or the primary structure of the protein. However, the folding is guided by some general rules: the overall shape is spherical, polar groups are on the surface while hydrophobic groups are buried in the interior, and the interior is closely packed. Guided by these rules, the protein folds into a unique structure. The forces that may stabilize the folded structure of a protein may be covalent or noncovalent. Disulfide bridges between cysteine residues provide a covalent linkage that can hold together two chains or two parts of the same chain. The noncovalent forces that can stabilize the protein include hydrogen bonding, salt bridges, and hydrophobic interactions between the side chains of amino acid residues. Unlike proteins, peptides may display several conformations in solution. This is because they lack the hydrogen bonds and disulfide bridges that stabilize the three-dimensional structure of proteins.

8.3 PROTEASE INHIBITORS AND PERMEATION ENHANCERS FOR TRANSDERMAL PROTEIN DELIVERY

8.3.1 PROTEOLYTIC ACTIVITY OF SKIN

Though skin is low in proteolytic activity compared to mucosal routes, it still retains substantial proteolytic activity to present a major barrier to delivery of peptides and proteins. The proteolytic enzymes in skin were reviewed (17,18). Of the exopeptidases in the skin, aminopeptidases are the best known, while carboxypeptidases are in relatively smaller quantity or are absent. The endopeptidases include the proteinases such as the caseinolytic enzymes, chymotrypsin- and trypsin-substrate hydrolyzing enzymes, thiol proteinase, and carboxyl proteinases. In addition, proteinases attacking specific substrates such as collagenases, elastase, fibrinolysins, and enzymes of the kallikrein–kinin system are present. Most of these enzymes have not yet been isolated, purified, or characterized. It must be realized that studies with skin homogenates may not be good predictors of proteolytic degradation actually encountered during transport. This is because if proteolysis in skin is discussed based on studies with skin homogenates, the structural complexity of skin is ignored. The subcellular compartmentalization of the proteolytic enzymes may be such that these enzymes do not encounter the peptide or protein drug, as the pathways of penetration may not pass through the cells that hold these enzymes. Once a skin homogenate has been made, the origins of the individual enzymes in the homogenate cannot be determined. Furthermore, the harsh techniques required to disrupt the cells during homogenization may destroy some enzymes. The enzymatic activity also differs between the stratum corneum side and the dermal side. When aqueous luteinizing hormone-releasing hormone (LHRH) solutions were exposed to stratum corneum side on excised hairless mouse skin, they were stable over a period of 40 hours, but solutions exposed to the dermal side degraded by about 43% after 40 hours (19). Degradation of proteins may also be pH dependent. For example, LHRH was stable when exposed to skin at pH 3 but was degraded at pH 5 and 7 (20). Hair follicles and sebaceous glands may have higher metabolic activity than keratinocytes. Because the distribution of these appendages varies with the anatomical site, this could be one contributing factor for the enzymatic activity being different at various anatomical locations. Another factor could be the varying ratio of epidermis to dermis at different sites (18). Degradation of bovine serum albumin (BSA) during iontophoresis experiments with hairless mouse skin was investigated by distribution of radioactivity in selected molecular weight ranges. Although little changes were seen in monomeric BSA (70 to 50 kD) after contact with stratum corneum or dermis for 24 hours, it was converted to lower molecular weight species (15 to 1.5 and <1.5 kD) during transport (21). Stratum corneum is known to have a chymotryptic enzyme that is a partially glycosylated 25 kD molecule with an isoelectric point of about 10. It has an optimum pH of neutral to alkaline but is also active at pH 5.5, the pH of the stratum corneum (22). It should be noted that the proteolytic activity of skin from many animal species that have been used in transdermal studies may typically be higher than that of human skin (Section 2.2.1).

8.3.2 Protease Inhibitors

Protease inhibitors provide a viable means to circumvent the enzymatic barrier for the delivery of peptides and proteins. Numerous agents could act as protease inhibitors by a variety of mechanisms (e.g., by tightly binding to or modifying covalently the active sites of proteases, or by chelating the metal ions essential for proteolytic activity). The selection of an appropriate protease inhibitor could be guided by studying the principal proteases responsible for the degradation of the peptide/protein to be delivered, their subcellular compartmentalization, and the mechanism for the transport of the peptide/protein. Protease inhibitors such as puromycin and amastatin were found to inhibit the degradation of leucine-enkephalin in skin homogenates of hairless mouse skin. Interestingly, these inhibitors were not effective in skin diffusion experiments (23). As discussed, this could be because the proteolytic enzymes may be present in membranes outside the actual diffusion path. Other potentially useful protease inhibitors include aprotinin, bestatin, boroleucine, p-chloromercuribenzoate, leupeptin, pepstatin, and phenylmethylsulfonyl fluoride. Sodium glycocholate, a penetration enhancer, may also act as a protease inhibitor. Protease inhibitors can be used to minimize the proteolytic degradation of peptides in skin during iontophoretic transport. This approach was found to be useful during delivery of calcitonin (see Section 8.5.5).

8.3.3 Permeation Enhancers

Permeation enhancers are generally not as effective as physical enhancement techniques to enable transdermal delivery of polypeptides, especially for larger protein molecules. However, enhancers have been investigated for delivery of many small peptides. A nonionic surfactant, n-decylmethyl sulfoxide, was found to increase the permeability of two amino acids, tyrosine and phenylalanine, through hairless mouse skin. This enhancer also increased the permeability of the dipeptide, Phe-Phe, and the pentapeptide, enkephalin (23). The percutaneous absorption of a tetrapeptide, hisetal (α-MSH) across hairless mouse skin, was found to be enhanced about 28-fold when oleic acid was used as a penetration enhancer. Oleic acid was the most effective enhancer, but Azone and dodecyl *N,N*-dimethylamino acetate (DDAA), a relatively new penetration enhancer, were also found to be effective (24). Compared to hairless mouse skin, human skin was much less permeable to hisetal. Also, the effectiveness of penetration enhancers was much greater with mouse skin as compared to human skin (25). The combined use of iontophoresis and penetration enhancers may allow greater amounts of drug to be delivered than either technique alone. In a study on transport of LHRH through porcine epidermis, iontophoresis was found to synergize with enhancers, such as 10% oleic acid in combination with ethanol, and 10% oleic acid in combination with propylene glycol. Fourier transform infrared (FT-IR) spectroscopic study showed that the combination of enhancer and iontophoresis increased the lipid fluidity, suggesting that the synergism in enhancement could be due to the greater fluidization of the stratum corneum lipids (26). A somewhat similar observation was made for the *in vitro* transport of cholecystokinin-8 through porcine epidermis. Iontophoresis was found to further increase the permeability of the drug through enhancer-pretreated porcine epidermis in comparison to the control (27).

8.4   IONTOPHORETIC DELIVERY OF PEPTIDES

Analogous to amino acids, the isoelectric point (pI) of a protein occurs at the pH where the positive and negative charges are balanced. At their isoelectric point, proteins behave like zwitterions. At any pH below its pI, the protein has a positive charge and should be delivered under anode. At pH above its pI, it has a negative charge and may be delivered under cathode. If a protein has an equal number of acidic and basic groups, its pI is close to 7. If it has more acidic than basic groups, the pI is low, and vice versa. For peptide drugs, the pH of the buffer used relative to the isoelectric point of the drug will determine whether the drug will have a positive or negative charge. The pH of the buffer should be at least one and preferably two pH units away from the pI of the peptide to ensure a high charge density on the peptide. Peptides with a high isoelectric point such as vasopressin (pI = 10.9) or salmon calcitonin (pI = 10.4) are perhaps the ideal candidates for iontophoretic delivery. This is because they will have a positive charge with high charge density at physiological or lower pH values. Because these will be delivered under anode, electroosmotic flow that occurs from anode to cathode (Section 4.7) will assist delivery in addition to direct electrostatic repulsion. In contrast, for peptides with very low pI (<3) delivered from a pH 7.4 buffer, the electroosmotic flow may hinder their transport. Thus, it will have to be established if optimal delivery takes place under the cathode or anode depending on whether direct electric repulsion or electroosmosis is the predominant mechanism of transport. If the formulation has a low pH (<4), the charge on the skin may be neutralized, and a diminished or even reversed electroosmotic flow is possible (Section 8.5). Iontophoretic delivery of peptides with a pI between 4 and 7.3 is faced with severe challenges (28). For a peptide to be iontophoretically delivered through the skin, it should ideally not encounter any environment where its charge may reverse. The pH of stratum corneum can range from 4 to 6, with a lowest pH of 3.5 to 4.5 at some small distance below the surface of the stratum corneum. The pH of the hydrated tissue just under the basement membrane is about 7.3. Thus, a peptide with a pI of, say 5, will be delivered into the skin up to some distance before it encounters a pH equal to its pI. At this point, the peptide will become uncharged, and the iontophoretic force will no longer apply. As it diffuses a little farther due to the concentration gradient (or electroosmotic flow), it will reverse its charge and may be pulled back toward the delivery electrode until it is again uncharged. Thus, the peptide will concentrate at some depth below the skin to form a depot or drug reservoir and may even precipitate in the skin (28).

The pH of the receiver fluid can also be an important variable for *in vitro* studies, because it will affect the pH of the viable epidermis (29). In order for the study to represent physiological conditions, a physiological receiver fluid such as pH 7.4 phosphate buffer should be used. It has also been suggested that electrically assisted transdermal delivery can be increased by nonionic and zwitterionic surfactants for peptides and proteins that have at least one hydrophobic site. Apparently, the hydrophobic site impedes transport due to interactions with the lipophilic regions of skin. The surfactant helps to shield the hydrophobic site, thus minimizing these interactions. It was shown that while cytochrome C, lysozyme, and ribonuclease have similar molecular weight, their *in vitro* skin flux was different and correlated with their hydrophobicity (30). The salt form of the peptide can also be very important. The chemically enhanced *in vitro* permeability of leuprolide in the base form was reported to be 10 times higher than in the acetate form due to improved lipophilicity of the base form (31). For iontophoretic delivery, salt forms are more likely to be permeable than the base. If the base is not water soluble, then salt forms may be the only choice. However, soluble peptides may also be available only in salt form due to procedures used in bulk synthesis. Selection of proper salt form is important. If hydrochloride salt is available and silver/silver chloride electrodes are being used, then it will at least partly obviate the necessity of adding chloride to drive the electrochemistry, thus improving efficiency of delivery by reducing the competitive ions. Because a peptide in aqueous solution is unlikely to have significant passive permeation, any comparison of the advantage of iontophoresis of aqueous solution of a salt of a lipophilic peptide should be with passive permeation of the base from a nonaqueous solution, rather than the aqueous salt formulation used in iontophoresis. The same is true for conventional drugs that may have been solubilized in aqueous solution for iontophoretic delivery. The technique of capillary electrophoresis, which is normally used as an analytical tool, can also be used to screen promising peptide candidates as well as to optimize the formulation and delivery conditions for iontophoresis. This technique has been used to estimate the electrophoretic mobility of thyrotropin-releasing hormone as a function of pH and ionic strength. The resulting data compared well with reported literature values for iontophoretic flux of the molecule under similar formulation conditions (32). With a similar rationale, the techniques of isoelectric focusing and capillary zone electrophoresis were used as tools to predict the ability of a peptide to be iontophoresed. Based on such studies, native LHRH was predicted to be better suited for iontophoretic delivery than its free-acid analog. The pI of LHRH was determined to be 9.6, while that of the analog was 6.9. Furthermore, native LHRH was more mobile at pH > 2.5, though the two compounds were chemically similar (33).

Liposomes have also been investigated for iontophoretic delivery of peptides. A charge could be imparted to neutral drugs by encapsulating them in charged liposomes, followed by their iontophoretic delivery. For preparation of charged liposomes, stearylamine (SA) can be added to induce positive charge, while phosphotidylserine (PS) can be added to induce negative charge. For drugs that carry a charge, such as peptides, liposomes could prevent or minimize their proteolytic breakdown in the skin during delivery. We reported the combined use of liposomes and iontophoresis for transdermal delivery (34). In this report, we investigated the iontophoretic delivery of a liposomal formulation of a pentapeptide, Leu-enkephalin (Tyr-Gly-Gly-Phe-Leu), across human cadaver skin. Leu-enkephalin was transported across human cadaver skin using a current density of 0.5 mA/cm^2. For liposome formulations, large unilamellar vesicles of dimyristoyl phosphatidyl choline and cholesterol were prepared. The mean particle diameter of liposomes was 110 ± 10 nm. Following iontophoretic delivery, liposomes or their constituents were found to be present in the skin. By HPLC analysis, the percent intact enkephalin, as determined by the radiochemical detector, starts to decline after a few hours. This could be due to the increased degradation of enkephalin as a function of time, as more and more proteolytic enzymes are leached from the skin. For liposomal formulation, the degradation is less, indicating that liposomes are preventing enkephalin against degradation. This study had several complicating variables operating at the same time, because the peptide had a charge (as a function of pH) in addition to the charge imparted on the liposomes. A simpler system would use a neutral drug in charged liposomes. We thus followed up the combined use of liposomes and iontophoresis using a model neutral drug. Delivery of the neutral drug in positively charged liposomes was four to five times greater than that of plain drug. Composition of the lipid used to make liposomes was also found to play a role (35).

8.5 TRANSDERMAL DELIVERY OF PEPTIDES AND PROTEINS: EXAMPLES

In addition to the peptides/proteins discussed here, extensive investigations have been done on delivery of insulin by iontophoresis and microneedles, and these are discussed in detail in Chapter 7 as a case study. Also, commercial development of a transdermal product for parathyroid hormone using coated microneedles is discussed in Chapter 10, and delivery of proteins and other macromolecules by sonophoresis is discussed in Chapter 6. There is extensive literature available on transdermal delivery of peptides by iontophoresis and chemical enhancers. Chemical enhancers are somewhat outside the scope of this book, but given the importance of peptide delivery via skin, these are also discussed in this chapter.

8.5.1 AMINO ACIDS

Passive permeability of amino acids through excised rat skin was found to vary with donor pH and the type of amino acids, with the permeability coefficient being highest for dication, followed by monocation, positively charged, uncharged, and negatively charged zwitterions (36). For *in vitro* passive permeation through porcine skin, the lipophilicity of amino acids was found to be a more dominant factor than the molecular weight (37). It was shown that the binding of a series of amino acids in excised abdominal skin of hairless rat decreased with an increase in the alkyl side chain (38). This suggests that binding is likely to be polar or electrostatic in nature. Iontophoresis of amino acids was explored in an attempt to predict or understand the delivery of peptides. Amino acids may also have a direct benefit of moisturizing the skin, as the presence of amino acids in the skin causes an increase in the penetration of aqueous solutions into lipids (39). Amino acids have also been used as model compounds to understand mechanisms of and factors affecting iontophoresis (40). Extraction of amino acids from the stratum corneum and subdermal tissue by reverse iontophoresis was reported (41). We investigated the iontophoretic delivery of amino acids by selecting one weakly charged amino acid, glycine (pI = 6), and one strongly charged amino acid, arginine (pI = 11.2). At the pH used (7.2), arginine would carry a strong positive charge, while glycine would carry a relatively weak negative charge. As expected, arginine delivery was significantly more under anode than under cathode or under passive conditions. For glycine, a higher delivery under cathode was expected as it carries a negative charge. Instead, a higher delivery was found under the anode. The reason for this behavior could be that electroosmotic flow from anode to cathode overshadows the effect of polarity on the ionic molecule, as the charge on the ion is relatively weak. It is also possible that as glycine enters the skin, the pH of the environment is closer to 6, so the molecule is essentially neutral in the solution (42).

Another study delivered a series of amino acids across excised hairless mouse skin to investigate the effects of permeant charge (neutral, +1 or −1), lipophilicity, and vehicle pH. Iontophoretic flux of zwitterions was significantly greater than their passive transport, and delivery from the anode was greater than from the cathode, presumably due to electroosmotic flow. For zwitterions (unlike the negatively charged amino acids), the iontophoretic/electroosmotic flux did not reach steady state under the experimental conditions used and was inversely proportional to the permeant octanol/pH 7.4 buffer distribution coefficient (43). Amino acids and their derivatives with blocked amino or carboxyl groups were used to investigate the relative importance of the electroosmotic flow in comparison to passive and electrical flow in transdermal iontophoresis across excised porcine skin. As expected, electroosmotic flow enhanced anodic delivery of amino acids, and flux decreased with increasing salt concentration in buffer. However, the negative effect of electroosmotic flow on anionic and neutral solutes was reduced as buffer concentration increased (44).

8.5.2 SMALL PEPTIDES

Thyrotropin-releasing hormone (TRH) is a tripeptide with a molecular weight of 362 Da and a pK_a of 6.2. A prodrug approach has been used for transdermal delivery of TRH. Results of diffusion experiments using excised human skin indicate that the *N*-octyloxycarbonyl derivative showed an enhanced permeability. The prodrug that penetrated into the receptor phase was found to exist primarily as TRH (45).

Iontophoretic transport of TRH across excised dorsal nude mouse skin was found to be directly proportional to the applied current density. In the absence of current, the flux of TRH across skin was undetectable. However, even uncharged TRH was transported with iontophoresis, presumably by the electroosmotic or convective flow that accompanies iontophoresis (46). Using response surface methodology to optimize the transdermal delivery of TRH, the maximum rate of permeation across rabbit inner pinna skin was achieved at low ionic strength, moderate pH, and large current duty cycle under the constraint of a pulsed current density of less than 1 mA/cm^2 (47).

In another study, flux of TRH across excised rabbit pinna skin in steady state was found to be proportional to the applied current density. At low ionic strength, the protonation of TRH was greater, and its flux at pH 4 was greater than at pH 8. At higher ionic strength, the trend was reversed, presumably due to the enhanced flux of unprotonated TRH due to electroosmotic flow (48). The stability of TRH under an electric field was also investigated. Using platinum wires as electrodes, the degradation increased when current density was increased >0.32 mA/cm^2, with rapid degradation at 0.64 mA/cm^2. After iontophoresis for 10 hours, more than 75% of the drug was degraded (49). Another tripeptide, threonine-lysine-proline (Thr-Lys-Pro), was successfully delivered across nude rat skin by iontophoresis under both *in vitro* and *in vivo* conditions. Transport of the tripeptide, which was positively charged at pH 7.4, was significantly enhanced by iontophoresis as compared to passive transport. The delivery was found to be directly proportional to the applied current density over the range 0.18 to 0.36 mA/cm^2. Following 6 hours of iontophoresis, 98.4% of the radioactivity in the donor was still the intact peptide, while 94% of the radioactivity penetrated in the receptor phase was the parent Thr-Lys-Pro, suggesting that degradation in the skin was not significant (50). Delivery of eight other tripeptides with the general structure alanine-X-alanine was investigated across hairless mouse skin, *in vitro*. The tripeptides exhibited very little degradation under the conditions of the experiment. The normalized iontophoretic flux was found to be independent of lipophilicity but inversely related to molecular weight. Steady-state fluxes were not achieved, suggesting that time-dependent changes in the properties of the skin barrier may be occurring (51). Iontophoretic delivery of another tripeptide, enalaprilat, was also investigated across hairless guinea pig skin (52).

8.5.3 VASOPRESSIN AND ANALOGS

Vasopressin, an antidiuretic hormone, and its analogs were investigated for transdermal delivery. Effects of enhancers such as DMSO, sodium lauryl sulfate, Azone, and others, on the transdermal permeation of vasopressin across excised rat skin were investigated (53,54). Sodium lauryl sulfate had no significant effect on permeation. Azone was found to be the most effective enhancer and increased flux about 15 times. These findings were confirmed by *in vivo* studies on Brattleboro rats. These rats are genetically deficient in vasopressin and secrete large volumes of urine with low osmolality. A significant reduction in urine volume and increase in osmolality over a 24-hour period resulted in enhancement of vasopressin absorption in the presence of Azone.

In another study, the potential of azacycloheptan-2-ones (azones) as penetration enhancers for desglycinamide arginine vasopressin (DGAVP) was investigated. Using human stratum corneum under *in vitro* conditions, it was found that the hydrocarbon chain length of Azone determined its effectiveness as a penetration enhancer. Pretreatment with hexyl- or octyl-Azone did not enhance the flux, but the permeability increased 1.9, 3.5, or 2.5-fold after pretreatment with decyl-, dodecyl-, or tetradecyl-Azone (55).

A nonapeptide with a molecular weight of 1084 Da, vasopressin makes a good model peptide for iontophoretic investigation because of its ideal molecular size and isoelectric point. Arginine vasopressin has a hexapeptide ring, in which two cysteine residues form a disulfide bridge, and a tripeptide tail. Due to the presence of basic arginine and blocking of the C-terminus with NH_2, arginine vasopressin has an isoelectric point of about 10.9. Thus, more than 99% of the drug is protonated in a buffer solution with pH lower than 9. Vasopressin also has therapeutic use in the treatment of diabetes insipidus and emergency treatment of bleeding from esophagogastric varices.

Furthermore, vasopressin may require a pulsed delivery for therapeutic effect rather than steady-state levels. This is due to the possibility of tolerance or desensitization of receptors by its continual presence at a receptor site (16). Iontophoretic delivery can provide for such pulsed delivery by controlling the current. Transdermal iontophoretic delivery of vasopressin across excised hairless rat skin was investigated (56–58). Delivery was found to be reversible upon termination of iontophoresis, with flux returning to the levels as seen for passive diffusion alone (56). It required about 1.5 to 2 hours of iontophoresis treatment for vasopressin permeation to reach steady-state levels (57). Transdermal iontophoretic transport and degradation of vasopressin (spiked with labeled peptide) across human cadaver skin was investigated by the author using a current density of 0.5 mA/cm^2 and analyzing results by an HPLC attached to a radio-chromatography detector. Because a radiochemical detector was used for analysis, any free tritium or degradation products would not show up at the vasopressin retention time on the chromatogram. A higher degradation was observed in the receptor where the peptide was in contact with the dermal side of the skin. Higher enzymatic degradation in the receptor solution is most likely because more proteolytic enzymes would be present on the dermal side of the skin, as compared to the stratum corneum, which was in contact with the donor solution. This particular experiment was designed so that no transdermal transport took place during this experiment. Because vasopressin was mostly protonated at the pH (7.20) of HEPES buffer during the study, only anodal delivery was investigated. The cumulative amount of intact vasopressin permeated during 8 hours of iontophoretic transport was 15.37 ± 5.31 µg/cm^2. Of the total radioactivity permeated, only about 40% was intact vasopressin at 12 hours, and several degradation peaks could be seen in the chromatogram. This suggests that the human cadaver skin still had enough enzymatic activity left to cause enzymatic cleavage of vasopressin into several products. The number of degradation products could be even higher than what would appear from such a chromatogram, because any cleaved fragment that did not carry the tracer would not show up on the chromatogram. No intact vasopressin was found to permeate under passive conditions in this study. The flux of vasopressin across skin was observed to stop as soon as the current was terminated, suggesting reversibility of delivery. Thus, modulation of peptide delivery by multiple applications of current should be possible. This can be used to achieve pulsatile delivery of peptides, which would be desirable because many physiological peptides are released in a pulsatile manner (59). A vasopressin analog, 9-desglycinamide, 8 arginine vasopressin (DGAVP) is a neuropeptide drug shown to improve memory process in men and is also used as a model peptide that is more potent and more resistant to metabolism as compared to vasopressin. Iontophoretic delivery of DGAVP across human and snake skin has been compared. In another study, DGAVP was iontophoretically delivered through dermatomed human skin. The transport was observed to be predominantly controlled by electroosmosis, and the flux appeared to be controlled by the applied voltage rather than by the current density (60). Iontophoretic delivery of desmopressin acetate (DDAVP) was investigated *in vivo* with diabetes insipidus model in rats. Repeated short iontophoretic treatments with low current density were found to be best to maintain a constant response. Using pulsed current, the antidiuretic response was dependent on current duty (on/off ratio) but not frequency (61,62). In another study, the prolongation of antidiuretic response to desmopressin acetate in diabetes insipidus rats was compared with other routes of administration. Delivery by iontophoresis was comparable to the nasal route when a dose about five times higher than the nasal route dose was used. In comparison to oral route, iontophoresis was two to three times as effective (63).

8.5.4 Luteinizing Hormone-Releasing Hormones and Analogs

Luteinizing hormone-releasing hormone (LHRH), also known as gonadotropin-releasing hormone (GnRH), is a naturally occurring decapeptide that stimulates the release of pituitary gonadotropins. LHRH is secreted in a pulsatile manner, and its replacement for treatment of primary infertility of hypothalamic origin requires pulsatile administration every 60 to 90 minutes through programmable pumps (64). LHRH is readily degraded in the body by proteolytic enzymes and has a half-life of only 8 minutes. Several agonists and antagonists of LHRH have been synthesized and are promising agents for a range of clinical applications. Compared to LHRH, these agonists have relatively longer half-lives, and they can mimic the effects of continuous LHRH infusion if administered twice daily. Such administration stimulates the release of pituitary gonadotropins initially, but repeated doses abolish the stimulatory effects. Due to receptor desensitization and downregulation, a stage of chemical castration results in about 4 weeks. Chronic application of LHRH agonists in controlled release formulations such as microspheres has proved effective in treating hormone-dependent cancers, prostate and breast cancer, and sex-hormone-dependent disorders (e.g., endometriosis and uterine fibroids) (16).

Transdermal delivery provides a noninvasive route for the delivery of LHRH and its analogs. Chemical enhancers have been reported to significantly enhance the permeability of LHRH (65) and its analogs, leuprolide acetate through nude mouse skin (31), and nafarelin acetate through human cadaver skin (66). A combination of penetration enhancers and iontophoresis has been reported to act synergistically to enhance delivery of LHRH across human epidermal membrane (67). LHRH and an antagonist have been shown to be rapidly iontophoresed through hairless mouse skin, but metabolism by the skin to fragment peptides was observed (19). Degradation of LHRH by iontophoretic transport across porcine skin has also been reported (26). The ability to transdermally deliver LHRH by iontophoresis was demonstrated using the isolated perfused porcine flap model (for details of the model, see Section 2.2.3). A current of 0.2 mA/cm^2 was applied for 3 hours, and the mean LHRH delivered in 5 hours using 21 skin flaps was 959 ± 444 ng. LHRH concentrations increased throughout the current application period and then decreased when current was stopped. The flux data from isolated perfused porcine skin flap (IPPSF) tests (see Section 2.2.3) were used as input into a two-compartment mammillary model, and the predictions were checked by *in vivo* studies in pigs. It was found that the model was able to predict the *in vivo* serum concentration of iontophoretically delivered LHRH. Increases in follicle-stimulating hormone (FSH) and luteinizing hormore (LH) concentrations were observed with increasing LHRH concentrations, indicating that iontophoretically delivered LHRH is both immunologically and biologically active (68).

In another study using the IPPSF model, a drug depot was identified in the skin underlying the electrode, which was approximately two times as large as the entire mass of drug delivered systemically. A small portion of this skin depot could be delivered with just a saline electrode during a second application period (69). Delivery of LHRH by electroporation was also investigated (see Section 5.7.1). Another LHRH analog, leuprolide, was delivered across skin by iontophoresis (70,71). Therapeutic doses of leuprolide were delivered in 13 normal men using a double-blind, randomized crossover study conducted under an investigational new drug (IND) granted by the U.S. Food and Drug Administration (FDA). The patches used were about 70 cm^2 and contained two reservoirs and an intrinsic power source supplied by small batteries. The electrical circuit was completed when the patch was applied to the skin. Passive patches were identical except that the wiring to complete the circuit was missing. Data analysis by analysis of variance (ANOVA) showed significant differences between the active and passive patches. The magnitude of elevation of LH produced by the active patches was in therapeutic range and comparable to that achieved by subcutaneous administration. The only adverse effect reported was a mild erythema at the site of the active patch in 6 of the 13 subjects. This erythema resolved rapidly without sequelae (70). Further evaluation compared the acute biological effect seen after subcutaneous and iontophoretic delivery of leuprolide to 18 volunteers. It was observed that the two techniques were similar in the magnitude and duration of response achieved. The onset of LH response was more rapid after subcutaneous administration, because the period of absorption required for transdermal absorption is not a factor in this case (72). A potentiometric titration curve of leuprolide shows that it has a net charge of +1 at a pH of 7.5 (protonation of Arg) and +2 at a pH of 4.5 (protonation of His). The enhancement factor for leuprolide across synthetic membrane as a function of applied voltage at pH 4.5 was thus found to be double that at pH 7.5, directly proportional to the charge as discussed (73).

Nafarelin is another potent analog of LHRH which is commercially available as a nasal solution for the treatment of endometriosis and central precocious puberty. Transdermal delivery could provide an alternative route, and use of chemical enhancers to facilitate *in vitro* permeation of nafarelin acetate through human cadaver and monkey skin has been investigated. Flux through human cadaver skin from a propylene glycol/Azone vehicle was 0.14 $\mu g/cm^2$ per hour. The steady-state lag time ranged from 24 to 40 hours. Monkey skin was slightly more permeable than human cadaver skin (66). Iontophoretic transport of nafarelin across hairless mouse skin (74,75) and human skin (76) was also investigated. Anodal delivery of positively charged nafarelin was found to result in a strong association of the lipophilic cation with the fixed negative charges on the skin, resulting in neutralization of the skin charge and reversal of electroosmotic flow. Due to this phenomenon, the peptide was found to inhibit its own delivery as a function of increasing concentration. An increased iontophoretic flux across hairless mouse skin was observed when the donor concentration was doubled from 0.5 to 1 mg/ml, but further elevation in concentration to 2 or 3 mg/ml did not increase the flux (75). In the case of human skin, the cumulative amount of nafarelin delivered iontophoretically (0.3 mA/cm^2) in 24 hours actually decreased from 27.3 to 5.9 $nmol/cm^2$ as the donor concentration was increased from 0.1 to 1 mg/ml. The average delivery rate was about 1 $\mu g/cm^2/hour$ and will allow delivery of target dose with a patch of 10 to 20 cm^2. Some metabolism of nafarelin during electrotransport was also observed (76). Similarly, iontophoretic delivery of triptorelin ([D-Trp[6]]LHRH) was reported to be reduced twofold when the concentration is doubled from 3 to 6 mM, though therapeutically relevant flux was still obtained (77). However, the parent peptide (LHRH) does not impair electroosmosis, leading investigators to search for a structural motif that may be responsible for this behavior. It was observed that the inhibition of electroosmosis by oligopeptides is reduced as the lipophilicity of the molecules is reduced. Also, the presence of a positive charge in close proximity to the hydrophobic portion of the molecule was considered essential. The carboxylate groups that give skin its negative charge probably interact with the positive charge of the oligopeptide, while the adjacent lipophilic group on the peptide anchors the molecule somewhere along the pathway for iontophoresis. This changes the permselectivity of the skin and progressively reduces the electroosmotic flow (78). Permeation of another analog, buserelin, through isolated human stratum corneum following iontophoretic treatment was also reported. In this study, it was found that passive permeation of buserelin through human stratum corneum was not feasible. However, permeation was achieved and controlled by iontophoresis (79).

8.5.5 CALCITONIN

Calcitonin is a polypeptide hormone secreted by the parafollicular cells of the thyroid gland. It is a 32-amino-acid polypeptide chain with a disulfide bridge and a molecular weight of about 3600 Da. It is commercially available as human calcitonin and salmon calcitonin, and it is used in the management of Paget's disease, hypercalcemia, and postmenopausal osteoporosis. Another protein useful for osteoporosis, PTH(1-34), is discussed in Chapter 10. Salmon calcitonin is more potent, with a dose about 50 times less than that of human calcitonin. Human calcitonin has a marked tendency to aggregate in aqueous solutions (80–82), but salmon calcitonin is more resistant to aggregation, though it can also undergo base-catalyzed dimerization and sulfide disproportionation (83). A nasal product was commercialized, but treatment with calcitonin is usually prolonged; thus, repeated administrations would be required. Transdermal delivery could offer an alternative treatment option, allowing continuous input. Electrical enhancement mechanisms such as iontophoresis and electroporation can modulate drug delivery by adjusting the current/voltage profiles so that pulsatile delivery is also feasible, when required. This continuous input could be used for chronic conditions such as Paget's disease and osteoporosis, while bolus dose could be given for control of hypercalcemic emergencies. Salmon calcitonin (pI = 10.4) and human calcitonin (pI = 9) have very high isoelectric points and so are good candidates for iontophoretic delivery.

Iontophoretic delivery of salmon calcitonin, *in vivo*, has been investigated using male Wistar rats. A hypocalcemic effect was noticed when delivered under anode from a pH 4 formulation. This was enhanced by the proteolytic enzyme inhibitors, aprotinin and camostat mesilate. No enhancement was seen with soybean trypsin inhibitor. This could be because aprotinin is a peptide (MW 6500, pI = 10.5) that is positively charged under most pH conditions and may be iontophoretically driven into the skin to provide protection. On the other hand, soybean trypsin inhibitor (MW 8000, pI – 4.1) was possibly neutral at the pH used and was not delivered into skin in significant amounts (84). Iontophoretic delivery of human calcitonin was also investigated both *in vitro* and *in vivo* using hairless rats, and a hypocalcemic effect was observed (85). We investigated the iontophoretic delivery of salmon calcitonin *in vitro* across human epidermal membrane (86), as well as *in vivo* in the hairless rat model (87). *In vitro* delivery was best under the anode at pH 4, where the molecule was strongly ionized with four positive charges, and electroosmotic flow from anode to cathode will help electromigration (86). *In vivo* delivery under similar conditions delivered 177.9 ± 58.7 ng/(min.kg) from an iontophoretic patch and produced an average steady-state concentration of 7.58 ± 1.35 ng/ml. In addition to a direct measurement of the serum calcitonin levels by enzyme-linked immunosorbent assay (ELISA), the pharmacodynamic response was measured as reflected by lowering of serum calcium levels and was found to follow a similar profile as that achieved by subcutaneous or intravenous injection (87) (Figure 8.1). The fraction (F) of dose absorbed into systemic circulation was calculated as:

$$F \times \text{dose delivered} = AUC_{\text{iontophoretic}} \times \text{Clearance}_{IV}$$

and the rate of infusion was then calculated as

$$R_0 = \frac{F \times \text{dose delivered}}{\text{duration of patch application}}$$

Pharmacokinetic parameters calculated by noncompartmental analysis were modeled by a one-compartment model with zero-order infusion. We also investigated the transdermal delivery of salmon calcitonin by a combination of iontophoresis and electroporation and reported that electroporation reduced the lag time and enhanced the flux for iontophoretic delivery of calcitonin across human epidermal membrane (88).

In an *in vivo* study with hairless guinea pigs, a significant increase in the transdermal electrotransport of salmon calcitonin was seen in the presence of a surfactant,

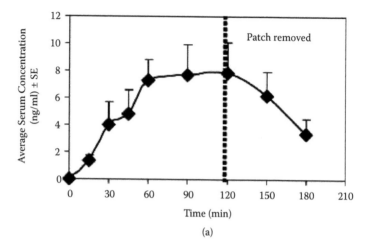

(a)

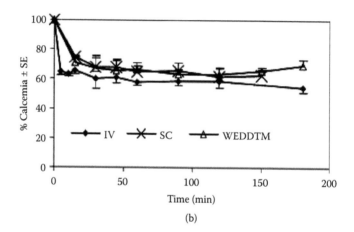

(b)

FIGURE 8.1 Serum concentrations of salmon calcitonin after application of iontophoresis patch (WEDD) in hairless rats (a), and the corresponding drop in blood calcium as compared to intravenous (IV) or subcutaneous (SC) administration (b). (Reprinted from *Int. J. Pharm.*, A. Chaturvedula et al., *In vivo* Iontophoretic Delivery and Pharmacokinetics of Salmon Calcitonin, 297:190–196, 2005 [87]. Copyright 2005, with permission from Elsevier.)

polysorbate 20 (30). Another study delivered a calcitonin analog to human volunteers by iontophoresis. A plasma concentration-time profile resulting from 6 hours of iontophoresis was found to have a similar profile as intravenous infusion, but the appearance of the peptide in plasma following iontophoresis had a higher lag time than that during intravenous infusion (89). Another published study found that transdermal iontophoresis of sCT under anode caused a decrease in serum calcium levels in rabbits, while no significant fall was measured in the absence of electric current. Substantial degradation of the polypeptide under the effect of electric current was observed. Due to stability concerns in solution, a solid formulation was designed as a thin, dry disc made by gentle compression of freeze-dried mixture of salmon calcitonin and gelatin. When required, the disc was placed directly on moistened skin and covered with a wetted pad that allowed immediate dissolution. The pad, in turn, was fixed to the skin by means of an adhesive patch containing the electrode (90). We also investigated the stability of salmon calcitonin formulation under conditions relevant for iontophoretic transdermal drug delivery (91).

8.5.6 LARGE PROTEINS

Iontophoretic delivery generally works best for molecules up to around 10 kDa, so in that sense, delivery of somewhat larger proteins has also been recently reported, including cytochrome c (92), daniplestim (93), and ribonuclease A (4). We reported on the delivery of daniplestim (12.76 kDa; pI 6.2), an IL-3 receptor agonist with 112 amino acid residues, and showed that it was delivered *in vitro* across hairless rat skin by iontophoresis at pH 4 (93). Ribonuclease A (13.6 kDa; pI 8.64) has physicochemical and structural similarities to cytochrome c (12.4 kDa; pI 10.2) and was delivered *in vitro* across porcine and human skin in clinically relevant amounts. As this is an enzyme, its functionality could be easily tested, and it was shown that the protein was functionally active following iontophoretic delivery. Using confocal microscopy and rhodamine-tagged RNAse, it was also shown that fluorescence was visible throughout the skin following iontophoresis but was localized at the skin surface following passive delivery (4). In contrast to iontophoresis, the pathways (microchannels) that can carry drug molecules into skin following microporation are several microns in dimensions; therefore, there is no size limit for delivery of macromolecules dissolved in a vehicle, as the hydrodynamic radii of even macromolecules such as monoclonal antibodies are typically in the nanometer size range. We investigated the delivery of large protein molecules across skin by microporation technologies alone or in combination with iontophoresis or sonophoresis (see also Chapters 3 and 8). Using immunohistochemical staining, we showed IgG (approximately 150 kDa) transporting across microchannels created by maltose microneedles (94). We also investigated the *in vivo* delivery of daniplestim in the hairless rat model by iontophoresis, sonophoresis, and microneedles and reported that a combination of microneedles and iontophoresis gave a C_{max} of 9.4 ± 2.5 ng/ml and was the most effective approach to maximize delivery (5).

8.5.7 OTHER PEPTIDE/PROTEIN CANDIDATES

In this section, some miscellaneous studies with various peptides are discussed. Iontophoretic delivery of a model dipeptide, H-Tyr-D-Arg-OH [D-(Arg)-kyotorphin or YdR; MW 337 Da], has been investigated *in vitro* across porcine skin, and a twofold increase in current resulted in a proportionate increase in both the electroosmotic and electromigration components of iontophoretic delivery (95). Iontophoretic delivery of octreotide acetate in rabbits was investigated. Octreotide, a synthetic octapeptide and an analog of somatostatin, is commercially available (Sandostatin®, Novartis, Basel, Switzerland) for treatment of acromegaly and carcinoid tumors. Plasma levels of octreotide were negligible without application of current. However, they increased proportional to current density in the range 50 to 150 $\mu A/cm^2$ and declined rapidly upon removal of the device. Plasma levels also increased proportionally by increasing drug concentration in the device from 2.5 to 5 mg/ml, but not beyond this concentration (96). Other polypeptides delivered iontophoretically include an analogue of growth hormone–releasing factor with a molecular weight of 3929 Daltons. No permeability was observed *in vitro* across hairless guinea pig skin under passive conditions, but delivery was achieved by iontophoresis (97).

REFERENCES

1. Z. Antosova, M. Mackova, V. Kral, and T. Macek. Therapeutic application of peptides and proteins: Parenteral forever? *Trends Biotechnol.*, 27:628–635 (2009).
2. A. K. Banga. New technologies to allow transdermal delivery of therapeutic proteins and small water-soluble drugs, *Am. J. Drug Del.*, 4:221–230 (2006).
3. H. A. Benson and S. Namjoshi. Proteins and peptides: Strategies for delivery to and across the skin, *J. Pharm. Sci.*, 97:3591–3610 (2008).
4. S. Dubey and Y. N. Kalia. Non-invasive iontophoretic delivery of enzymatically active ribonuclease A (13.6 kDa) across intact porcine and human skins, *J. Control. Release*, 145:203–209 (2010).
5. S. Katikaneni, G. Li, A. Badkar, and A. K. Banga. Transdermal delivery of a approximately 13 kDa protein—An *in vivo* comparison of physical enhancement methods, *J. Drug Target.*, 18:141–147 (2010).
6. N. Celebi, N. Erden, B. Gonul, and M. Koz. Effects of epidermal growth factor dosage forms on dermal wound strength in mice, *J. Pharm. Pharmacol.*, 46:386–387 (1994).
7. K. Okumura, Y. Kiyohara, F. Komada, S. Iwakawa, M. Hirai, and T. Fuwa. Improvement in wound healing by epidermal growth factor (EGF) ointment. I. Effect of Nafamostat, Gabexate, or Gelatin on stabilization and efficacy of EGF, *Pharm. Res.*, 7:1289–1293 (1990).
8. Y. Kiyohara, F. Komada, S. Iwakawa, T. Fuwa, and K. Okumura. Systemic effects of epidermal growth factor (EGF) ointment containing protease inhibitor or gelatin in rats with burns or open wounds, *Biol. Pharm. Bull.*, 16:73–76 (1993).
9. B. Matuszewska, M. Keogan, D. M. Fisher, K. A. Soper, C. Hoe, A. C. Huber, and J. V. Bondi. Acidic fibroblast growth factor: Evaluation of topical formulations in a diabetic mouse wound healing model, *Pharm. Res.*, 11(1):65–71 (1994).
10. M. Okumura, T. Okuda, T. Nakamura, and M. Yajima. Acceleration of wound healing in diabetic mice by basic fibroblast growth factor, *Biol. Pharm. Bull.*, 19:530–535 (1996).
11. N. Weiner, N. Williams, G. Birch, C. Ramachandran, C. Shipman, and G. Flynn. Topical delivery of liposomally encapsulated interferon evaluated in a cutaneous herpes guinea pig model, *Antimicrob. Agents Chemother.*, 33:1217–1221 (1989).
12. J. Ophir, S. Brenner, R. Bali, S. Krissleventon, Z. Smetana, and M. Revel. Effect of topical interferon-beta on recurrence rates in genital herpes: A double-blind, placebo-controlled, randomized study, *J. Interferon Cytokine Res.*, 15:625–631 (1995).
13. Y. W. Chien, O. Siddiqui, W. M. Shi, P. Lelawongs, and J. C. Liu. Direct current iontophoretic transdermal delivery of peptide and protein drugs, *J. Pharm. Sci.*, 78:376–383 (1989).
14. C. Cullander and R. H. Guy. (D) Routes of delivery: Case studies (6): Transdermal delivery of peptides and proteins, *Adv. Drug Del. Rev.*, 8:291–329 (1992).
15. D. Parasrampuria and J. Parasrampuria. Percutaneous delivery of proteins and peptides using iontophoretic techniques, *J. Clin. Pharm. Ther.*, 16:7–17 (1991).
16. A. K. Banga. *Therapeutic peptides and proteins: Formulation, processing, and delivery systems*, Taylor and Francis, London, 2006.

17. V. K. Hopsu-Havu, J. E. Fraki, and M. Jarvinen. Proteolytic enzymes in the skin. In A. J. Barrett (ed), *Proteinases in mammalian cells and tissues*, North-Holland Biomedical, 1977, pp. 547–581.
18. I. Steinstrasser and H. P. Merkle. Dermal metabolism of topically applied drugs: Pathways and models reconsidered, *Pharmaceut. Acta Helvetiae*, 70:3–24 (1995).
19. L. L. Miller, C. J. Kolaskie, G. A. Smith, and J. Rivier. Transdermal iontophoresis of gonadotropin releasing hormone (LHRH) and two analogues, *J. Pharm. Sci.*, 79:490–493 (1990).
20. L. H. Chen and Y. W. Chien. Development of a skin permeation cell to simulate clinical study of iontophoretic transdermal delivery, *Drug Develop. Ind. Pharm.*, 20:935–945 (1994).
21. M. J. Pikal and S. Shah. Transport mechanisms in iontophoresis. III. An experimental study of the contributions of electroosmotic flow and permeability change in transport of low and high molecular weight solutes, *Pharm. Res.*, 7:222–229 (1990).
22. T. Egelrud, A. Lundstrom, and B. Sondell. Stratum corneum cell cohesion and desquamation in maintenance of the skin barrier. In F. N. Marzulli and H. I. Maibach (eds), *Dermatotoxicology*, Taylor and Francis, London, 1996, pp. 19–27.
23. H. Choi, G. L. Flynn, and G. L. Amidon. Transdermal delivery of bioactive peptides: The effect of n- decylmethyl sulfoxide, pH, and inhibitors on enkephalin metabolism and transport, *Pharm. Res.*, 7:1099–1106 (1990).
24. A. Ruland, J. Kreuter, and J. H. Rytting. Transdermal delivery of the tetrapeptide hisetal (melanotropin (6-9)). I. Effect of various penetration enhancers: *In vitro* study across hairless mouse skin, *Int. J. Pharm.*, 101:57–61 (1994).
25. A. Ruland, J. Kreuter, and J. H. Rytting. Transdermal delivery of the tetrapeptide hisetal (Melanotropin (6-9)): II. Effect of various penetration enhancers—*In vitro* study across human skin, *Int. J. Pharm.*, 103:77–80 (1994).
26. K. S. Bhatia, S. Gao, and J. Singh. Effect of penetration enhancers and iontophoresis on the FT-IR spectroscopy and LHRH permeability through porcine skin, *J. Control. Release*, 47:81–89 (1997).
27. K. S. Bhatia, S. Gao, T. P. Freeman, and J. Singh. Effect of penetration enhancers and iontophoresis on the ultrastructure and cholecystokinin-8 permeability through porcine skin, *J. Pharm. Sci.*, 86:1011–1015 (1997).
28. B. H. Sage, C. R. Bock, J. D. Denuzzio, and R. A. Hoke. Technological and developmental issues of iontophoretic transport of peptide and protein drugs. In V. H. L. Lee, M. Hashida, and Y. Mizushima (eds), *Trends and future perspectives in peptide and protein drug delivery*, Harwood Academic, Chur, Switzerland, 1995, pp. 111–134.
29. J. H. Kou, S. D. Roy, J. Du, and J. Fujiki. Effect of receiver fluid pH on *in vitro* skin flux of weakly ionizable drugs, *Pharm. Res.*, 10:986–990 (1993).
30. L. A. Holladay, L. G. Treat-Clemons, and P. M. Bassett. Composition and method for enhancing electrotransport agent delivery. Assignee: Alza Corp. [WO 96/15826], 1–51. 1996.
31. M. F. Lu, D. Lee, and G. S. Rao. Percutaneous absorption enhancement of leuprolide, *Pharm. Res.*, 9:1575–1579 (1992).
32. B. V. Huff, G. G. Liversidge, and G. L. Mcintire. The electrophoretic mobility of tripeptides as a function of pH and ionic strength: Comparison with iontophoretic flux data, *Pharm. Res.*, 12:751–755 (1995).
33. M. C. Heit, A. McFarland, R. Bock, and J. E. Riviere. Isoelectric focusing and capillary zone electrophoretic studies using luteinizing hormone releasing hormone and its analog, *J. Pharm. Sci.*, 83:654–656 (1994).
34. N. B. Vutla, G. V. Betageri, and A. K. Banga. Transdermal iontophoretic delivery of enkephalin formulated in liposomes, *J. Pharm. Sci.*, 85:5–8 (1996).

35. S. B. Kulkarni, A. K. Banga, and G. V. Betageri. Transdermal iontophoretic delivery of colchicine encapsulated in liposomes, *Drug Delivery*, 3:245–250 (1996).
36. T. Hatanaka, T. Kamon, C. Uozumi, S. Morigaki, T. Aiba, K. Katayama, and T. Koizumi. Influence of pH on skin permeation of amino acids, *J. Pharm. Pharmacol.*, 48:675–679 (1996).
37. R. Y. Lin, C. W. Hsu, and W. Y. Chen. A method to predict the transdermal permeability of amino acids and dipeptides through porcine skin, *J. Control. Release*, 38:229–234 (1996).
38. L. L. Wearley, K. Tojo, and Y. W. Chien. A numerical approach to study the effect of binding on the iontophoretic transport of a series of amino acids, *J. Pharm. Sci.*, 79:992–998 (1990).
39. L. Coderch, M. Oliva, L. Pons, and J. L. Parra. Percutaneous penetration *in vivo* of amino acids, *Int. J. Pharm.*, 111:7–14 (1994).
40. J. Hirvonen, F. Hueber, and R. H. Guy. Current profile regulates iontophoretic delivery of amino acids across the skin, *J. Control. Release*, 37:239–249 (1995).
41. C. C. Bouissou, J. P. Sylvestre, R. H. Guy, and M. B. Delgado-Charro. Reverse iontophoresis of amino acids: Identification and separation of stratum corneum and subdermal sources *in vitro*, *Pharm Res.*, 26:2630–2638 (2009).
42. A. K. Banga, S. Kulkarni, and R. Mitra. Selection of electrode material and polarity in the design of iontophoresis experiments, *Int. J. Pharm. Adv.*, 1:206–215 (1995).
43. P. G. Green, R. S. Hinz, C. Cullander, G. Yamane, and R. H. Guy. Iontophoretic delivery of amino acids and amino acid derivatives across the skin *in vitro*, *Pharm. Res.*, 8:1113–1120 (1991).
44. R. Y. Lin, Y. C. Ou, and W. Y. Chen. The role of electroosmotic flow on *in vitro* transdermal iontophoresis, *J. Control. Release*, 43:23–33 (1997).
45. J. Moss and H. Bundgaard. Prodrugs of peptides. 7. Transdermal delivery of thyrotropin-releasing hormone (TRH) via prodrugs, *Int. J. Pharm.*, 66:39–45 (1990).
46. R. R. Burnette and D. Marrero. Comparison between the iontophoretic and passive transport of thyrotropin releasing hormone across excised nude mouse skin, *J. Pharm. Sci.*, 75:738–743 (1986).
47. Y. Y. Huang, S. M. Wu, and C. Y. Wang. Response surface method: A novel strategy to optimize iontophoretic transdermal delivery of thyrotropin-releasing hormone, *Pharm. Res.*, 13:547–552 (1996).
48. Y. Y. Huang and S. M. Wu. Transdermal iontophoretic delivery of thyrotropin-releasing hormone across excised rabbit pinna skin, *Drug Develop. Ind. Pharm.*, 22:1075–1081 (1996).
49. Y. Y. Huang and S. M. Wu. Stability of peptides during iontophoretic transdermal delivery, *Int. J. Pharm.*, 131:19–23 (1996).
50. P. Green, B. Shroot, F. Bernerd, W. R. Pilgrim, and R. H. Guy. *In vitro* and *in vivo* iontophoresis of a tripeptide across nude rat skin, *J. Control. Release*, 20:209–218 (1992).
51. P. G. Green, R. S. Hinz, A. Kim, F. C. Szoka, and R. H. Guy. Iontophoretic delivery of a series of tripeptides across the skin *in vitro*, *Pharm. Res.*, 8:1121–1127 (1991).
52. S. K. Gupta, S. Kumar, S. Bolton, C. R. Behl, and A. W. Malick. Optimization of iontophoretic transdermal delivery of a peptide and a non-peptide drug, *J. Control. Release*, 30:253–261 (1994).
53. P. S. Banerjee and W. A. Ritschel. Transdermal permeation of vasopressin. I. Influence of pH, concentration, shaving and surfactant on *in vitro* permeation, *Int. J. Pharm.*, 49:189–197 (1989).
54. P. S. Banerjee and W. A. Ritschel. Transdermal permeation of vasopressin. II. Influence of Azone on *in vitro* and *in vivo* permeation, *Int. J. Pharm.*, 49:199–204 (1989).

55. A. J. Hoogstraate, J. Verhoef, J. Brussee, A. P. Ijzerman, F. Spies, and H. E. Bodde. Kinetics, ultrastructural aspects and molecular modelling of transdermal peptide flux enhancement by N-alkylazacycloheptanones, *Int. J. Pharm.*, 76:37–47 (1991).

56. P. Lelawongs, J. Liu, O. Siddiqui, and Y. W. Chien. Transdermal iontophoretic delivery of arginine-vasopressin (I): Physicochemical considerations, *Int. J. Pharm.*, 56:13–22 (1989).

57. P. Lelawongs, J. Liu, and Y. W. Chien. Transdermal iontophoretic delivery of arginine-vasopressin (II): Evaluation of electrical and operational factors, *Int. J. Pharm.*, 61:179–188 (1990).

58. Y. W. Chien, P. Lelawongs, O. Siddiqui, Y. Sun, and W. M. Shi. Facilitated transdermal delivery of therapeutic peptides and proteins by iontophoretic delivery devices, *J. Control. Release*, 13:263–278 (1990).

59. A. K. Banga, M. Katakam, and R. Mitra. Transdermal iontophoretic delivery and degradation of vasopressin across human cadaver skin, *Int. J. Pharm.*, 116:211–216 (1995).

60. W. H. M. Craanevanhinsberg, L. Bax, N. H. M. Flinterman, J. Verhoef, H. E. Junginger, and H. E. Bodde. Iontophoresis of a model peptide across human skin *in vitro*: Effects of iontophoresis protocol, pH, and ionic strength on peptide flux and skin impedance, *Pharm. Res.*, 11:1296–1300 (1994).

61. M. Nakakura, M. Terajima, Y. Kato, E. Hayakawa, K. Ito, and T. Kuroda. Effect of iontophoretic patterns on *in vivo* antidiuretic response to desmopressin acetate administered transdermally, *J. Drug Target.*, 2:487–492 (1995).

62. M. Nakakura, Y. Kato, E. Hayakawa, K. Ito, and T. Kuroda. Effect of pulse on iontophoretic delivery of desmopressin acetate in rats, *Biol. Pharm. Bull.*, 19:738–740 (1996).

63. M. Nakakura, Y. Kato, and K. Ito. Prolongation of antidiuretic response to desmopressin acetate by iontophoretic transdermal delivery in rats, *Biol. Pharm. Bull.*, 20:537–540 (1997).

64. B. H. Vickery. Biological actions of synthetic analogs of luteinizing hormone-releasing hormone. In P. D. Garzone, W. A. Colburn, and M. Mokotoff (eds), *Pharmacokinetics and pharmacodynamics. Vol. 3: Peptides, peptoids and proteins*, Harvey Whitney Books, Cincinnati, OH, 1991, pp. 41–49.

65. K. S. Bhatia and J. Singh. Percutaneous absorption of LHRH through porcine skin: Effect of N-methyl 2-pyrrolidone and isopropyl myristate, *Drug Develop. Ind. Pharm.*, 23:1111–1114 (1997).

66. S. D. Roy and J. S. Degroot. Percutaneous absorption of nafarelin acetate, an LHRH analog, through human cadaver skin and monkey skin, *Int. J. Pharm.*, 110:137–145 (1994).

67. H. D. Smyth, G. Becket, and S. Mehta. Effect of permeation enhancer pretreatment on the iontophoresis of luteinizing hormone releasing hormone (LHRH) through human epidermal membrane (HEM), *J. Pharm. Sci.*, 91:1296–1307 (2002).

68. M. C. Heit, P. L. Williams, F. L. Jayes, S. K. Chang, and J. E. Riviere. Transdermal iontophoretic peptide delivery—*In vitro* and *in vivo* studies with luteinizing hormone releasing hormone, *J. Pharm. Sci.*, 82:240–243 (1993).

69. M. C. Heit, N. A. Monteiroriviere, F. L. Jayes, and J. E. Riviere. Transdermal iontophoretic delivery of luteinizing hormone releasing hormone (LHRH): Effect of repeated administration, *Pharm. Res.*, 11:1000–1003 (1994).

70. B. R. Meyer, W. Kreis, and J. Eschbach. Successful transdermal administration of therapeutic doses of a polypeptide to normal human volunteers, *Clin. Pharmacol. Ther.*, 44:607–612 (1988).

71. C. Kochhar and G. Imanidis. *In vitro* transdermal iontophoretic delivery of leuprolide under constant current application, *J. Control. Release*, 98:25–35 (2004).

72. B. R. Meyer, W. Kreis, J. Eschbach, V. O'Mara, S. Rosen, and D. Sibalis. Transdermal versus subcutaneous leuprolide: A comparison of acute pharmacodynamic effect, *Clin. Pharmacol. Ther.*, 48:340–345 (1990).
73. A. J. Hoogstraate, V. Srinivasan, S. M. Sims, and W. I. Higuchi. Iontophoretic enhancement of peptides: Behaviour of leuprolide versus model permeants, *J. Control. Release*, 31:41–47 (1994).
74. M. B. Delgadocharro and R. H. Guy. Iontophoretic delivery of nafarelin across the skin, *Int. J. Pharm.*, 117:165–172 (1995).
75. M. B. Delgadocharro, A. M. Rodriguezbayon, and R. H. Guy. Iontophoresis of nafarelin: Effects of current density and concentration on electrotransport *in vitro, J. Control. Release*, 35:35–40 (1995).
76. A. M. R. Bayon and R. H. Guy. Iontophoresis of nafarelin across human skin *in vitro, Pharm. Res.*, 13:798–800 (1996).
77. Y. B. Schuetz, A. Naik, R. H. Guy, E. Vuaridel, and Y. N. Kalia. Transdermal iontophoretic delivery of triptorelin *in vitro, J. Pharm. Sci.*, 94:2175–2182 (2005).
78. J. Hirvonen, Y. N. Kalia, and R. H. Guy. Transdermal delivery of peptides by iontophoresis, *Nat. Biotechnol.*, 14:1710–1713 (1996).
79. P. Knoblauch and F. Moll. *In vitro* pulsatile and continuous transdermal delivery of buserelin by iontophoresis, *J. Control. Release*, 26:203–212 (1993).
80. T. Arvinte, A. Cudd, and A. F. Drake. The structure and mechanism of formation of human calcitonin fibrils, *J. Biol. Chem.*, 268:6415–6422 (1993).
81. H. H. Bauer, U. Aebi, M. Haner, R. Hermann, T. Muller, T. Arvinte, and H. P. Merkle. Architecture and polymorphism of fibrillar supramolecular assemblies produced by *in vitro* aggregation of human calcitonin, *J. Str. Biol.*, 115:1–15 (1995).
82. A. Cudd, T. Arvinte, R. E. G. Das, C. Chinni, and A. Macintyre. Enhanced potency of human calcitonin when fibrillation is avoided, *J. Pharm. Sci.*, 84:717–719 (1995).
83. V. Windisch, F. DeLuccia, L. Duhau, F. Herman, J. J. Mencel, S. Y. Tang, and M. Vuilhorgne. Degradation pathways of salmon calcitonin in aqueous solution, *J. Pharm. Sci.*, 86:359–364 (1997).
84. K. Morimoto, Y. Iwakura, E. Nakatani, M. Miyazaki, and H. Tojima. Effects of proteolytic enzyme inhibitors as absorption enhancers on the transdermal iontophoretic delivery of calcitonin in rats, *J. Pharm. Pharmacol.*, 44:216–218 (1992).
85. S. Thysman, C. Hanchard, and V. Preat. Human calcitonin delivery in rats by iontophoresis, *J. Pharm. Pharmacol.*, 46:725–730 (1994).
86. S. Chang, G. A. Hofmann, L. Zhang, L. J. Deftos, and A. K. Banga. Transdermal iontophoretic delivery of salmon calcitonin. *Int. J. Pharmaceut.*, 200:107–113 (2000).
87. A. Chaturvedula, D. P. Joshi, C. Anderson, R. L. Morris, W. L. Sembrowich, and A. K. Banga. *In vivo* iontophoretic delivery and pharmacokinetics of salmon calcitonin, *Int. J. Pharm.*, 297:190–196 (2005).
88. S. Chang, G. A. Hofmann, L. Zhang, L. J. Deftos, and A. K. Banga. The effect of electroporation on iontophoretic transdermal delivery of calcium regulating hormones. *J. Control. Release*, 66:127–133 (2000).
89. P. Green. Iontophoretic delivery of peptide drugs, *J. Control. Release*, 41:33–48 (1996).
90. P. Santi, P. Colombo, R. Bettini, P. L. Catellani, A. Minutello, and N. M. Volpato. Drug reservoir composition and transport of salmon calcitonin in transdermal iontophoresis, *Pharm. Res.*, 14:63–66 (1997).
91. S. L. Chang, G. A. Hofmann, L. Zhang, L. J. Deftos, and A. K. Banga. Stability of a transdermal salmon calcitonin formulation, *Drug Deliv.*, 10:41–45 (2003).
92. J. Cazares-Delgadillo, A. Naik, A. Ganem-Rondero, D. Quintanar-Guerrero, and Y. N. Kalia. Transdermal delivery of cytochrome C—A 12.4 kDa protein—across intact skin by constant-current iontophoresis, *Pharm. Res.*, 24:1360–1368 (2007).

93. S. Katikaneni, A. Badkar, S. Nema, and A. K. Banga. Molecular charge mediated transport of a 13 kD protein across microporated skin, *Int. J. Pharm.*, 378:93–100 (2009).

94. G. Li, A. Badkar, S. Nema, C. S. Kolli, and A. K. Banga. *In vitro* transdermal delivery of therapeutic antibodies using maltose microneedles, *Int. J. Pharm.*, 368:109–115 (2009).

95. N. Abla, A. Naik, R. H. Guy, and Y. N. Kalia. Contributions of electromigration and electroosmosis to peptide iontophoresis across intact and impaired skin, *J. Control. Release*, 108:319–330 (2005).

96. D. T. W. Lau, J. W. Sharkey, L. Petryk, F. A. Mancuso, Z. L. Yu, and F. L. S. Tse. Effect of current magnitude and drug concentration on iontophoretic delivery of octreotide acetate (Sandostatin(R)) in the rabbit, *Pharm. Res.*, 11:1742–1746 (1994).

97. S. Kumar, H. Char, S. Patel, D. Piemontese, K. Iqbal, A. W. Malick, E. Neugroschel, and C. R. Behl. Effect of iontophoresis on *in vitro* skin permeation of an analogue of growth hormone releasing factor in the hairless guinea pig model, *J. Pharm. Sci.*, 81:635–639 (1992).

9 Transcutaneous Immunization via Physical Methods

9.1 INTRODUCTION

Immunization using vaccines has played a vital role in eradicating and limiting the spread of several dangerous infections over the years. However, the threat of bioterrorism that rose in 2001 from the anthrax spores sent via the postal service and the pandemic caused in 2009 due to the H1N1 virus necessitated the need to investigate faster and more efficient vaccine delivery systems. This has brought a renewed interest to develop methods suited for mass immunizations. Currently, vaccines are typically administered by parenteral, intranasal, or oral routes. The majority of vaccines are administered by intramuscular or subcutaneous route using needle sticks with hypodermic needles. However, this mode of administration has several inherent disadvantages. These include possible pain and tissue reactions at site of administration resulting in poor patient compliance. Furthermore, the involvement of trained medical personnel adds to the cost of the health-care system. To add to the complications, a cold chain is often required for storage of the vaccines, and reuse of nonsterile syringes is a major source of hepatitis B infections. Noninvasive routes of administration have therefore been explored and, in many cases, provide the further advantage of inducing mucosal immune responses in addition to systemic immune responses (1,2). Some of the needle-free approaches that have been explored include jet injectors, mucosal immunization, and transcutaneous immunization (3,4).

Skin is an immune competent organ and provides an attractive target for vaccine delivery. Langerhans cells occupy only about 1% of the cell population of the viable epidermis, but because of their long protrusions and horizontal orientation, they cover nearly 20% of the surface area to effectively take up antigens. Upon uptake, these antigen-laden cells then migrate to dermal lymphatic channels and present the information to T lymphocytes in lymph nodes. Therefore, delivery of vaccines into skin is usually very efficient and produces good immune response at less dosage as compared to intramuscular (IM) or subcutaneous (SC) injections. Furthermore, the latter delivery routes deliver antigens to areas that do not have significant resident populations of antigen-presenting cells (5). For example, this dose-sparing ability of microneedle-mediated vaccine delivery has been demonstrated in many studies (6,7). Using three different types of influenza vaccines, Alarcon et al. demonstrated that microneedle-based intradermal delivery is at least as effective as intramuscular delivery at 5- to 100-fold lower doses (8). This dose-sparing ability can be critically important to respond to a pandemic when production of the antigen can be a rate-limiting step. This chapter will discuss the various approaches for vaccination via skin and uses the term *transcutaneous immunization* somewhat loosely to include various approaches, including intradermal injections. Related approaches for delivering plasmid DNA or oligonucleotide for vaccination or gene delivery are also discussed in this chapter, because there are similarities in the intended application (e.g., delivery of plasmid DNA by one of the physical approaches may be intended either for DNA vaccination or for gene delivery).

Improved transcutaneous immunization through intact skin has been shown to be facilitated by adjuvants or by using enhancement techniques such as iontophoresis, electroporation, or sonophoresis. Low-frequency ultrasound has also been reported as a transcutaneous immunization adjuvant by itself, and this is attributed partly to enhanced delivery of the antigen and partly to activation of the immune cells by application of ultrasound (9–11). Also, vaccines can be delivered intradermally by jet injectors or by using hypodermic needles. Other unique approaches such as bioneedles composed of hollow mini-implants made of biodegradable polymers have also been investigated. The bioneedles were filled with liquid formulation of hepatitis B surface antigen, freeze dried, and then implanted under the skin by injecting with compressed air (12). More recently, microneedles (discussed in Chapter 3) are being investigated for delivering vaccines to skin. Initial investigations have been very promising, and there is currently a tremendous amount of interest and excitement about microneedle-based vaccine delivery systems. The primary focus of this book is on skin enhancement technologies and not on vaccines per se. Nevertheless, this is a very important application, and this chapter will provide a brief overview of the efforts in this area along with a listing of some recent literature to provide an understanding of these emerging applications. The economics of drug and vaccine delivery and how enhancement technologies such as skin microporation will have an impact on the development of drug delivery systems are discussed in Chapter 10.

9.2 IMMUNOLOGY OF THE SKIN

Skin performs a complex defense function that may be described as "immunological." The capacity of skin to distinguish self from nonself is remarkable if one considers the vast variety of exogenous substances to which the skin is continuously exposed. Our understanding of skin immunology has now advanced tremendously. The immunological environment of skin includes the humoral and cellular components and is given the acronym SIS (skin immune system). Dysregulations of this system can manifest with several immunodermatological diseases, including atopic eczema, psoriasis, cutaneous lupus erythematosus, scleroderma, and autoimmune bullous disease (13,14). When allergens penetrate skin, they can, in some cases, lead to allergic contact dermatitis, which is characterized by redness and vesicles, followed by scaling and dry skin. Additional relevant compounds in skin immunology are the eicosanoids. Eicosanoids, which are oxygenated metabolites of 20 carbon fatty acid, especially arachidonic acid, are a class of compounds that have a role in the pathophysiology of inflammatory and immunologic skin disorders. For example, leukotrienes play a central role in the pathogenesis of psoriasis, a chronic, scaly, inflammatory skin disorder (15). Thus, the immunology of skin needs to be carefully considered to design transdermal systems, especially for delivery of vaccines. It is known that Langerhans cells reside in the epidermis and express a high level of major histocompatibility complex (MHC) class II molecules and strong stimulatory functions for the activation of T lymphocytes. Langerhans cells comprise 2% to 4% of the cells of the epidermis and are also found in lymph nodes. Langerhans cells are dendritic-shaped cells located in the basal parts of the epidermis. They act on antigens and present them to lymphocytes, thus providing immune surveillance for viruses, neoplasms, and nonautologous grafts. In the skin-associated lymphoid tissue or "SALT," Langerhans cells in the epidermis are believed to act as antigenic traps, and the antigen-laden cells then migrate into dermal lymphatic channels to present the information to T lymphocytes in lymph nodes. Keratinocytes also play a role in immunity.

9.3 TRANSCUTANEOUS IMMUNIZATION BY PASSIVE APPROACHES

Starting from the late 1990s, several reports started suggesting that contrary to intuitive thinking, antigenic macromolecules can be delivered into skin to generate immune responses. Since then, interest in transcutaneous immunization has increased significantly (16–21). This may possibly be because vaccines/adjuvants might be able to diffuse through hair follicles (see Section 3.2.6). However, these studies reported that the presence of an adjuvant is usually needed to get an immune response. Bacterial products such as heat-labile enterotoxin from *Escherichia coli* (LT) and cholera toxin (CT) have been commonly used as adjuvants, and antigens formulated with these adjuvants were applied to intact skin for achieving immune responses. These adjuvants most likely have structural elements similar to bacteria, and these are recognized by the defense mechanisms of the body (1). A high-throughput screening method (see Section 1.4.1) has been reported to be useful to screen formulations and chemicals for their adjuvanticity potential (22). One study that used a hydrogel patch on intact skin reported that antigen-specific Ig antibodies were produced, and immunofluorescence histochemical analysis revealed that Langerhans cells in the epidermal layer captured the antigens and migrated to draining lymph nodes. The authors hypothesized that the antigenic proteins (45 to 150 kDa) were delivered across intact skin due to factors related to concentration gradient, hydration effects on skin, and transport via hair follicles (21). Even though there are reports of vaccine delivery via application of the formulation to intact skin, it is generally believed that enhancement methods are needed to be able to deliver macromolecules efficiently into the skin, even though only miniscule amounts of antigen are required to generate an immune response. These enhancement methods for delivery of vaccines into the skin is the focus of this chapter.

9.4 MICRONEEDLE-BASED INTRADERMAL INJECTION

Intradermal injection, sometimes also known as the Mantoux method, as currently administered typically uses a 27-gauge 10-mm needle that is inserted parallel to the skin surface about 3 mm until the entire bevel is covered. Intradermal injection has been validated as a safe and effective method of immunization based on our experience with smallpox, tuberculosis (TB), and other vaccines. The smallpox vaccination is still accomplished today by administering the *Vaccinia* virus to skin by scarification using a bifurcated needle that can pick up a minute drop (2 μL) of the vaccine solution between its two prongs. The needle is jabbed into the papillary (upper) dermis for delivery of the vaccine into skin. However, these administration techniques are painful, and the Mantoux technique requires training and practice. In contrast, a microneedle-based intradermal injection typically uses a 30-gauge 1- to 1.5-mm needle that is inserted perpendicularly in the skin and is reported to be painless (23,24). Two studies in human subjects reported that the microneedle-based intradermal injection can be performed without any training or practice. Injections were made in the deltoid, suprascapular, thigh, and waist areas, and these sites all have skin thickness greater than the needle length (1.5 mm) used irrespective of age, gender, and ethnic origin or body mass index (BMI). This ensures that the needle is reaching into the dermis. Skin thickness was 2.54 mm at the suprascapular, 2.02 mm at the upper arm (deltoid area), 1.91 mm at the waist (lower quadrant of the abdomen), and 1.55 mm at the thigh (anterolateral area) as measured by echography using a 20-MHz probe in 342 adult subjects. These sites have efficient lymphatic drainage, and despite the small differences in skin thickness, a 1.5-mm needle is expected to reach the papillary and reticular dermis, with the possible exception of the thigh area. The differences in skin thickness between people of different BMI, age, gender, or ethnic origin were even less than the difference seen between different body sites on the same individual (23,25). As stated, a 1.5-mm needle is expected to reach the papillary and reticular dermis, and this has taken into consideration the long bevel of the microneedle (Figure 9.1). In contrast, a 3-mm needle will reach the deep reticular dermis, while a 1-mm needle will mostly reach the papillary dermis. In a study by Laurent et al., these three lengths (1 to 3 mm) were used to deliver a rabies vaccine to 66 volunteers, and it was found that a good immune response was obtained with intradermal administration at one fourth the dose used for intramuscular injection control, though there was no difference between the groups using needles of different lengths. However, another control that used a skin microabrader (Onvax™) composed of an array of 200-μm long plastic projections that were used to disrupt the stratum corneum by rubbing on the skin, did not produce an immune response. Draize scoring was used in this study to show that skin irritation was higher with intradermal injection relative to intramuscular administration, but the injections were well tolerated, and no medical intervention was needed (26). Intradermal injection can reduce the dosage requirements for the antigen by as much as 80% by reducing the volume injected from 0.5 to 0.1 mL, while still getting the same immune response (27). In a study that delivered influenza vaccine intradermally to 1107 elderly volunteers (>60 years of age), immune responses superior to those achieved after intramuscular vaccination were observed (28). Another study that used the same device delivered the influenza vaccine to 1150 volunteers who were <60 years of age and reported comparable results to a licensed intramuscular vaccine (29). A marketed device based on this and other studies uses a single 30-gauge 1.5-mm microneedle (Soluvia™, BD, Franklin Lakes, New Jersey) for intradermal delivery of influenza vaccine (Intanza®/ID Flu®, Sanofi-pasteur, Lyon, France) and is discussed in Section

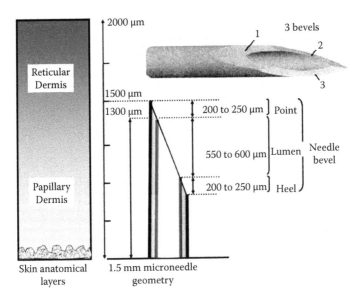

FIGURE 9.1 Microneedle geometry of a hollow 1.5-mm microneedle for intradermal injection showing how the length of the microneedle and the bevel relate to anatomical placement in skin. (Reprinted from *Vaccine*, P. E. Laurent et al., Safety and Efficacy of Novel Dermal and Epidermal Microneedle Delivery Systems for Rabies Vaccination in Healthy Adults, 28:5850–5856, 2010 [26]. *In vivo* delivery of nucleic acid-based agents with electroporation, 10:37–41, 2010 [101]. Copyright 2010, with permission from Elsevier.) **(See color insert.)**

10.5. Microneedle-based intradermal delivery was also reported to be effective for delivery of anthrax vaccine (30,31), delivery of a live attenuated chimeric vaccine for Japanese encephalitis (32), and delivery of a Yersinia pestis–based vaccine for plague as a proactive guard against bioterrorism (33).

9.5 MICRONEEDLE-MEDIATED VACCINATION

Microneedles have been reported to efficiently deliver antigens into skin in a reproducible manner, to achieve consistent and enhanced antibody titers for improved vaccination (34,35). Vaccine delivery using this technology is anticipated to offer several additional benefits. Because skin is immunocompetent, the dosage required to produce an immune response may be reduced (dose sparing). Microneedles also have the potential for delivering combination vaccines (36). Microneedles are simple in design, can be produced at low cost, and can be self-administered without the need for any activation or other devices. They can be coated with stabilized antigen formulations that do not need cold storage. They can also be designed to fit on existing standard syringes if desired. Also, the risk of pain, needlestick injuries or transmission of blood-borne pathogens is reduced, as microneedles do not penetrate deep into the skin (see also Chapter 3). Therefore, this technology seems to be appropriate to provide a cost-efficient means for mass immunization programs, especially in developing countries. Using focus groups composed of public and health-care professionals, a recent survey study found that 100% of the public and 74% of the professional group participants had an overall positive opinion of the microneedle technology to bring this technology from the bench to bedside (37). Safety of this technology is further evident from the fact that in thousands of microneedle insertions in human and animal subjects reported to date, no infections or serious adverse effects have been reported (5). However, the treatment site will have to be cleaned with an alcohol swab, and the microneedles should have a low bioburden or should be sterile (also see Section 3.2.5) to eliminate even the least possible risk of infection that may occur as the barrier is disrupted by these needles.

In general, the term *microneedle* is used for lengths <1 mm, with a more typical length being about 500 µm. However, microneedle-based intradermal injection discussed in Section 9.4 uses a needle that is typically around 1.5 mm in length. Because the entire microneedle may not enter the skin due to its elasticity (see Chapter 3), the needle may sometimes reach into the epidermis rather than the dermis. In fact, very short (65 to 110 µm) microneedles have also been designed, coated with antigen, adjuvant, or DNA payloads, and assembled in a patch with dense packing (over 3000 microprojections per device). The rationale has been that these densely packed microprojections can directly target the thousands of antigen-presenting cells that are present in the basal parts of the epidermis. When DNA payloads were delivered by these needles, expression of encoded proteins was observed within 24 hours (38,39). A densely packed microprojection array of short dissolving microneedles was also recently reported (40). These studies were done on mouse ear skin; penetration of such short microneedles into human skin may need specially designed insertion devices (see Section 3.2.4). Longer microneedles have also been shown to provide an immune response, perhaps due to the presence of dendritic cells throughout the dermis. Also, any antigen delivered to the epidermis past the stratum corneum is likely to diffuse into the dermis as well. For example, Zhu et al. used 500-µm long microneedles to demonstrate immune response to influenza virus antigen coated on the microneedles (41). Longer microneedles may also allow coating of larger quantities of vaccine on the microneedle surface. In another study using ovalbumin-coated titanium microneedles (approximately 1300 microneedles per array) in the 225- to 600-µm length range, immune response was reported to be dose dependent but independent of the depth of delivery (35).

The use of adjuvants has also been found helpful when using microneedles to deliver vaccines into the skin. Ding et al. (42) reported that an immune response was observed to influenza vaccine administered to mice, when formulated with cholera toxin (CT; 84 kDa). In the same study, the authors reported that for diphtheria toxoid, microneedle treatment of skin was required before an immune response could be observed (42). Other adjuvants that increased the immune response to delivery of diphtheria toxoid by microneedles include lipopolysaccharide (LPS), Quil A, and CpG oligo deoxynucleotide (CpG) (43). Another adjuvant, *N*-trimethyl chitosan (TMC), was also found to enhance the immune response when diphtheria toxoid was delivered to skin by microneedles. A solution formulation was more effective than a nanoparticle-based formulation, due to limited transport of the nanoparticles across the microneedle conduits (44). Other studies demonstrated that TMC

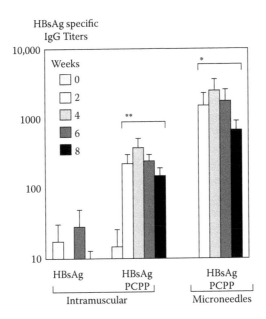

FIGURE 9.2 Serum IgG specific HBsAg titers after single-dose intramuscular (IM) or intradermal immunization of pigs with HBsAg or HBsAg-PCPP microneedles; results were significantly higher ($p < 0.05$; ANOVA) than in both IM groups (*) and the nonadjuvanted IM group (**). (Reproduced with permission from A. K. Andrianov et al., *Proc. Natl. Acad. Sci. USA*, 106:18936–18941, 2009 [47].)

functions as an immune potentiator for antigens (45,46). Mikszta et al. reported that recombinant protective antigen of *Bacillus anthracis* formulated with aluminum and administered by a microneedle-based device is as effective as intramuscular administration and required less dosage (31). Apogee Technology (Norwood, Massachusetts) is developing microneedles coated with PCPP, a polyphosphazene, that was found to be a potent adjuvant for intradermal immunization (47,48). In a study where 600-μm long titanium microneedle arrays coated with PCPP and hepatitis B surface antigen (HBsAg) were inserted in pig skin, immune responses (Figure 9.2) obtained for the microneedle group were three orders of magnitude higher than those obtained for a nonadjuvanted intramuscular control group, and were 10 times higher when the HBsAg-PCPP formulation was used for intramuscular administration (47).

Several researchers used microneedles for investigating delivery of influenza vaccine (6,28,41,42,49–53), including those based on virus-like particles (7,54,55). Influenza is a major public health threat, with over 200,000 hospitalizations and 36,000 deaths annually in the United States alone. Using coated metal microneedles, induction of strong humoral and cellular immune responses was demonstrated, and the immune responses conferred protection against virus challenge (51). Stability of the inactivated influenza virus vaccine has been shown to be enhanced by trehalose as monitored by hemagglutination activity and virus particle aggregation (50,53,56). Such stabilization of vaccine-coated microneedles can eliminate the need for refrigeration, providing significant cost savings by eliminating the cold chain. Mice vaccinated with a single dose of trehalose-stabilized influenza vaccine, delivered by 700-μm long metal microneedles, developed strong and long-lived antibody titers. A five-needle array of microneedles was coated with 0.4 μg of inactivated influenza virus by dip coating in a 1 mg/ml solution. Protection against a lethal challenge infection was demonstrated, and the stabilized coated vaccine formulation provided better protective immunity as compared to intramuscular administration or unstabilized microneedle formulation (52,53). Another recent study used 650-μm polymeric microneedles that encapsulated 3 μg of inactivated influenza virus vaccine per patch and generated robust antibody and cellular immune responses in mice when applied to skin. It was shown that the microneedles dissolved almost completely within 5 minutes in *ex vivo* pig skin or within 15 minutes when applied *in vivo* to mice skin (Figure 9.3). Processing of the vaccine by lyophilization or changes in formulation by adding PVP did not affect the IgG titers (57).

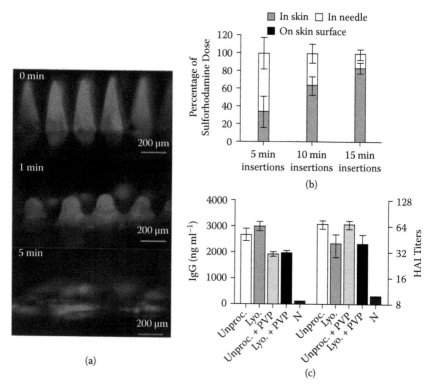

FIGURE 9.3 Dissolution of polymeric microneedles (a) *ex vivo* in pig skin or (b) *in vivo* in mice skin, and (c) effect of processing (lyophilization) or formulation (PVP) on immunogenicity in mice following intramuscular administration. (Reprinted by permission from Macmillan Publishers Ltd., *Nat. Med.*, 16:915–920, S. P. Sullivan et al. [57]. Copyright 2010.) **(See color insert.)**

9.6 JET INJECTORS

Jet injectors deliver a liquid vaccine or drug formulation at a very high velocity such that the formulation can enter the skin by creating superficial micropathways in the skin. Jet injectors hold the liquid in a dose chamber until it is later delivered from a tiny orifice, at a very high velocity exceeding 100 meters per second, to directly enter the skin. The peak pressures generated within the dose chamber are typically in the range of 14 to 35 MPa. The liquid then spreads in a conical distribution, and the depth of delivery depends on several factors including velocity, orifice size, formulation viscosity, and skin parameters such as thickness and tautness. Liquid jet injectors have been around for a long time and have been widely used for mass vaccination programs in the last 50 years (27). However, they are constrained by cost, occasional pain, or local adverse reactions as well as by the potential risk of transfer of blood-borne pathogens when multiuse nozzle jet injectors are used (58). Several new, improved injectors protected by several patents are now available to overcome these disadvantages and are finding applications for delivery of insulin, vaccines, and other molecules (59). Liquid jets have been reported to create a hole around 100 μm in the skin (60), which is much smaller than the diameter of standard hypodermic needles (0.4 to 0.7 mm) and may therefore be considered a type of microporation technology. In a study with human subjects, it was shown that increasing gas mass in the device (chamber pressure) and orifice size, injection performance also increased (61). The depth of penetration in skin has been reported to increase to an asymptotic limit with increasing jet volume at a constant jet exit velocity and nozzle diameter. Further, it was found to correlate with a parameter based on the hypothesis that jet penetration occurs by skin puncture followed by jet dispersion. At an ejection volume of 0.13 mL, the hole depth in skin was around 0.8 mm (60). A somewhat different approach, where solid particles are directly delivered into the skin by a similar technology, is discussed in Section 9.6.1.

9.6.1 PARTICLE-MEDIATED IMMUNIZATION

Particle-mediated immunization, also referred to as gene-gun (62–64), particle-mediated epidermal delivery (PMED), epidermal powder immunization (EPI), or biolistic particle delivery, typically refers to the use of a needle-free device to deliver DNA or protein-coated microscopic gold particles or a powder formulation directly into the cells of the epidermis (65,66). The technique typically uses a gas (often helium) to propel or deliver powdered proteins or vaccines directly into the epidermis at supersonic speeds. This technique takes advantage of the immune competence of the skin organ. Skin has lymphoid tissue with specialized cells that enhance immune responses. It was shown that a single intradermal injection of <1 μg of specific DNA was sufficient to induce an immune response against the influenza virus. Gene expression in epidermis and dermis was greater at 10 days than at 3 days after plasmid DNA (pDNA) injection. As the epidermal cells (keratinocytes) are sloughed off, the expression remained mostly in the dermis at 30 days after injection (67). Intradermal injection seems to work, and the amount of DNA required for immunization can be reduced and immune responses can be improved if the DNA can be targeted directly to the cells, such as by using a gene gun to deliver DNA-coated gold beads to the epidermis. Once delivered to the cells, DNA dissolves and can then be expressed. In fact, it has been found that this route is more efficient as compared to other parenteral or mucosal routes and needs as little as 0.4 µg of DNA, which is 250 to 2500 times less DNA amount than required for direct parenteral inoculation of purified DNA in saline. In this study, expression was found to be transient, with most being lost within 2 to 3 days due to the normal sloughing of the epidermis. However, short-term expression of antigen is sufficient to raise long-term immune responses, because epidermal Langerhans cells are capable of presenting transfected antigens to the T-helper component of the immune system, priming both T-helper and B-cell memory (68).

PowderJect Pharmaceuticals, which was investigating vaccine delivery, was acquired by Chiron Corporation for its conventional flu shots, and Chiron, in turn, was acquired by Novartis Vaccine and Diagnostics. However, DNA vaccine delivery research was spun out of PowderJect as PowderMed, which is now owned by Pfizer, Inc. (New York), mostly for its interest in the DNA vaccine–based flu program. PowderJect Vaccines now operates as a subsidiary of Pfizer, Inc., and is investigating use of compressed helium gas to deliver DNA-coated microscopic gold particles into skin. This system was reportedly used to deliver hepatitis B virus coated on gold particles to healthy volunteers (69), and in a different study, gold particles were coated with vaccine powder formulations of diphtheria toxoid for delivery into skin (70). Other studies using this technology have been reviewed elsewhere (71).

9.7 DELIVERY OF DNA VACCINES, GENES, AND OLIGONUCLEOTIDES

Introduction of DNA into cells, called transfection, allows cells to produce proteins continuously, thus avoiding potential toxic effects that may result from high-dose bolus delivery of recombinant proteins. Delivery of proteins is also faced with the problem of short half-life, thus requiring repeated administrations. Furthermore, proteins produced by transfected eukaryotic cells will undergo post-translational modifications such as glycosylation, unlike those produced in *Escherichia coli* (72). When the transfected cells are producing protein antigens, they can provide a means of vaccination. DNA vaccination by electroporation has been particularly promising and is discussed in Section 9.7.1 but other approaches such as liposomes (73) and microneedles (74) have also been used for DNA vaccination. Gene-based drugs and gene therapy are expected to revolutionize therapeutics in the coming years. However, clinical application of these agents is limited by several problems, such as limited physical and chemical stability or undesirable attributes for adequate absorption and distribution. Thus, as these macromolecules are made available, it will be essential to formulate these drugs into safe, stable, and efficacious delivery systems. Iontophoresis, electroporation, and other skin enhancement techniques provide viable opportunities for the delivery of these agents.

9.7.1 DNA Vaccination by Electroporation

DNA vaccination or DNA-mediated immunization is based on the use of direct inoculation of plasmid DNA (pDNA) to raise immune responses by expression of an antigenic protein directly within the transfected cells. Using cloned genes in the form of plasmids allows one to introduce only the genes required for antigen production. In addition to providing coding sequences for the antigenic protein, the plasmid serves as the physical vector carrying the genes. This innovative approach to immunization holds true promise for the development of needed vaccines (75). The promise of DNA vaccines was first shown in the 1990s (76). Since then, it has been demonstrated by several scientists and includes the possibility of vaccination against genital infections (77), hepatitis B virus (78), murine cytomegalovirus (79), rotavirus (80), tuberculosis (81), and HIV (82–85). A strong immune response involving both the humoral and cellular arms of the immune system is involved. Also, DNA vaccines are easier to manufacture than an inactivated pathogen or recombinant protein vaccine. Furthermore, the DNA is very stable even under high-temperature conditions so that storage, transport, and distribution will be easy (86). Currently, intramuscular injection is commonly used, but it is now being realized that skin may also provide a viable route for administration. Skin and mucosal membranes are the anatomical sites where most exogenous antigens are commonly encountered. Following inoculation of DNA into skin, transfected keratinocytes will be lost in a few days due to the normal sloughing of the epidermis. However, it seems that relatively short-term expression of antigen is sufficient to raise long-term immune responses. However, expression of the gene in the DNA vaccine must be high because DNA does not replicate in the mammalian cell. Thus, all expression comes from the small amount of DNA internalized by the cells. Langerhans' cells of skin carry the antigen from the skin to the draining lymph nodes. These antigen-loaded Langerhans' cells are potent activators of T lymphocytes, and a type of lymphocytes called epidermal lymphocytes also play a key role in skin immunity (67). pDNA may be administered naked (dissolved in saline), complexed with lipids, or dried on the surface of particles.

Introduction of plasmid DNA into bacteria represents an important application of electroporation (87–89). Contact between plasmid DNA and bacteria must be present during the pulse, and the proportion of electrotransformed cells is dependent on DNA concentration (90). Other factors determining the electrotransformation of bacteria by plasmid DNA were investigated, including the use of electroporation buffer (91), temperature (92), physical factors (93), molecular form of DNA (94) and other factors (95). Electroporation has also been widely used to transform mammalian cells with plasmids (96,97). Electroporation offers a nonviral method for delivery of DNA or other nucleic acid–based drugs for gene therapy, DNA vaccination, or other applications. Nucleic acid–based drugs delivered into cells by electroporation can express to produce an endogenous supply of a therapeutic protein as an alternative to repeated exogenous administrations (98–101). Electroporation can also be helpful to downregulate gene expression by delivery of interfering RNA (101). Currently, delivery of pDNA into skin cells is typically accomplished by direct intradermal injection or particle bombardment. The use of gene gun technology to deliver DNA into the skin is discussed in Section 9.6.1. It would be preferable to have a noninvasive means to introduce DNA into skin cells. The technique of electroporation (see also Chapter 5) can be used for DNA vaccination. Electroporation pulses can allow the uptake of DNA by cells and thereby result in an effective immune response that cannot be achieved by injecting the DNA into skin (102,103). One study has shown that *in vivo* electroporation can be used for the introduction of plasmid DNA into skin cells of mouse. A mixture of two supercoiled pDNAs was introduced subcutaneously to newborn mice, and after 10 to 60 minutes, the pleat of skin was exposed to two high-voltage pulses. Fibroblast primary cell cultures were obtained from the treated skin, and after 2 to 3 weeks of selection, clones of stable transformed mouse fibroblasts were obtained (104). In this study, DNA was first injected subcutaneously before electroporation pulse was applied. Another study has used the Easy Vax™ delivery system (see Section 5.3) to deliver an experimental smallpox DNA vaccine in mice and demonstrate the generation of protective immune responses (105). In more recent studies, DNA vaccination by *in vivo* electroporation of muscle following injection of the plasmid has been investigated in mice and pig animal models (106–109). Several studies have also been done in rhesus macaques as nonhuman primates (110–115). Macaques immunized in combination with electroporation have reported a 10- to 40-fold higher HIV-specific enzyme-linked immunospot assay response as compared to those receiving a fivefold higher dose of vaccine but without *in vivo* electroporation (112). DNA vaccines administered by injection have often been less effective in large animals; therefore, electroporation offers a significant advantage for DNA vaccination in large animals (116) and humans. DNA vaccination studies were performed more for muscle and less for skin; one likely reason is because larger volumes and therefore higher plasmid doses can be administered intramuscularly. However, in a study with pigs and rhesus macaques, gene expression for plasmid GFP delivered by intradermal/subcutaneous route was demonstrated. Microelectrodes were used in this study to prevent skin discomfort or damage during administration of the plasmid (115). In an earlier study, intradermal injection of plasmid DNA followed by electroporation was found to improve transfection in skin of a mouse model (117). DNA vaccination studies have now also been performed on healthy human volunteers (86,101,118).

9.7.2 DELIVERY OF GENES

The ability to target genes to the various layers, cell types, and appendages of skin is of interest and could be used to correct skin disorders including inherited skin disorders, wound healing, and changes due to aging, such as wrinkles (119,120). Several melanoma and nonmelanoma skin cancers have their origin in mutations caused by several risk factors, including ultraviolet radiation (121,122), and perhaps may be treated by gene therapy. Similarly, the damage induced in skin by ultraviolet light involves DNA damage (123,124) and might possibly be treated by gene therapy. Topical diseases and conditions such as nonmelanoma skin cancers, damage by ultraviolet light, lupus vulgaris, fungal infections, warts, and herpes simplex virus infections, can benefit from electroporation of nucleotide-based or other drugs. Skin is also an attractive target tissue due to its accessibility. While most of the work in the field of gene therapy has focused on viral gene transfer (i.e., using viral vectors), there is a need to develop nonviral approaches, as these would offer several advantages, the primary advantage being that nonviral gene therapy will eliminate the risk of generating novel infectious agents. Also, nonviral methods can introduce large molecules of DNA into the cell, unlike viral systems that are limited by the carrying capacity of the virus. In addition, nonviral methods generally are associated with a lower risk of immunogenicity as compared to viral methods (125,126). Liposomes and lipids have been extensively investigated as carriers for gene delivery. Cationic liposomes or lipids can complex DNA, RNA, or short-stranded antisense sequences, and results have been very promising (127–132). A DNA/liposome complex applied *in vivo* to mouse skin was found to rapidly penetrate the skin and exhibit its β-galactosidase gene expression in epidermis, dermis, and hair follicles. Expression was seen as early as 6 hours after application and was high for 24 to 48 hours but reduced significantly by 7 days following application (127). Similarly, even synthetic cationic polymers were investigated as a vector for the design of *in vivo* and *ex vivo* gene transfection systems (133). The use of electroporation for gene therapy of skin has been shown in an *in vivo* study where *lacZ* DNA was delivered to hairless mice using three exponential decay pulses of amplitude 120 V and pulse length of 10 to 20 ms. Expression of the *lacZ* gene as indicated by blue staining was observed extensively in the dermis, including the hair follicles (119,134). Microneedle arrays have also been used, and it was reported that they can enable topical gene transfer with a 2800-fold increase in reporter gene activity relative to a topical control (135).

9.7.3   Delivery of Oligonucleotides

Antisense oligonucleotides are short lengths of single-stranded RNA or DNA which have base sequences complementary to a specific gene or its mRNA. The length is typically 15 to 30 nucleotide bases, usually denoted by "mer" (i.e., 15 to 30 mers). They can inhibit gene expression, primarily by binding to the target gene mRNA, and thus have the ability to selectively block disease-causing genes from producing disease-associated proteins. Antisense technology offers a straightforward approach to take advantage of the elucidation of genomic sequences; thus, collaboration between antisense and genomic companies is developing. Conventional delivery routes such as oral delivery are not suitable for oligonucleotides due to their large size, high negative charge, biological instability toward intra- and extracellular nucleases (136), rapid *in vivo* plasma elimination kinetics, poor cellular uptake, and ineffective delivery to target site. Some of these drawbacks may be overcome by the use of controlled-release delivery systems such as biodegradable polymer matrices or by using cationic lipids or liposomes that may enhance cellular uptake of oligonucleotides due to their attraction to the negatively charged surfaces of most cells (137–139). Transdermal delivery offers one possibility for the delivery of antisense oligonucleotides. Also, some diseases may be best treated via topical application of these agents. Recently, delivery of antisense oligonucleotides into skin has been reported using several techniques such as cationic elastic liposomes (140), tape stripping (141), Er:YAG laser (142), and iontophoresis (143). Iontophoretic delivery of the representative bases (uracil and adenine), nucleosides (uridine and adenosine), and nucleotides (AMP, ATP, GTP, and imido-GTP) across mammalian skin *in vitro* was demonstrated (144). For the delivery of gene-based drugs, the type of challenges faced may be somewhat different than those faced by conventional or peptide drugs. Charge and delivery considerations need to be carefully considered. The pK_a value of the phosphate backbone of oligonucleotides is very low (approximately 2.2), so the molecules are negatively charged over a wide range of pH. Differences in pH, therefore, have little or no effect on the charge/mass ratio of the molecule and are not likely to affect iontophoretic delivery (145) from this point of consideration. However, other considerations are involved, such as the pH-dependent electroosmotic flow and the type of backbone in the oligonucleotide so that pH would still be a major factor (146). Oligonucleotides and DNA are negatively charged, so their iontophoretic delivery would normally be under cathode. Because electroosmotic flow is from anode to cathode (see Section 4.7), it will hinder transport, and this needs to be taken into consideration. This electroosmotic flow will be pH-dependent, and its overall contribution will be dependent on the size of the molecule being transported. In Section 8.3.1, it was discussed that the enzymatic activity of skin presents a barrier to the delivery of peptides. Skin also contains nuclease and phosphatase activity, and oligonucleotides and DNA penetrating the skin are susceptible to the activity of these enzymes. Because many nucleases require bivalent cations for their activity, they can be inhibited by ethylenediaminetetraacetic (EDTA) while phosphatase activity can be inhibited by inorganic phosphate. Thus, appropriate choice of buffer systems may be able to minimize degradation of oligonucleotides during transport through skin. Topical delivery of small interfering RNA (siRNA) can help to target genes that are involved in cutaneous disorders, viral infections, skin cancer, and atopic dermatitis. Because siRNA is a hydrophilic macromolecule with no passive permeation, it has been delivered by iontophoresis to an atopic dermatitis rat model and was reported to accumulate in the epidermis and demonstrate significant silencing of the target gene (147).

Larger oligonucleotides may be delivered by iontophoresis or electroporation, which also offer means to avoid the use of penetration enhancers. Iontophoretic *in vitro* delivery of oligonucleotides across excised full-thickness hairless mouse skin was investigated. Flux was found to decrease with increasing size, with the 10-mer oligonucleotide being transported at a rate about sixfold faster than the 30-mer and twofold faster than the 20-mer. Skin was first pretreated with a buffer containing EDTA and K_3PO_4, which were also added to the transport buffer, as these are known to be effective inhibitors of nuclease and phosphatase activity, respectively. The buffer also contained 2.5 mg/ml of salmon testes DNA to prevent nonspecific binding of the oligonucleotide to surfaces, during studies with transdermal diffusion cells. EDTA or K_3PO_4 alone were not very effective, but when present together, they resulted in oligonucleotide delivery, mostly as an intact molecule in the receptor, by iontophoresis. Dephosphorylation and degradation were observed only during transport of oligonucleotides through skin and were found to occur at a pH of 5.5 or below. However, virtually no degradation was seen at pH 9.5. Incubation of the oligonucleotides in contact with the skin resulted in no degradation or dephosphorylation of the drug, suggesting that both nuclease and phosphatase activities are located within the tissue. A steady-state lag time of about

30 to 60 minutes was observed before the oligonucleotides appeared in the receptor chamber following the start of current, using a current density of 0.3 mA/cm^2 (145). In contrast to this study, a small six-base-sequence oligonucleotide, TAG-6 (MW 1927), was reported to cross hairless mouse skin with little or no degradation in an *in vitro* study. Cathodal iontophoretic delivery for 12 hours using either the 5′-FITC or ^{35}S-labeled oligonucleotide resulted in a substantial flux, with steady-state levels of 273 ± 65 or 285 ± 71 ng/cm^2-hour respectively, suggesting that FITC labeling does not alter TAG-6 transport (146). Electroporation has been shown to be useful to improve intracellular delivery of synthetic antisense oligonucleotides to suppress a target protein (148,149). The electroporation technique has been successfully used to transport 15-mer (4.8 kDa) and 24-mer (7.0 kDa) antisense oligonucleotides through human skin *in vitro* with fluxes of 6.4 pM/cm^2-hour and 11.5 pmol/cm^2-hour, respectively. The transport of fluorescein-labeled oligonucleotides was determined by measuring the fluorescence of the receptor compartment solution by spectrofluorimeter. To visualize sites of local transport, a 2% agarose gel was prepared in the receptor, and lack of staining at sites corresponding to sweat ducts and hair follicles indicated that the transappendageal route was not a primary pathway. Transport was found to increase significantly with transdermal voltages greater than about 70 V, but then it formed a plateau soon afterward. Transport also increased as the pulse length increased from 1.1 to 2.2 ms (150). In another study, the 15-mer phosphodiesters were delivered to hairless rat skin by electroporation. It was shown that the 3′ protected phosphodiesters were better for topical delivery to skin, due to good stability against skin nucleases as compared to phosphorothiolate oligonucleotides (151). Phosphorothiolate oligonucleotides complementary to TGF-β mRNA were designed to eliminate scars, which can be caused by excessive collagen deposition due to overexpression of TGF-β in wounded skin. Partially modified forms of these oligonucleotides have been shown to penetrate normal and tape-stripped damaged rat skin during *in vitro* studies, with higher amounts permeating through damaged skin. For normal skin, the oligonucleotide did not penetrate through skin into the receptor medium, presumably due to its high molecular weight (MW 8000) and polyanionic charge (152). Penetration enhancers have been investigated *in vitro* for skin penetration and retention of a series of antisense methyl phosphonate oligonucleotides using hairless mouse or human cadaver skin. As the molecular weight of the oligonucleotide increased, the penetration rate was found to decrease, though the study was limited to a 18-mer (MW 5500) oligonucleotide as the largest size (153).

REFERENCES

1. A. K. Banga. *Therapeutic peptides and proteins: Formulation, processing, and delivery systems*, Taylor and Francis, London, 2006.
2. G. Kersten and H. Hirschberg. Antigen delivery systems, *Expert. Rev. Vaccines*, 3:453–462 (2004).
3. E. L. Giudice and J. D. Campbell. Needle-free vaccine delivery, *Adv. Drug Deliv. Rev.*, 58:68–89 (2006).
4. M. A. Kendall. Needle-free vaccine injection, *Handb. Exp. Pharmacol.*, 197:193–219 (2010).

234   Transdermal and Intradermal Delivery of Therapeutic Agents

5. M. R. Prausnitz, J. A. Mikszta, M. Cormier, and A. K. Andrianov. Microneedle-based vaccines, *Curr. Top. Microbiol. Immunol.*, 333:369–393 (2009).
6. P. Van Damme, F. Oosterhuis-Kafeja, W. M. Van der, Y. Almagor, O. Sharon, and Y. Levin. Safety and efficacy of a novel microneedle device for dose sparing intradermal influenza vaccination in healthy adults, *Vaccine*, 27:454–459 (2009).
7. F. S. Quan, Y. C. Kim, A. Vunnava, D. G. Yoo, J. M. Song, M. R. Prausnitz, R. W. Compans, and S. M. Kang. Intradermal vaccination with influenza virus-like particles by using microneedles induces protection superior to that with intramuscular immunization, *J. Virol.*, 84:7760–7769 (2010).
8. J. B. Alarcon, A. W. Hartley, N. G. Harvey, and J. A. Mikszta. Preclinical evaluation of microneedle technology for intradermal delivery of influenza vaccines, *Clin. Vaccine Immunol.*, 14:375–381 (2007).
9. A. Tezel, S. Paliwal, Z. Shen, and S. Mitragotri. Low-frequency ultrasound as a transcutaneous immunization adjuvant, *Vaccine*, 23:3800–3807 (2005).
10. S. Paliwal and S. Mitragotri. Therapeutic opportunities in biological responses of ultrasound, *Ultrasonics*, 48:271–278 (2008).
11. A. Dahlan, H. O. Alpar, P. Stickings, D. Sesardic, and S. Murdan. Transcutaneous immunisation assisted by low-frequency ultrasound, *Int. J. Pharm.*, 368:123–128 (2009).
12. H. J. Hirschberg, G. G. van de Wijdeven, H. Kraan, J. P. Amorij, and G. F. Kersten. Bioneedles as alternative delivery system for hepatitis B vaccine, *J. Control. Release*, 147:211–217 (2010).
13. M. Zierhut, T. Bieber, E. B. Brocker, J. V. Forrester, C. S. Foster, and J. W. Streilein. Immunology of the skin and the eye, *Immunol. Today*, 17:448–450 (1996).
14. J. D. Bos. The skin as an organ of immunity, *Clin. Exp. Immunol.*, 107:3–5 (1997).
15. J. D. Bos, Ed. *Skin immune system: Cutaneous immunology and clinical immunodermatology*, CRC Press, Boca Raton, FL, 1997.
16. G. M. Glenn, T. Scharton-Kersten, and C. R. Alving. Advances in vaccine delivery: Transcutaneous immunization, *Exp. Opin. Invest. Drugs*, 8:797–805 (2003).
17. S. A. Hammond, D. Walwender, C. R. Alving, and G. M. Glenn. Transcutaneous immunization: T cell responses and boosting of existing immunity. *Vaccine*, 19:2701–2707 (2001).
18. S. A. Hammond, M. Guebre-Xabier, J. Yu, and G. M. Glenn. Transcutaneous immunization: An emerging route of immunization and potent immunostimulation strategy, *Crit. Rev. Ther. Drug Carrier Syst.*, 18:503–526 (2001).
19. S. Babiuk, M. Baca-Estrada, L. A. Babiuk, C. Ewen, and M. Foldvari. Cutaneous vaccination: The skin as an immunologically active tissue and the challenge of antigen delivery. *J. Control. Rel.*, 66:199–214 (2000).
20. C. D. Partidos, A. S. Beignon, F. Brown, E. Kramer, J. P. Briand, and S. Muller. Applying peptide antigens onto bare skin: Induction of humoral and cellular immune responses and potential for vaccination, *J. Control. Release*, 85:27–34 (2002).
21. Y. Ishii, T. Nakae, F. Sakamoto, K. Matsuo, K. Matsuo, Y. S. Quan, F. Kamiyama, T. Fujita, A. Yamamoto, S. Nakagawa, and N. Okada. A transcutaneous vaccination system using a hydrogel patch for viral and bacterial infection, *J. Control. Release*, 131:113–120 (2008).
22. P. Karande, A. Arora, T. K. Pham, D. Stevens, A. Wojicki, and S. Mitragotri. Transcutaneous immunization using common chemicals, *J. Control. Release*, 138:134–140 (2009).
23. P. E. Laurent, S. Bonnet, P. Alchas, P. Regolini, J. A. Mikszta, R. Pettis, and N. G. Harvey. Evaluation of the clinical performance of a new intradermal vaccine administration technique and associated delivery system, *Vaccine*, 25:8833–8842 (2007).

24. A. J. Harvey, S. A. Kaestner, D. E. Sutter, N. G. Harvey, J. A. Mikszta, and R. J. Pettis. Microneedle-based intradermal delivery enables rapid lymphatic uptake and distribution of protein drugs, *Pharm. Res.*, [Epub] (2010).

25. A. Laurent, F. Mistretta, D. Bottigioli, K. Dahel, C. Goujon, J. F. Nicolas, A. Hennino, and P. E. Laurent. Echographic measurement of skin thickness in adults by high frequency ultrasound to assess the appropriate microneedle length for intradermal delivery of vaccines, *Vaccine*, 25:6423–6430 (2007).

26. P. E. Laurent, H. Bourhy, M. Fantino, P. Alchas, and J. A. Mikszta. Safety and efficacy of novel dermal and epidermal microneedle delivery systems for rabies vaccination in healthy adults, *Vaccine*, 28:5850–5856 (2010).

27. B. G. Weniger and M. J. Papania. Alternative vaccine delivery methods. In S. A. Plotkin, W. A. Orenstein, and P. A. Offit (eds), *Vaccines*, Saunders (Elsevier), Philadelphia, 2008, pp. 1357–1392.

28. D. Holland, R. Booy, F. De Looze, P. Eizenberg, J. McDonald, J. Karrasch, M. McKeirnan, H. Salem, G. Mills, J. Reid, F. Weber, and M. Saville. Intradermal influenza vaccine administered using a new microinjection system produces superior immunogenicity in elderly adults: A randomized controlled trial, *J. Infect. Dis.*, 198:650–658 (2008).

29. J. Beran, A. Ambrozaitis, A. Laiskonis, N. Mickuviene, P. Bacart, Y. Calozet, E. Demanet, S. Heijmans, P. Van Belle, F. Weber, and C. Salamand. Intradermal influenza vaccination of healthy adults using a new microinjection system: A 3-year randomised controlled safety and immunogenicity trial, *BMC. Med.*, 7:13 (2009).

30. J. A. Mikszta, V. J. Sullivan, C. Dean, A. M. Waterston, J. B. Alarcon, J. P. Dekker, III, J. M. Brittingham, J. Huang, C. R. Hwang, M. Ferriter, G. Jiang, K. Mar, K. U. Saikh, B. G. Stiles, C. J. Roy, R. G. Ulrich, and N. G. Harvey. Protective immunization against inhalational anthrax: A comparison of minimally invasive delivery platforms, *J. Infect. Dis.*, 191:278–288 (2005).

31. J. A. Mikszta, J. P. Dekker, III, N. G. Harvey, C. H. Dean, J. M. Brittingham, J. Huang, V. J. Sullivan, B. Dyas, C. J. Roy, and R. G. Ulrich. Microneedle-based intradermal delivery of the anthrax recombinant protective antigen vaccine, *Infect. Immun.*, 74:6806–6810 (2006).

32. C. H. Dean, J. B. Alarcon, A. M. Waterston, K. Draper, R. Early, F. Guirakhoo, T. P. Monath, and J. A. Mikszta. Cutaneous delivery of a live, attenuated chimeric flavivirus vaccine against Japanese encephalitis (ChimeriVax)-JE) in non-human primates, *Hum. Vaccin.*, 1:106–111 (2005).

33. J. Huang, A. J. D'Souza, J. B. Alarcon, J. A. Mikszta, B. M. Ford, M. S. Ferriter, M. Evans, T. Stewart, K. Amemiya, R. G. Ulrich, and V. J. Sullivan. Protective immunity in mice achieved with dry powder formulation and alternative delivery of plague F1-V vaccine, *Clin. Vaccine Immunol.*, 16:719–725 (2009).

34. J. A. Matriano, M. Cormier, J. Johnson, W. A. Young, M. Buttery, K. Nyam, and P. E. Daddona. Macroflux microprojection array patch technology: A new and efficient approach for intracutaneous immunization, *Pharm. Res.*, 19:63–70 (2002).

35. G. Widera, J. Johnson, L. Kim, L. Libiran, K. Nyam, P. E. Daddona, and M. Cormier. Effect of delivery parameters on immunization to ovalbumin following intracutaneous administration by a coated microneedle array patch system, *Vaccine*, 24:1653–1664 (2006).

36. G. L. Morefield, R. F. Tammariello, B. K. Purcell, P. L. Worsham, J. Chapman, L. A. Smith, J. B. Alarcon, J. A. Mikszta, and R. G. Ulrich. An alternative approach to combination vaccines: Intradermal administration of isolated components for control of anthrax, botulism, plague and staphylococcal toxic shock, *J. Immune. Based. Ther. Vaccines*, 6:5 (2008).

37. J. C. Birchall, R. Clemo, A. Anstey, and D. N. John. Microneedles in clinical practice—An exploratory study into the opinions of healthcare professionals and the public, *Pharm Res.*, [Epub] (2010).

38. X. Chen, T. W. Prow, M. L. Crichton, D. W. Jenkins, M. S. Roberts, I. H. Frazer, G. J. Fernando, and M. A. Kendall. Dry-coated microprojection array patches for targeted delivery of immunotherapeutics to the skin, *J. Control. Release*, 139:212–220 (2009).

39. T. W. Prow, X. Chen, N. A. Prow, G. J. Fernando, C. S. Tan, A. P. Raphael, D. Chang, M. P. Ruutu, D. W. Jenkins, A. Pyke, M. L. Crichton, K. Raphaelli, L. Y. Goh, I. H. Frazer, M. S. Roberts, J. Gardner, A. A. Khromykh, A. Suhrbier, R. A. Hall, and M. A. Kendall. Nanopatch-targeted skin vaccination against West Nile virus and Chikungunya virus in mice, *Small*, 6:1776–1784 (2010).

40. A. P. Raphael, T. W. Prow, M. L. Crichton, X. Chen, G. J. Fernando, and M. A. Kendall. Targeted, needle-free vaccinations in skin using multilayered, densely-packed dissolving microprojection arrays, *Small*, 6:1785–1793 (2010).

41. Q. Zhu, V. G. Zarnitsyn, L. Ye, Z. Wen, Y. Gao, L. Pan, I. Skountzou, H. S. Gill, M. R. Prausnitz, C. Yang, and R. W. Compans. Immunization by vaccine-coated microneedle arrays protects against lethal influenza virus challenge, *Proc. Natl. Acad. Sci. USA*, 106:7968–7973 (2009).

42. Z. Ding, F. J. Verbaan, M. Bivas-Benita, L. Bungener, A. Huckriede, D. J. van den Berg, G. Kersten, and J. A. Bouwstra. Microneedle arrays for the transcutaneous immunization of diphtheria and influenza in BALB/c mice, *J. Control. Release*, 136:71–78 (2009).

43. Z. Ding, E. van Riet, S. Romeijn, G. F. Kersten, W. Jiskoot, and J. A. Bouwstra. Immune modulation by adjuvants combined with diphtheria toxoid administered topically in BALB/c mice after microneedle array pretreatment, *Pharm. Res.*, 26:1635–1643 (2009).

44. S. M. Bal, Z. Ding, G. F. Kersten, W. Jiskoot, and J. A. Bouwstra. Microneedle-based transcutaneous immunisation in mice with N-trimethyl chitosan adjuvanted diphtheria toxoid formulations, *Pharm Res.*, 27:1837–1847 (2010).

45. S. M. Bal, B. Slutter, E. van Riet, A. C. Kruithof, Z. Ding, G. F. Kersten, W. Jiskoot, and J. A. Bouwstra. Efficient induction of immune responses through intradermal vaccination with N-trimethyl chitosan containing antigen formulations, *J. Control. Release*, 142:374–383 (2010).

46. B. Slutter, P. C. Soema, Z. Ding, R. Verheul, W. Hennink, and W. Jiskoot. Conjugation of ovalbumin to trimethyl chitosan improves immunogenicity of the antigen, *J. Control. Release*, 143:207–214 (2010).

47. A. K. Andrianov, D. P. Decollibus, H. A. Gillis, H. H. Kha, A. Marin, M. R. Prausnitz, L. A. Babiuk, H. Townsend, and G. Mutwiri. Poly[di(carboxylatophenoxy)phosphazene] is a potent adjuvant for intradermal immunization, *Proc. Natl. Acad. Sci. USA*, 106:18936–18941 (2009).

48. A. K. Andrianov, A. Marin, and D. P. Decollibus. Microneedles with intrinsic immunoadjuvant properties: microfabrication, protein stability, and modulated release, *Pharm. Res.*, (2010).

49. Y. C. Kim, F. S. Quan, D. G. Yoo, R. W. Compans, S. M. Kang, and M. R. Prausnitz. Improved influenza vaccination in the skin using vaccine coated microneedles, *Vaccine*, 27:6932–6938 (2009).

50. Y. C. Kim, F. S. Quan, R. W. Compans, S. M. Kang, and M. R. Prausnitz. Stability kinetics of influenza vaccine coated onto microneedles during drying and storage, *Pharm. Res.*, [Epub] (2010).

51. D. G. Koutsonanos, M. M. Del Pilar, V. G. Zarnitsyn, S. P. Sullivan, R. W. Compans, M. R. Prausnitz, and I. Skountzou. Transdermal influenza immunization with vaccine-coated microneedle arrays, *PLoS. One.*, 4:e4773 (2009).

52. Y. C. Kim, F. S. Quan, D. G. Yoo, R. W. Compans, S. M. Kang, and M. R. Prausnitz. Enhanced memory responses to seasonal H1N1 influenza vaccination of the skin with the use of vaccine-coated microneedles, *J. Infect. Dis.*, 201:190–198 (2010).

53. Y. C. Kim, F. S. Quan, R. W. Compans, S. M. Kang, and M. R. Prausnitz. Formulation and coating of microneedles with inactivated influenza virus to improve vaccine stability and immunogenicity, *J. Control. Release*, 142:187–195 (2010).

54. M. Pearton, S. M. Kang, J. M. Song, Y. C. Kim, F. S. Quan, A. Anstey, M. Ivory, M. R. Prausnitz, R. W. Compans, and J. C. Birchall. Influenza virus-like particles coated onto microneedles can elicit stimulatory effects on Langerhans cells in human skin, *Vaccine*, 28:6104–6113 (2010).

55. J. M. Song, Y. C. Kim, P. G. Barlow, M. J. Hossain, K. M. Park, R. O. Donis, M. R. Prausnitz, R. W. Compans, and S. M. Kang. Improved protection against avian influenza H5N1 virus by a single vaccination with virus-like particles in skin using microneedles, *Antiviral Res.*, [Epub] (2010).

56. F. S. Quan, Y. C. Kim, D. G. Yoo, R. W. Compans, M. R. Prausnitz, and S. M. Kang. Stabilization of influenza vaccine enhances protection by microneedle delivery in the mouse skin, *PLoS. One.*, 4:e7152 (2009).

57. S. P. Sullivan, D. G. Koutsonanos, M. M. Del Pilar, J. W. Lee, V. Zarnitsyn, S. O. Choi, N. Murthy, R. W. Compans, I. Skountzou, and M. R. Prausnitz. Dissolving polymer microneedle patches for influenza vaccination, *Nat. Med.*, 16:915–920 (2010).

58. P. H. Lambert and P. E. Laurent. Intradermal vaccine delivery: Will new delivery systems transform vaccine administration? *Vaccine* 26:3197–3208 (2008).

59. T. R. Singh, M. J. Garland, C. M. Cassidy, K. Migalska, Y. K. Demir, S. Abdelghany, E. Ryan, A. D. Woolfson, and R. F. Donnelly. Microporation techniques for enhanced delivery of therapeutic agents, *Recent Pat. Drug Deliv. Formul.*, 4:1–17 (2010).

60. J. Baxter and S. Mitragotri. Jet-induced skin puncture and its impact on needle-free jet injections: Experimental studies and a predictive model, *J. Control. Release*, 106:361–373 (2005).

61. L. Linn, B. Boyd, H. Iontchev, T. King, and S. J. Farr. The effects of system parameters on *in vivo* injection performance of a needle-free injector in human volunteers, *Pharm. Res.*, 24:1501–1507 (2007).

62. A. M. Bennett, R. J. Phillpotts, S. D. Perkins, S. C. Jacobs, and E. D. Williamson. Gene gun mediated vaccination is superior to manual delivery for immunisation with DNA vaccines expressing protective antigens from *Yersinia pestis* or Venezuelan Equine Encephalitis virus, *Vaccine*, 18:588–596 (2000).

63. M. T. Lin, L. Pulkkinen, J. Uitto, and K. Yoon. The gene gun: Current applications in cutaneous gene therapy, *Int. J. Dermatol.*, 39:161–170 (2000).

64. R. Han, C. A. Reed, N. M. Cladel, and N. D. Christensen. Immunization of rabbits with cottontail rabbit papillomavirus E1 and E2 genes: Protective immunity induced by gene gun-mediated intracutaneous delivery but not by intramuscular injection, *Vaccine*, 18:2937–2944 (2000).

65. D. Chen, K. F. Weis, Q. Chu, C. Erickson, R. Endres, C. R. Lively, J. Osorio, and L. G. Payne. Epidermal powder immunization induces both cytotoxic T-lymphocyte and antibody responses to protein antigens of influenza and hepatitis B viruses, *J. Virol.*, 75:11630–11640 (2001).

66. H. J. Dean, D. Fuller, and J. E. Osorio. Powder and particle-mediated approaches for delivery of DNA and protein vaccines into the epidermis, *Comp. Immunol. Microbiol. Infect. Dis.*, 26:373–388 (2003).

67. E. Raz, D. A. Carson, S. E. Parker, T. B. Parr, A. M. Abai, G. Aichinger, S. H. Gromkowski, M. Singh, D. Lew, M. A. Yankauckas, S. M. Baird, and G. H. Rhodes. Intradermal gene immunization: The possible role of DNA uptake in the induction of cellular immunity to viruses, *Proc. Natl. Acad. Sci.*, 91:9519–9523 (1994).

68. E. F. Fynan, R. G. Webster, D. H. Fuller, J. R. Haynes, J. C. Santoro, and H. L. Robinson. DNA Vaccines: Protective immunizations by parenteral, mucosal, and gene-gun inoculations, *Proc. Natl. Acad. Sci.*, 90:11478–11482 (1993).

69. M. J. Roy, M. S. Wu, L. J. Barr, J. T. Fuller, L. G. Tussey, S. Speller, J. Culp, J. K. Burkholder, W. F. Swain, R. M. Dixon, G. Widera, R. Vessey, A. King, G. Ogg, A. Gallimore, J. R. Haynes, and D. H. Fuller. Induction of antigen-specific CD8+ T cells, T helper cells, and protective levels of antibody in humans by particle-mediated administration of a hepatitis B virus DNA vaccine, *Vaccine*, 19:764–778 (2001).

70. D. Chen, C. A. Erickson, R. L. Endres, S. B. Periwal, Q. Chu, C. Shu, Y. F. Maa, and L. G. Payne. Adjuvantation of epidermal powder immunization, *Vaccine*, 19:2908–2917 (2001).

71. M. Kendall. Engineering of needle-free physical methods to target epidermal cells for DNA vaccination, *Vaccine*, 24:4651–4656 (2006).

72. I. Wicks. Human gene therapy, *Aust. N.Z. J. Med.*, 25:280–283 (1995).

73. G. Gregoriadis, R. Saffie, and J. B. deSouza. Liposome-mediated DNA vaccination, *FEBS Lett.*, 402:107–110 (1997).

74. H. S. Gill, J. Soderholm, M. R. Prausnitz, and M. Sallberg. Cutaneous vaccination using microneedles coated with hepatitis C DNA vaccine, *Gene Ther.*, 17:811–814 (2010).

75. H. L. Robinson, S. Lu, D. M. Feltquate, C. T. Torres, J. Richmond, C. M. Boyle, M. J. Morin, J. C. Santoro, R. G. Webster, D. Montefiori, Y. Yasutomi, N. L. Letvin, K. Manson, M. Wyand, and J. R. Haynes. DNA vaccines, *AIDS Res. Hum. Retroviruses*, 12:455–457 (1996).

76. J. J. Donnelly, J. B. Ulmer, and M. A. Liu. DNA vaccines, *Life Sci.*, 60:163–172 (1997).

77. J. J. Donnelly, D. Martinez, K. U. Jansen, R. W. Ellis, D. L. Montgomery, and M. A. Liu. Protection against papillomavirus with a polynucleotide vaccine, *J. Infect. Dis.*, 173:314–320 (1996).

78. H. L. Davis. DNA-based vaccination against hepatitis B virus, *Advan. Drug Del. Rev.*, 21:33–47 (1996).

79. J. C. G. Armas, C. S. Morello, L. D. Cranmer, and D. H. Spector. DNA immunization confers protection against murine cytomegalovirus infection, *J. Virol.*, 70:7921–7928 (1996).

80. J. E. Herrmann, S. C. Chen, E. F. Fynan, J. C. Santoro, H. B. Greenberg, S. X. Wang, and H. L. Robinson. Protection against rotavirus infections by DNA vaccination, *J. Infect. Dis.*, 174:S93–S97 (1996).

81. R. E. Tascon, M. J. Colston, S. Ragno, E. Stavropoulos, D. Gregory, and D. B. Lowrie. Vaccination against tuberculosis by DNA injection, *Nature Med.*, 2:888–892 (1996).

82. S. Lu, J. Arthos, D. C. Montefiori, Y. Yasutomi, K. Manson, F. Mustafa, E. Johnson, J. C. Santoro, J. Wissink, J. I. Mullins, J. R. Haynes, N. L. Letvin, M. Wyand, and H. L. Robinson. Simian immunodeficiency virus DNA vaccine trial in macaques, *J. Virol.*, 70:3978–3991 (1996).

83. B. Wang, J. Boyer, V. Srikantan, K. Ugen, L. Gilbert, C. Phan, K. Dang, M. Merva, M. G. Agadjanyan, M. Newman, R. Carrano, D. McCallus, L. Coney, W. V. Williams, and D. B. Weiner. Induction of humoral and cellular immune responses to the human immunodeficiency type 1 virus in nonhuman primates by *in vivo* DNA inoculation, *Virology*, 211:102–112 (1995).

84. J. W. Shiver, J. B. Ulmer, J. J. Donnelly, and M. A. Liu. Humoral and cellular immunities elicited by DNA vaccines: Application to the human immunodeficiency virus and influenza, *Adv. Drug Del. Rev.*, 21:19–31 (1996).

85. J. W. Shiver, M. E. Davies, H. C. Perry, D. C. Freed, and M. A. Liu. Humoral and cellular immunities elicited by HIV-1 DNA vaccination, *J. Pharm. Sci.*, 85:1317–1324 (1996).

86. M. P. Fons. Next generation vaccines: Antigen-encoding DNA delivered via electropo-ration, *Drug Del. Technol.*, 9:32–35 (2009).

87. E. L. Rosey, M. J. Kennedy, D. K. Petrella, R. G. Ulrich, and R. J. Yancey. Inactivation of *Serpulina hyodysenteriae* flaA1 and flaB1 periplasmic flagellar genes by electropora-tion-mediated allelic exchange, *J. Bacteriol.*, 177:5959–5970 (1995).

88. H. Li and H. K. Kuramitsu. Development of a gene transfer system in *Treponema denti-cola* by electroporation, *Oral Microbiol. Immunol.*, 11:161–165 (1996).

89. S. H. Yang, T. S. Song, and Y. M. Kim. Transformation of methylotrophic bacteria by electroporation of pRO1727 plasmid from *Pseudomonas aeruginosa*, *Mol. Cells*, 6:225–228 (1996).

90. B. Rittich and A. Spanova. Electrotransformation of bacteria by plasmid DNAs: Statistical evaluation of a model quantitatively describing the relationship between the number of electrotransformants and DNA concentration, *Bioelectrochem. Bioenerg.*, 40:233–238 (1996).

91. Z. J. Li, R. L. Jarret, M. Cheng, and J. W. Demski. Improved electroporation buffer enhances transient gene expression in *Arachis hypogaea* protoplasts, *Genome*, 38:858–863 (1995).

92. B. J. Wards and D. M. Collins. Electroporation at elevated temperatures substantially improves transformation efficiency of slow-growing mycobacteria, *FEMS Microbiol. Lett.*, 145:101–105 (1996).

93. B. Rittich, K. Manova, A. Spanova, and L. Pribyla. Electrotransformation of bacteria by plasmid DNA: Effect of serial electroporator resistor, *Chem. Papers*, 50:245–248 (1996).

94. H. Kimoto and A. Taketo. Studies on electrotransfer of DNA into *Escherichia coli*: Effect of molecular form of DNA, *Biochim. Biophys. Acta*, 1307:325–330 (1996).

95. F. Berthier, M. Zagorec, M. ChampomierVerges, S. D. Ehrlich, and F. MorelDeville. Efficient transformation of *Lactobacillus* sake by electroporation, *Microbiology*, 142:1273–1279 (1996).

96. M. Tatsuka, N. Yamagishi, M. Wada, H. Mitsui, T. Ota, and S. Odashima. Electroporation-mediated transfection of mammalian cells with crude plasmid DNA preparations, *Genet. Anal. Biomol. Eng.*, 12:113–117 (1995).

97. H. Melkonyan, C. Sorg, and M. Klempt. Electroporation efficiency in mammalian cells is increased by dimethyl sulfoxide (DMSO), *Nucleic Acids Res.*, 24:4356–4357 (1996).

98. M. L. Lucas, L. Heller, D. Coppola, and R. Heller. IL-12 plasmid delivery by *in vivo* electroporation for the successful treatment of established subcutaneous B16.F10 mela-noma, *Mol. Ther.*, 5:668–675 (2002).

99. R. Draghia-Akli, K. M. Ellis, L. A. Hill, P. B. Malone, and M. L. Fiorotto. High-efficiency growth hormone-releasing hormone plasmid vector administration into skeletal muscle mediated by electroporation in pigs, *FASEB J.*, 17:526–528 (2003).

100. T. E. Tjelle, A. Corthay, E. Lunde, I. Sandlie, T. E. Michaelsen, I. Mathiesen, and B. Bogen. Monoclonal antibodies produced by muscle after plasmid injection and elec-troporation, *Mol. Ther.*, 9:328–336 (2004).

101. K. E. Dolter, C. F. Evans, and D. Hannaman. *In vivo* delivery of nucleic acid-based agents with electroporation, *Drug Del. Technol.*, 10:37–41 (2010).

102. G. Widera, M. Austin, D. Rabussay, C. Goldbeck, S. W. Barnett, M. Chen, L. Leung, G. R. Otten, K. Thudium, M. J. Selby, and J. B. Ulmer. Increased DNA vaccine deliv-ery and immunogenicity by electroporation *in vivo*, *J. Immunol.*, 164:4635–4640 (2000).

103. R. Heller, J. Schultz, M. L. Lucas, M. J. Jaroszeski, L. C. Heller, R. A. Gilbert, K. Moelling, and C. Nicolau. Intradermal delivery of interleukin-12 plasmid DNA by *in vivo* electroporation, *DNA Cell Biol.*, 20:21–26 (2001).

104. A. V. Titomirov, S. Sukharev, and E. Kistanova. *In vivo* electroporation and stable trans-formation of skin cells of newborn mice by plasmid DNA, *Biochim. Biophys. Acta*, 1088:131–134 (1991).

105. J. W. Hooper, J. W. Golden, A. M. Ferro, and A. D. King. Smallpox DNA vaccine deliv-ered by novel skin electroporation device protects mice against intranasal poxvirus chal-lenge, *Vaccine*, 25:1814–1823 (2007).

106. S. Buchan, E. Gronevik, I. Mathiesen, C. A. King, F. K. Stevenson, and J. Rice. Electroporation as a "prime/boost" strategy for naked DNA vaccination against a tumor antigen, *J. Immunol.*, 174:6292–6298 (2005).

107. R. Draghia-Akli, A. S. Khan, M. A. Pope, and P. A. Brown. Innovative electroporation for therapeutic and vaccination applications, *Gene Ther. Mol. Biol.*, 9:329–338 (2005).

108. C. Curcio, A. S. Khan, A. Amici, M. Spadaro, E. Quaglino, F. Cavallo, G. Forni, and R. Draghia-Akli. DNA immunization using constant-current electroporation affords long-term protection from autochthonous mammary carcinomas in cancer-prone transgenic mice, *Cancer Gene Ther.*, 15:108–114 (2008).

109. R. Draghia-Akli, A. S. Khan, P. A. Brown, M. A. Pope, L. Wu, L. Hirao, and D. B. Weiner. Parameters for DNA vaccination using adaptive constant-current electropora-tion in mouse and pig models, *Vaccine*, 26:5230–5237 (2008).

110. Z. Li, H. Zhang, X. Fan, Y. Zhang, J. Huang, Q. Liu, T. E. Tjelle, I. Mathiesen, R. Kjeken, and S. Xiong. DNA electroporation prime and protein boost strategy enhances humoral immunity of tuberculosis DNA vaccines in mice and non-human primates, *Vaccine*, 24:4565–4568 (2006).

111. S. Capone, I. Zampaglione, A. Vitelli, M. Pezzanera, L. Kierstead, J. Burns, L. Ruggeri, M. Arcuri, M. Cappelletti, A. Meola, B. B. Ercole, R. Tafi, C. Santini, A. Luzzago, T. M. Fu, S. Colloca, G. Ciliberto, R. Cortese, A. Nicosia, E. Fattori, and A. Folgori. Modulation of the immune response induced by gene electrotransfer of a hepatitis C virus DNA vaccine in nonhuman primates, *J. Immunol.*, 177:7462–7471 (2006).

112. A. Luckay, M. K. Sidhu, R. Kjeken, S. Megati, S. Y. Chong, V. Roopchand, D. Garcia-Hand, R. Abdullah, R. Braun, D. C. Montefiori, M. Rosati, B. K. Felber, G. N. Pavlakis, I. Mathiesen, Z. R. Israel, J. H. Eldridge, and M. A. Egan. Effect of plasmid DNA vaccine design and *in vivo* electroporation on the resulting vaccine-specific immune responses in rhesus macaques, *J. Virol.*, 81:5257–5269 (2007).

113. A. D. Cristillo, D. Weiss, L. Hudacik, S. Restrepo, L. Galmin, J. Suschak, R. Draghia-Akli, P. Markham, and R. Pal. Persistent antibody and T cell responses induced by HIV-1 DNA vaccine delivered by electroporation, *Biochem. Biophys. Res. Commun.*, 366:29–35 (2008).

114. M. Rosati, A. Valentin, R. Jalah, V. Patel, A. von Gegerfelt, C. Bergamaschi, C. Alicea, D. Weiss, J. Treece, R. Pal, P. D. Markham, E. T. Marques, J. T. August, A. Khan, R. Draghia-Akli, B. K. Felber, and G. N. Pavlakis. Increased immune responses in rhesus macaques by DNA vaccination combined with electroporation, *Vaccine*, 26:5223–5229 (2008).

115. L. A. Hirao, L. Wu, A. S. Khan, A. Satishchandran, R. Draghia-Akli, and D. B. Weiner. Intradermal/subcutaneous immunization by electroporation improves plasmid vaccine delivery and potency in pigs and rhesus macaques, *Vaccine*, 26:440–448 (2008).

116. S. Babiuk, M. E. Baca-Estrada, M. Foldvari, M. Storms, D. Rabussay, G. Widera, and L. A. Babiuk. Electroporation improves the efficacy of DNA vaccines in large animals, *Vaccine*, 20:3399–3408 (2002).

117. L. Zhang, E. Nolan, S. Kreitschitz, and D. P. Rabussay. Enhanced delivery of naked DNA to the skin by non-invasive *in vivo* electroporation, *Biochim. Biophys. Acta*, 1572:1–9 (2002).

118. N. A. Charoo, Z. Rahman, M. A. Repka, and S. N. Murthy. Electroporation: An avenue for transdermal drug delivery, *Curr. Drug Deliv.*, 7:125–136 (2010).

119. L. Zhang, L. N. Li, G. A. Hoffmann, and R. M. Hoffman. Depth-targeted efficient gene delivery and expression in the skin by pulsed electric fields: An approach to gene therapy of skin aging and other diseases, *Biochem. Biophys. Res. Commun.*, 220:633–636 (1996).
120. A. H. Trainer and M. Y. Alexander. Gene delivery to the epidermis, *Hum. Mol. Genet.*, 6:1761–1767 (1997).
121. S. Kanjilal, S. S. Strom, G. L. Clayman, R. S. Weber, A. K. Elnaggar, V. Kapur, K. K. Cummings, L. A. Hill, M. R. Spitz, M. L. Kripke, and H. N. Ananthaswamy. p53 mutations in nonmelanoma skin cancer of the head and neck: Molecular evidence for field cancerization, *Cancer Res.*, 55:3604–3609 (1995).
122. R. S. Camplejohn. DNA damage and repair in melanoma and non-melanoma skin cancer, *Cancer Surv.*, 26:193–206 (1996).
123. X. S. Qin, S. M. Zhang, K. Oda, Y. Nakatsuru, S. Shimizu, Y. Yamazaki, O. Nikaido, and T. Ishikawa. Quantitative detection of ultraviolet light-induced photoproducts in mouse skin by immunohistochemistry, *Jpn. J. Cancer Res.*, 86:1041–1048 (1995).
124. P. V. Bennett, R. W. Gange, H. Hacham, V. S. Hejmadi, M. Moran, S. Ray, and B. M. Sutherland. Isolation of high-molecular-length DNA from human skin, *Biotechniques*, 21:458–461 (1996).
125. D. A. Treco and R. F. Selden. Non-viral gene therapy, *Mol. Med. Today*, 1:314–321 (1995).
126. K. Slavikova and E. Massouridou. DNA-mediated gene transfer into mammalian cells and cancer, *Neoplasma*, 42:293–297 (1995).
127. M. Y. Alexander and R. J. Akhurst. Liposome-mediated gene transfer and expression via the skin, *Hum. Mol. Genet.*, 4:2279–2285 (1995).
128. O. Zelphati and F. C. Szoka. Liposomes as a carrier for intracellular delivery of antisense oligonucleotides: A real or magic bullet? *J. Control. Release*, 41:99–119 (1996).
129. D. D. Lasic and N. S. Templeton. Liposomes in gene therapy, *Adv. Drug Del. Rev.*, 20:221–266 (1996).
130. C. J. Wheeler, P. L. Felgner, Y. J. Tsai, J. Marshall, L. Sukhu, S. G. Doh, J. Hartikka, J. Nietupski, M. Manthorpe, M. Nichols, M. Plewe, X. W. Liang, J. Norman, A. Smith, and S. H. Cheng. A novel cationic lipid greatly enhances plasmid DNA delivery and expression in mouse lung, *Proc. Natl. Acad. Sci. USA*, 93:11454–11459 (1996).
131. M. B. Bally, Y. P. Zhang, F. M. P. Wong, S. Kong, E. Wasan, and D. L. Reimer. Lipid/DNA complexes as an intermediate in the preparation of particles for gene transfer: An alternative to cationic liposome/DNA aggregates, *Adv. Drug Del. Rev.*, 24:275–290 (1997).
132. R. I. Mahato, A. Rolland, and E. Tomlinson. Cationic lipid-based gene delivery systems: Pharmaceutical perspectives, *Pharm. Res.*, 14:853–859 (1997).
133. J. Y. Cherng, P. van de Wetering, H. Talsma, D. J. A. Crommelin, and W. E. Hennink. Effect of size and serum proteins on transfection efficiency of poly((2-dimethylamino) ethyl methacrylate)-plasmid nanoparticles, *Pharm. Res.*, 13:1038–1042 (1996).
134. L. Zhang, L. N. Li, Z. L. An, R. M. Hoffman, and G. A. Hofmann. *In vivo* transdermal delivery of large molecules by pressure-mediated electroincorporation and electroporation: A novel method for drug and gene delivery, *Bioelectrochem. Bioenerg.*, 42:283–292 (1997).
135. J. A. Mikszta, J. B. Alarcon, J. M. Brittingham, D. E. Sutter, R. J. Pettis, and N. G. Harvey. Improved genetic immunization via micromechanical disruption of skin-barrier function and targeted epidermal delivery, *Nat. Med.*, 8:415–419 (2002).
136. A. J. Hudson, W. Lee, J. Porter, J. Akhtar, R. Duncan, and S. Akhtar. Stability of antisense oligonucleotides during incubation with a mixture of isolated lysosomal enzymes, *Int. J. Pharm.*, 133:257–263 (1996).

137. K. J. Lewis, W. J. Irwin, and S. Akhtar. Biodegradable poly(L-lactic acid) matrices for the sustained delivery of antisense oligonucleotides, *J. Control. Release*, 37:173–183 (1995).
138. S. T. Crooke. Delivery of oligonucleotides and polynucleotides, *J. Drug Target.*, 3:185–190 (1995).
139. Y. S. Jong, J. S. Jacob, K. P. Yip, G. Gardner, E. Seitelman, M. Whitney, S. Montgomery, and E. Mathiowitz. Controlled release of plasmid DNA, *J. Cont. Rel.*, 47:123–134 (1997).
140. S. T. Kim, K. M. Lee, H. J. Park, S. E. Jin, W. S. Ahn, and C. K. Kim. Topical delivery of interleukin-13 antisense oligonucleotides with cationic elastic liposome for the treatment of atopic dermatitis, *J. Gene Med.*, 11:26–37 (2009).
141. J. Inoue, S. Yotsumoto, T. Sakamoto, S. Tsuchiya, and Y. Aramaki. Changes in immune responses to antigen applied to tape-stripped skin with CpG-oligodeoxynucleotide in mice, *J. Control. Release*, 108:294–305 (2005).
142. W. R. Lee, S. C. Shen, C. R. Liu, C. L. Fang, C. H. Hu, and J. Y. Fang. Erbium:YAG laser-mediated oligonucleotide and DNA delivery via the skin: An animal study, *J. Control. Release*, 115:344–353 (2006).
143. I. I. Hashim, K. Motoyama, A. E. Abd-Elgawad, M. H. El Shabouri, T. M. Borg, and H. Arima. Potential use of iontophoresis for transdermal delivery of NF-kappaB decoy oligonucleotides, *Int. J. Pharm.*, 393:127–134 (2010).
144. R. van der Geest, F. Hueber, F. C. Szoka, and R. H. Guy. Iontophoresis of bases, nucleosides, and nucleotides, *Pharm. Res.*, 13:553–558 (1996).
145. K. R. Oldenburg, K. T. Vo, G. A. Smith, and H. E. Selick. Iontophoretic delivery of oligonucleotides across full thickness hairless mouse skin, *J. Pharm. Sci.*, 84:915–921 (1995).
146. R. M. Brand and P. L. Iversen. Iontophoretic delivery of a telomeric oligonucleotide, *Pharm. Res.*, 13:851–854 (1996).
147. K. Kigasawa, K. Kajimoto, S. Hama, A. Saito, K. Kanamura, and K. Kogure. Noninvasive delivery of siRNA into the epidermis by iontophoresis using an atopic dermatitis-like model rat, *Int. J. Pharm.*, 383:157–160 (2010).
148. R. Bergan, F. Hakim, G. N. Schwartz, E. Kyle, R. Cepada, J. M. Szabo, D. Fowler, R. Gress, and L. Neckers. Electroporation of synthetic oligodeoxynucleotides: A novel technique for *ex vivo* bone marrow purging, *Blood*, 88:731–741 (1996).
149. R. Bergan, Y. Connell, B. Fahmy, and L. Neckers. Electroporation enhances c-myc antisense oligodeoxynucleotide efficacy, *Nucleic Acids Res.*, 21:3567–3573 (1993).
150. T. E. Zewert, U. F. Pliquett, R. Langer, and J. C. Weaver. Transdermal transport of DNA antisense oligonucleotides by electroporation, *Biochem. Biophys. Res. Commun.*, 212:286–292 (1995).
151. V. Regnier, A. Tahiri, N. Andre, M. Lamaitre, V. Preat, and T. L. Doan. Delivery of 3′-protected phosphodiester oligodeoxynucleotides to the skin, *Pharm. Res.*, 14:S–640 (1997).
152. Y. M. Lee, K. Song, S. H. Lee, G. I. Ko, J. B. Kim, and D. H. Sohn. Percutaneous absorption of antisense phosphorothioate oligonucleotide *in vitro*, *Arch. Pharm. Res.*, 19:116–121 (1996).
153. H. W. Nolen, P. Catz, and D. R. Friend. Percutaneous penetration of methyl phosphonate antisense oligonucleotides, *Int. J. Pharm.*, 107:169–177 (1994).

10 Commercial Development of Devices and Products for Transdermal Physical Enhancement Technologies

10.1 INTRODUCTION

The drug delivery market is estimated to be about $80 billion, and one driving factor for its growth is the need for innovative delivery systems for biopharmaceuticals that typically cannot be administered by oral administration. They could be administered by the parenteral route, but repeated injections are often required due to their short half-lives. Noninvasive delivery systems are therefore highly desired. Furthermore, biopharmaceuticals may constitute about 25% of the total pharmaceutical market; therefore, interest in technologies that might enable noninvasive or minimally invasive delivery of biopharmaceuticals is on the rise. Pulmonary administration was one such promising system, but interest has diminished following the withdrawal of inhalable insulin (Exubera®) from the market. Transdermal delivery would be a very attractive route of administration, but most biopharmaceuticals are hydrophilic macromolecules that do not normally enter the skin (1). Physical enhancement techniques discussed in this book, in particular the microporation technologies, can enable the transdermal delivery of these macromolecules, irrespective of their size. These macromolecules include large proteins and vaccines. As discussed in Chapters 3 and 9, microneedles offer significant advantages for vaccine delivery. Therefore, microporation-based technologies may capture a good portion of the drug delivery market and perhaps, more importantly, they may capture some portion of the global vaccine market, which is around $10 billion. They may become an important class of the currently estimated $15 billion worldwide transdermal market. Furthermore, the global dermatologicals market is estimated to be around $18 billion (2), with sales of Duragesic® (fentanyl) patch alone exceeding $1 billion. Over $150 billion worth of brand-name drugs will come off patent in the next 5 years, creating significant opportunities for generic drug delivery systems, including transdermals. The focus of this chapter is on the commercial development of microneedles and iontophoresis, as these two technologies have perhaps seen the highest activity for commercial development efforts. Commercial development of sonophoresis is limited to a very small number of players and is discussed in Chapter 6. Electroporation is a promising method for delivery of DNA vaccines, and this was discussed in Chapter 5.

In recent years, it has become more difficult to get U.S. Food and Drug Administration (FDA) approval for new drug molecules, and pharmaceutical companies will look at drug delivery systems to extend the life of patents. Therefore, the future of these technologies looks very promising, especially after the world recovers from its current economic downturn. There should be a clear therapeutic rationale for the selection of the drug candidate for transdermal delivery. The drug must be potent, with dosage preferably being only a few milligrams per day, as these can be easily delivered. The drug should also have a low or zero skin irritation or sensitization potential. Local irritation and potential to induce allergic response have been implicated as the factors that were often discovered late in the development path for transdermal patches and caused failure and high financial loss (3). The drug must also have the desired physicochemical properties to qualify as a viable candidate for delivery by the physical method of enhancement being considered. The cost effectiveness of the delivery system will depend on several factors including the cost of the drug. In addition to scientific/technical considerations, commercial considerations include patent rights, collaboration agreements, and details of milestone and royalty payments, as well as other details such as who will be the first to market and so on. Two products based on microporation technologies have already been approved and several more are in preclinical and clinical development. Similarly, many iontophoresis devices are already on the market. For success of these technologies, it will be ideal if drug molecules with significant unmet therapeutic need are first identified and pursued for commercialization diligently through the various phases of clinical trials (4). For a small company, this, of course, gets more difficult unless a suitable bigger partner can be identified to take over or share the costs of more expensive Phase III clinical trials. For example, Zosano's ZP-PTH patch (Zosano Pharma, Fremont, California) has completed Phase II clinical trials and will enter Phase III clinical trials once a partner is identified (see Section 10.5).

10.2   WHAT CAN BE LEARNED FROM THE PASSIVE PATCHES ON THE MARKET?

As discussed in Chapter 1, over 15 different drug molecules in various brands are already on the market as passive patches. There is significant literature available on the design and manufacture of these passive patches, which is beyond the scope of this book but should nevertheless be read and understood, as it can be beneficial while designing an enhancement technique–based delivery system. However, there are also important differences with respect to passive versus active delivery of the drug into or through the skin which need understanding. Passive patches may be of reservoir or matrix type, though the recent trend has been toward a simple drug in an adhesive (DIA) patch. A DIA patch typically consists of a drug-loaded pressure-sensitive adhesive sandwiched between a backing layer and a release liner. Pressure-sensitive adhesives (PSAs) are typically acrylics, silicones, or polyisobutylene (PIB). Noven

Pharmaceuticals (Miami, Florida) uses a mix of acrylic and silicone adhesives in their DOT matrix technology, where the drug is dissolved in the acrylic mix while silicone allows adhesion to the skin. This allows more drug to be loaded onto the adhesive without losing its stickiness and enables development of smaller patches. The patches are typically tested for thickness, weight variation, drug content, moisture content, moisture uptake, peel adhesion, tack, flatness, and *in vitro* release (dissolution) studies (5,6). Besides these, design, usage, and safety concerns associated with the passive patches will have to be addressed for "active" patches as well, which are based on the enhancement technologies discussed in this book.

10.3   SAFETY ISSUES AND EFFECT OF HEAT

Safety and risk issues associated with patches also need to be considered. All patches should be applied to intact skin, as application of a patch to cut or irritated areas may lead to absorption of too much drug into the body. Mild erythema and irritation under the site of application are usually considered acceptable, and the application site is rotated to let the skin recover. It should be realized that skin is not a biologically inert membrane even though much of the transdermal literature treats it like one. Skin will respond to any damage to its barrier by initiating a series of biochemical events designed to repair the damage, and the resulting level of irritation is reflective of the extent of perturbation caused by the enhancement mechanism (7). In addition, biophysical considerations matter in safety issues. Authors from the FDA presented a risk analysis of transdermal drug delivery systems with emphasis on adequate adhesion of the patch to the skin (8,9). Failure of adhesive can lead to improper dosing of patient due to decreased patch/skin contact area. Poor adhesion can also lead to accidents resulting from fallen patches being picked up by children or being accidentally transferred to them by hugging and children sitting or lying on fallen patches (8). Accidental transfer can also happen with gels (e.g., with testosterone gel). Some patches have aluminum or other metal in the backing layer which can overheat and cause skin burns during magnetic resonance imaging (MRI) scans due to the generation of electric fields under the strong magnets used in MRI machines.

In addition to possible failure of the adhesive, another factor of increasing concern is the dramatic effect of heat on percutaneous absorption, attributed to increased blood circulation which can draw on any drug depot in the skin or be attributed to increased drug solubility (and concentration gradient) in the patch and increased kinetic energy of drug molecules (9). Exposure to heat can occur in everyday situations like physical exercise, hot climate, steam bath, saunas, or use of warming blankets and can cause increased absorption of drugs (10). Reports have been published to show the effect of heat on increased delivery of model penetrants (11) and on testosterone (12) and fentanyl (13) from transdermal patches. Over the past few years, many deaths have resulted from fentanyl patches due to increased delivery or abuse (14,15). A suicide attempt using nicotine patches was also reported (16). Leakage in case of reservoir patches can also result in safety issues. Increased temperature and resulting increased flux can also be taken advantage of to deliver drugs through skin if heat can be applied in a controlled manner. Skin surface temperatures up to around 42°C can be tolerated for periods up to 24 hours (2). Zars (Salt Lake City, Utah) is working with controlled heat assisted delivery (CHADD) for transdermal delivery. Zars CHADD technology has been commercialized for a topical patch (Synera®) recently launched by Endo Pharmaceuticals (Newark, Delaware). The lidocaine/tetracaine patch is approved for prevention of pain associated with superficial venous access and superficial dermatological procedures. The patch contains a formulation integrated with an oxygen-activated heating component. The company also designed a CHADD unit that can be placed on top of the patch rather than incorporating into the patch as in Synera (17). Use of short pulses of very high temperature to cause thermal ablation of the skin is a somewhat different phenomena and is discussed in Section 3.3.1.

Safety issues with microporation technologies are still in the process of being addressed, but issues related to perception of pain and some other considerations are discussed in Chapter 3. High-voltage electroporation pulses are generally too intense for drug delivery applications but are finding applications for electrochemotherapy (see Chapter 5) and for delivery of DNA vaccines (see Chapter 9). Skin damage caused by high-frequency sonophoresis is discussed in Chapter 6. In the remaining part of this section, some safety issues specifically related to iontophoretic delivery will be discussed. Because iontophoretic transport takes place via an appendageal pathway, immunology of the hair follicle may be important to understand (18). Several isolated literature reports exist on the adverse effects of application of current, but it should be realized that many of these early studies were not well controlled. In these studies, the electrochemical changes at electrodes were not controlled and might have led to pH shifts, or the pH of the drug formulation might have been low. In general, transient erythema under the skin is common and would be considered normal. In human studies done by the author in collaboration with coworkers (see Chapter 4), an erythema under the electrodes was noted which resolved within a few hours. A mild tingling sensation was reported as the current was being ramped up (19). Another report also describes a mild tingling sensation in subjects as the current was ramped up to 0.25 mA/cm^2, with more tingling perceived under the anode. The sensation diminished with time of current application and disappeared in less than 30 minutes. Erythema observed under the electrodes resolved within 10 to 60 minutes of turning the current off. On occasion, a few miniscule punctate lesions were observed which persisted for several days. Nevertheless, these side effects were similar to those observed with conventional transdermal drug delivery systems (20). The erythema associated with iontophoresis is most probably a result of nonspecific irritation such as from an irritant drug or due to microscopic cellular damage at sites of high current density. This microscopic damage would release cytokine or prostaglandin, resulting in local vasodilatation. Direct electrical stimulation of C-fibers, a specific class of nociceptors, may also be responsible for sensations of tingling and pricking and development of erythema. C-fibers contain substance P and calcitonin gene-related peptide. Because the latter is a potent vasodilator causes a localized erythema lasting over several hours. Substance P is involved in histamine-dependent wheal and flare responses (21). In a comprehensive study on 30 pigs *in vivo* and 112 IPPSFs, alterations in skin were seen following iontophoretic delivery of lidocaine hydrochloride. The change was characterized by light microscopy as the appearance of dark basophilic staining nuclei oriented parallel to the stratum corneum in the stratum granulosum and spinosum layers. However, the dose-dependent non-immune-mediated epidermal alteration was considered to have minimum toxicological significance (22). Probably the worst electric burn would result if the electrode metal was allowed to touch the skin directly, as this would produce very high localized current density. It should be realized that any metal under a properly placed electrode pad will also cause shunting of current and can lead to a burn. Electric burns have been reported at the site of contact with a metal ring worn during iontophoresis (23). Even if the current density is uniformly applied to the skin, it should be realized that significant current density may flow through appendageal pathways, as these are the primary pathways of transport. For the use of pilocarpine iontophoresis in a clinical setting for collection of sweat, it was suggested that patients should be informed that the procedure carries a small risk of minor burns (24).

The electrical resistance of skin decreases with time during iontophoresis, the effect being larger at higher current densities. This decreased skin resistance often reflects increased permeability of the skin due to changes caused by current flow. The postiontophoresis passive permeability is thus generally higher than the passive permeability before iontophoresis. In one study, the passive flux measured after iontophoresis was about a factor of 10 greater than the corresponding flux measured before the skin was exposed to electric current (25). Similarly, the passive permeability of water and mannitol after 10 hours of iontophoresis on hairless mouse skin was, respectively, 6 and 30 times greater than pretreatment values (26). The permeability increase factor may be evaluated by electrical conductivity (resistance) measurements. The time-dependent electrical resistance of porcine skin was found to initially drop quickly and then level off (27). Continuous membrane alteration has been shown to occur during *in vitro* iontophoresis of human epidermal membranes at high applied voltage drops (1000 mV) but does not occur when 250 mV is applied across the membrane. These changes were found to apparently reverse over time once the voltage drop was removed (28). A study investigated the effects of electric current on the fine structure of human stratum corneum lipids. The stratum corneum was exposed to pulsed constant current (0.013 to 13 mA/cm^2) for 1 hour and then rapidly frozen and processed for freeze-fracture electron microscopy or subjected to X-ray diffraction analysis. A disordering of intercellular lipids was observed and was attributed to mutual repulsion following polarization of the lipid head group induced by the electric field. This change in the ordering of the intercellular lipid lamellae is responsible for the observed changes in skin resistance, because both the resistance and capacitance properties of the skin are determined by the intercellular lipid lamellae. Other possible mechanisms involved include interactions of lipids with water or ions and displacement of structurally important ions such as Ca^{++} (29).

Safety features can be easily built into the circuitry of iontophoresis devices, and several patents have been issued in this area (30). In order to prevent exposure of patient to excessive current, an intermediary storage device can be used that can use energy from the power supply and later transfer a predetermined safe level to the patient's skin to guard against failure of any component of the circuit (31). Some patients may be sensitive to electricity, most likely due to a psychosomatic illness (32), and electrically assisted delivery of drugs may be excluded in these individuals. Although allergic contact dermatitis is rare in clinical applications of iontophoresis, it could develop, as all drugs can potentially be allergenic and the iontophoresis patch has a relatively complex construction (21). A case report describes a cutaneous allergic reaction to iontophoresis of 5-fluorouracil which was reactivated at a distant site upon a subsequent treatment with iontophoresis (33). Another potential concern may be dose dumping. However, solid-state circuits are quite reliable and can be set to avoid the discharge of excessive current during a rare but possible event of circuit failure. Similarly, if the current stops for some reason, means to alert the patient can be built into the device. In the event of an adverse reaction, the device can just be removed because the remaining dose resides in the patch outside the body (34).

10.4 SKIN IRRITATION TESTING

All transdermal products need to be tested for irritation during the development process. Much of the animal testing is based on a method developed by John Draize in the mid 1940s using rabbits (35). Dermal irritation studies are initially done in rabbits, and these studies are usually required to initiate Phase I studies in humans. If only transient erythema is observed, then human studies may be warranted. If more severe skin reactions are seen, the rabbit study will have to be repeated with reduced dosage of the drug. For example, Park et al. tested a buprenorphine patch for skin irritation after single or repeat application to backs of rabbits 24 hours after hair was clipped with an electric clipper. At 24 and 72 hours after application, test sites were scored for erythema and edema on a Draize dermal scoring criteria on a scale of 0 to 8 (36). Sensitization studies are also carried out in guinea pigs and involve an induction and challenge phase. Sensitization or "allergic contact dermatitis" involves an immunological response specific for the drug being used. The challenge phase occurs after a subsequent encounter of the sensitized person with the inducing substance.

Human skin equivalent tests and other nonanimal testing methods have also been widely used for determining the potential of drugs/chemicals to cause skin irritation or corrosion, especially for cosmeceuticals (35,37). EpiDerm™ (MatTek Corporation, Ashland, Massachusetts), for example, is a three-dimensional *in vitro* skin model tissue derived from human cells and is composed of *in vitro* cultures of human keratinocytes. It is metabolically active and mimics human epidermis, structurally as well as biochemically. Two testing protocols are available—one for irritation potency assessment and one for hazard identification. Skin irritation potential of the chemical/drug being tested is predicted based on whether the test material is able to penetrate to the underlying cell layers to affect cell viability below or above a defined threshold. Viability of epidermal cells can be measured by the release of interleukin-1-alpha and using the methyl thiazol tetrazolium (MTT) uptake assay (38). Similarly, Episkin™ (SkinEthic Laboratories, Lyon, France) has been widely used for skin irritation testing. It is obtained by culturing human keratinocytes on a collagen base under conditions that allow the differentiation of keratinocytes to reconstruct an epidermis with a functional stratum corneum. Guidelines for human skin equivalent tests are available from the Organization for Economic Cooperation and Development (OECD) and have been validated by the European Centre for the Validation of Alternative Methods (ECVAM). Instrumental tools like Chromameter and VapoMeter which can be used to monitor skin irritation are discussed in Section 2.5.

In 1999, the FDA issued a draft guidance document for skin irritation and sensitization testing of generic transdermal drug products in human subjects. The document was not finalized or revised but is still used by most as a guide for irritation testing when submitting abbreviated new drug applications for generic patches. Irritation studies involve comparison of the test patch with the innovator patch and the placebo control in 30 subjects. Patches are applied for 24 hours each day for 21 days to the same skin site, and the site is evaluated following each patch removal by a blinded trained observer. Dermal reactions are scored, typically on a 0 to 7 scale, though many companies use a 0 to 4 scale. At the extreme score, skin will have erythema and edema along with vesicles and bullae. Additionally, adhesion of the patch to the skin can be assessed using an adhesion score from 0 to 4, where a score of 0 signifies no lift off the skin and a score of 4 signifies a detached patch. Skin sensitization (allergic contact dermatitis) can also be tested using a modified Draize test which is carried out in 200 subjects over a period of 6 weeks. The induction phase involves nine applications of the patch for 48 to 72 hours each, over 3 weeks. After a rest phase of 2 weeks, patches are again applied to new skin sites for 48 hours during the challenge phase, and skin reactions are evaluated at 30 minutes and at 24, 48, and 72 hours after patch removal. Examples of such testing can be found in the literature (e.g., an acute and cumulative irritation patch test in 151 volunteers was described) (39).

Skin irritation testing is also required when physical enhancement technologies are employed for transdermal delivery. Bioengineering techniques (see Section 2.5) such as transepidermal water loss (TEWL), electrical capacitance moist determination (ECM), laser-Doppler flowmetry (LDF) or impedance spectroscopy (IS) can be used to assess changes in the skin during skin microporation or during application of iontophoresis currents or electroporation pulses. We have shown that TEWL values increase by more than 150% following treatment of skin by microneedles (40). Passage of an iontophoresis current through human skin *in vivo* causes a significant reduction in the magnitude of skin impedance, and it has been suggested that impedance spectroscopy can complement TEWL as a technique for measuring *in vivo* skin permeability (41). When the stratum corneum is physically damaged, both TEWL and IS are well correlated (42). Impedance spectroscopy, a noninvasive biophysical technique, can be used to study damage induced by chemical enhancers as well. IS has shown that enhancers increase the resistance of the skin because they open new penetration routes and increase ohmic resistance and capacitive properties, suggesting a rougher or more heterogeneous surface for excised human abdominal skin (43). This technique has also been used to investigate the extent to which the effect of iontophoresis on skin can be modulated by pretreatment with penetration enhancers (44). A study using Fourier-transform infrared (FTIR) and differential scanning calorimetry (DSC) techniques has suggested that electrical treatment of the skin is less damaging than treatment of the skin with chemical penetration enhancers (45). The postiontophoresis recovery of hairless mouse skin following a 2-hour exposure to 0.5 mA/cm^2 current has been investigated using impedance spectroscopy and laser scanning confocal microscopic images of permeability of calcein through preiontophoresed skin. The time for the recovery of skin impedance and permeability characteristics to preiontophoresis levels was found to be 18 to 24 hours (46). In a study with percutaneous absorption of leuprolide, it was observed that a combination of a high level of ethanol and chemical enhancer was the main cause of skin irritation. In this study, rabbits were shaved at an abdominal site by a clipper, and the skin was then allowed to recover for 24 hours before applying the formulation via Hill Top Chamber patches (Hill Top Research, Cincinnati, Ohio) (47). Using LDF, it has been shown that novamide, a synthetic analog of capsaicin, is much less irritating to humans than capsaicin upon topical application (48). The LDF technique was also used to assess the increased blood flow following iontophoresis of histamine in humans. When the flare disappeared, the level of LDF at the site of wheal formation was still higher than the basal value (49). Any irritation induced in humans can also be quantitatively and noninvasively measured based on changes in electrical impedance. The values obtained with impedance measurements seem to agree with those obtained using TEWL and visual readings (50,51). In a study with nine volunteers, the effect of iontophoresis on the stratum corneum barrier function and on development of erythema was measured using TEWL and LDF, respectively. Current was applied on the volar forearms of the subjects for 30 minutes. TEWL values measured after removal of patch were about twofold higher than the baseline value irrespective of the presence or absence of current. This suggests that iontophoresis had a negligible and indistinguishable effect on skin barrier as compared to the effect attributed to the occlusion. The erythematous response, however, differed from the controls and was 1.5 to 2.7 times higher at the anode than at the cathode and was considered moderate (52). In addition to TEWL, other techniques such as capacitance, skin temperature, skin color, and a visual scoring system have been used to assess any damage to skin barrier function following iontophoresis. The first stage of a multistage study investigated the effect of 10 minutes of iontophoresis to validate the experimental methodology. The long-term goal of the study was to assess the safety of 24 hours of saline iontophoresis. Using 36 volunteers, current was applied at 0.1 mA/ cm^2 on a 1-cm^2 patch or 0.2 mA/cm^2 on a 6.5 cm^2 patch, and results were compared to a control. No damage to skin barrier function was seen but a slight, subclinical short-lasting skin erythema was observed (53).

Another study used LDF and TEWL techniques to study the safety of iontophoresis using six healthy female volunteers. A current of 0.1 or 0.2 mA/cm^2 was applied for 30 minutes. It was observed that iontophoresis increased cutaneous blood flow, with a more pronounced effect at higher current density. However, the increase was reversible within 1 hour. TEWL values were not enhanced compared to control (no iontophoresis) except for a small Joule heating effect at higher current density. The lipid structure of the skin was not affected, and it was concluded that the drug delivery by iontophoresis should be safe (54). In this study, the TEWL and LDF measurements were the same at the anode and cathode sites. In contrast, another study with 26 healthy subjects suggests that the erythema was higher at the cathode as measured by LDF (55). Electrode details for this study are not provided, and pH changes could at least partly be responsible for the observed effects. Skin impedance has been measured *in vivo* in man for direct electrical evaluation of the effects of iontophoresis on tissue. At a current density of 0.5 mA/cm^2 applied for just a few minutes, skin resistance dropped within seconds to essentially that of the electrolyte contained in the electrode chambers. Recovery of skin resistance was slow, with only

35% of the pretreatment value recovering 4 hours after the termination of current. These decreases are much faster than those typically seen with *in vitro* studies. The reasons for this discrepancy are not clear but may be related to damage to current conducting pathways when skin is removed for *in vitro* studies, such as due to possible occlusion of sweat glands. For an equivalent current dose, short applications of high current were found to be more damaging to skin as compared to longer applications at lower current density (56). The degree of erythema of skin can be quantitated by a Chromameter, as described in the literature (57). X-ray diffraction technique has also been used and has shown that intercellular lipids are extensively reorganized by iontophoresis (58).

10.5 COMMERCIAL DEVELOPMENT OF MICROPORATION DEVICES

Microporation technologies such as microneedles, thermal or radio-frequency ablation, and laser ablation were discussed in detail in Chapter 3. In this chapter, commercial development of this technology will be discussed, with emphasis on the challenges and products in development.

10.5.1 CHALLENGES FOR COMMERCIAL DEVELOPMENT OF A MICRONEEDLE-BASED PATCH

As discussed in Chapter 3, some of the potential hurdles to overcome when developing a safe and effective product that can receive regulatory approval include considerations of skin elasticity and applicators, bioavailability, pore closure kinetics, sterilization, and immunogenicity. Fabrication methods that can be scaled up to produce inexpensive and biocompatible microneedles are also important. Because these microneedles will be for single usage only, methods of fabrication become important, and molding methods generally tend to be more economical (59). A detailed discussion of the microporation technologies and some of the challenges can be found in Chapter 3. Any variations in delivery from site to site or patient to patient will need to be investigated. These include any possible differences relating to site of administration due to structural or enzymatic factors. Other potential variables to be investigated include gender, age, and race. Microneedles in a patch must penetrate skin in a uniform manner up to a uniform depth and must not break off in the skin. The patch will most likely have many similar considerations as passive patches (e.g., to be applied at a non-hairy site on unbroken skin). Any potential immunogenicity resulting from delivery via skin will also need to be investigated, though currently there are no such reports of increased immunogenicity resulting solely due to administration via the skin route. In clinical studies that administered PTH by coated microneedles, no anti-PTH antibody formation was observed (60). More work needs to be done to gain a better understanding of pore closure following skin microporation (see Section 3.2.5). It is also not clear at this time if microneedle-based patches will need to be sterile or will just need to be manufactured under low bioburden specifications and applied after disinfecting the application site with an alcohol swab. Microneedles typically penetrate skin to only shallow depths, but sterilization may still be required. Sterilization of metal microneedles coated with the polypeptide PTH(1-34) has been investigated using gamma or e-beam irradiation. It was found that the effect of irradiation on the long-term stability of the polypeptide made it difficult to achieve the 95% purity specification required at the end of the shelf life. Aseptic assembly of patch components can be considered if it can be demonstrated that terminal sterilization is not feasible (61). Packaging considerations are also important to provide proper protection of microneedles, especially their tips, during the shelf life and later during usage and administration.

Potential transdermal delivery of monoclonal antibodies poses more challenges as, unlike other proteins, these tend to have high dosage requirements. Use of coated microneedles or soluble microneedles with incorporated drug may be difficult, as these approaches will typically work only for sub-milligram quantities. If a monoclonal antibody has some topical/local therapeutic indication for a skin disease, then dosage requirements may be lower, and it will be easier to develop a product. One promising approach may be an intradermal injection using a single hollow microneedle or a patch having an array of hollow microneedles. In a recent study from BD Technologies (Franklin Lakes, New Jersey), Harvey et al. (62) used a 34-gauge needle with a tissue penetration length of 1 mm to investigate *in vivo* delivery of four proteins in the swine model. The proteins included a 5.8 kDa (regular recombinant insulin), a 11.6 kDa (dimeric insulin lispro), a 22 kDa (recombinant human growth hormone), and a 132 kDa (TNFα-binding dimeric fusion protein etanercept) molecule. Intradermal delivery resulted in a unique pharmacokinetic profile characterized by rapid uptake, elevated peak concentrations, and systemic distribution relative to subcutaneous injection. Imaging studies with a near-infrared (NIR) fluorescent dye suggested that these results may at least be partly attributed to rapid lymphatic uptake. The authors reported that a volume of up to 250 μL (20 μL/min) was infused over a few minutes with a protein concentration as high as 80 mg/mL, thus enabling delivery of 20 mg of protein over a few minutes. Generally, when an array of hollow microneedles is used, there are limitations on the volumes that can be infused into skin owing to its densely packed structure. A volume of up to 2 mL was reportedly administered over 6 minutes, painlessly and efficiently, with good skin tolerability, though a bleb (<1 mm) may form in the skin which will resolve in about an hour or so (60). If any of the hollow microneedles does not enter the skin, it will result in leakage of the fluid at that point once the formulation is being infused into the skin under pressure or another driving force. The amount of fluid that can be delivered into the skin can possibly be increased by the use of the enzyme hyaluronidase that can temporarily degrade hyaluronan, which is a major component of skin. Halozyme Therapeutics, Inc. (San Diego, California) makes a recombinant human version of the enzyme (HYLENEX; hyaluronidase human injection) for use as an adjuvant to increase the absorption and dispersion of other injected drugs. Halozyme has partnerships with Roche and Baxter for their Enhanze technology to use the enzyme for specific applications.

Postmarketing questions also need to be anticipated along with development of the product. For example, one such question is, can a microneedle patch be cut in half to achieve half the dosage? The answer to this question involves several considerations.

First, a patch with metal microneedles may not be amenable for cutting simply based on mechanics if the needles are very strong. Similarly, integrity of a patch with an elaborate design will be lost upon cutting. However, assuming that the patch can be cut, a further complication may be that drug flux across microporated skin may not be linearly related to the number of microchannels in the skin. Furthermore, any contribution of passive diffusion through intact skin also needs to be considered. Theoretically, a patch with coated microneedles could be cut, because the amount of drug that will be deposited in the skin is directly related to the number of coated microneedles penetrating the skin. In reality, however, the company may not recommend such an approach due to liability reasons.

10.5.2 MICRONEEDLE-BASED TRANSDERMAL PRODUCTS IN DEVELOPMENT

In recent years, microporation technologies have received increasing attention as it offers several advantages for potential commercialization. This minimally invasive delivery can be accomplished by a simple inexpensive patch that does not need a power supply or advanced microelectronics. The microchannels created in skin do not reach the nerve endings in the dermis and are therefore painless if the depth is controlled carefully. Coated microneedles can potentially be used for mass immunization programs in developing countries, and also for delivery of very potent drug molecules such as PTH. As many of the macromolecules that are potential candidates for microneedle-based delivery tend to be expensive biopharmaceuticals, bioavailability considerations become important. Several companies including 3M, Apogee Technology, Becton Dickinson, Corium (acquired technology from Procter and Gamble), Debiotech, Elegaphy, Imtek, ISSYS, Kumetrix, Nanopass, Norwood Abbey, Silex Microsystems, Theraject, Valeritas (a subsidiary of BioValve), Zeopane, and Zosano (formerly Macroflux, and Alza/J&J spin off) are exploring or have recently explored the use of microneedles for drug delivery. However, not all of these companies have capabilities to scale up the process and give reliable and rapid turnaround times to work with clients for commercialization of this technology. Once a first approval is obtained, the regulatory pathway may be a little easier and cheaper for subsequent approvals. In addition, several organizations are involved in the vaccine development efforts including CDC, Gates and other foundations, NIH, and the PATH organization. Academic groups at Georgia Tech, MIT, Welsh School of Pharmacy, Tyndall National Institute, and several other places are also investigating fabrication of microneedles and their applications to drug delivery. Our transdermal delivery laboratory is also exploring applications of microneedles for drug delivery. Progress is being made on development of microporation devices, and the patent literature provides useful information on device development (63). Products on market and in clinical or preclinical development are discussed below and are listed in Table 10.1.

10.5.2.1 Marketed Products

A tuberculin tine test that uses four to six short needles coated with the tuberculin antigen has been used for a long time. The test is used for medical diagnosis of tuberculosis. A positive test is verified by the Mantoux test (see Section 9.4). A more recent product that has received approval is a device with a single 30-gauge 1.5-mm

TABLE 10.1

Some of the Microporation Products in Development

Microporation Principle	Company	Technology/Products
Microneedles	Apogee Technology	PyraDerm™ solution
	Corium	MicroCor™
	DermaRoller	DermaRoller®
	Elegaphy	Soluble microneedles
	Kumetrix	Silicon microneedle technology
	NanoPass Technologies	MicronJet needle (hollow); MicroPyramid platform (solid)
	Norwood Abbey	Microneedle technology
	TheraJect	TheraJectMAT and VaxMAT (with dissolving microneedles)
	Valeritas	Micro-Trans™
	Zeopane	Hollow microneedles (insulin patch)
	Zosano Pharma	Macroflux® technology (ZP-PTH patch for osteoporosis has completed Phase 2 clinical trials successfully)
	ISSYS	Microneedle technology
	3M	Solid microstructured transdermal system (sMTS); hollow microstructured transdermal system (hMTS)
	Debiotech	Nanoject (hollow microneedles)
	Becton Dickinson	Soluvia™ (Intanza®/ID Flu®)
	TransDerm	Protrusion array device (PAD) (with dissolvable PVA microneedles)
Thermal ablation	Altea Therapeutics	PassPort® Patch
Radio-frequency ablation	TransPharma Medical	ViaDerm™ (Completed Phase II clinical trials for hPTH(1-34) delivery)
Laser ablation	Norwood Abbey	Epiture Easytouch™ (received U.S. FDA approval for lidocaine delivery)
	Pantech Biosolutions AG	P.L.E.A.S.E.®

microneedle (Soluvia™, BD) that is currently on the market for intradermal delivery of influenza vaccine (Intanza®/ID Flu®, Sanofi Pasteur, Lyon, France) in Europe (62). The product was tested in clinical trials on over 7000 subjects and is approved for use in the European Union territory for prevention of seasonal influenza in both adult and elderly populations. Another company, Norwood Abbey (Victoria, Australia), received approval from the FDA for laser-assisted delivery of a lidocaine formulation (Epiture Easytouch™). In a controlled, randomized, multicenter study of 320 subjects, the company reported that laser ablation of stratum corneum facilitated the delivery of 4% lidocaine to achieve dermal anesthesia (64). There are several more products based on microporation technologies in clinical and preclinical development as discussed in the following sections.

10.5.2.2   Products in Clinical Development

If a patch delivers a fraction of its payload, say about 2%, without achieving therapeutic levels, then just the mere feasibility of transdermal delivery does not create any value for the pharmaceutical industry. Therefore, having products in clinical development is essential to establish the viability of any delivery technology. One major product in clinical development at this time using microporation technologies of Zosano and TransPharma is hPTH(1-34). TransPharma Medical (Lod, Israel) is using radiofrequency ablation technology to also develop a hPTH(1-34) product for osteoporosis, in collaboration with Eli Lilly (Indianapolis, Indiana). TransPharma also has two additional peptides currently in Phase I clinical development, including a GLP-1 analog that, if successful, will compete with Amylin/Lilly's Byetta® (Exenatide) for patients with type 2 diabetes who currently need two daily injections. Exenatide is also being developed into a product by Altea Therapeutics (Atlanta, Geogia) using their thermal ablation PassPort® patch in an agreement with Eli Lilly and Amylin Pharmaceuticals. Altea is also developing a basal insulin patch that is currently in clinical development, and the company carried out some clinical studies with interferon, parathyroid hormone, hepatitis B surface antigen, and hydromorphone HCl in the past. Clinical trials with Altea's device indicated its safety in a range of subjects even after 5500 applications overall (65). More scientific details of the thermal ablation process and the published literature are discussed in Section 3.3.1.

As discussed, one of the products in clinical development at this time using microporation technologies is hPTH(1-34). Currently, this is the only osteoporosis drug on the market with anabolic properties to stimulate new bone growth, and even though there are other osteoporosis drugs on the market, it has a huge growth potential due to the 10 million people in the United States with osteoporosis and another 34 million with low bone mass (65). Zosano Pharma (Fremont, California) is using coated microneedles technology in developing the ZP-PTH patch for treatment of osteoporosis as an alternative to the daily subcutaneous injections of PTH (Forteo®; Lilly). It was reported that ≥85% of the coated polypeptide is delivered from the patch in <30 minutes. T_{max} was achieved three times faster as compared to Forteo SC Injection, and the intra- and intersubject variability was similar. The patch has a storage stability time of >2 years at ambient temperature, and skin tolerability was good only with mild and transient erythema. Capabilities exist to scale up the manufacturing process to produce 20 to 50 million units per year (60). Delivery of parathyroid hormone (1-34) by these coated metal microneedles and sterilization of the patch were described in the literature. The microneedles have an arrowhead design to maximize surface area available for coating (61,66,67). The patch contains a 2 cm^2 metal (titanium) microprojection array with approx 1300 microprojections, each 190 μm in length. The penetration depth is around 80 μm. The array is attached to an adhesive backing to create a 5 cm^2 patch which in turn sits in a retainer ring and the patch is applied using an applicator (see Figure 10.1). The patch is then placed inside a nitrogen-purged, heat-sealed foil patch with a desiccant sachet. Compatibility of each component of the patch with the drug needs to be checked, because outgassing from any component in the packaged product can interact with the drug. Biopotency of the drug was maintained over the 2-year storage period. It

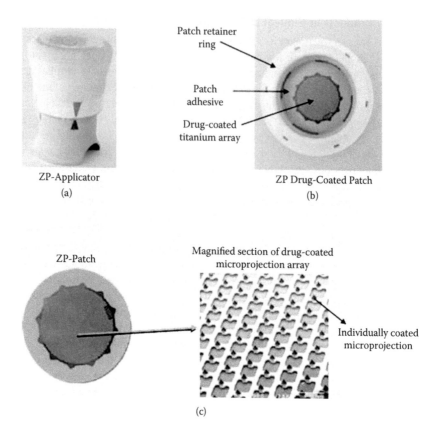

ZP-Applicator
(a)

Patch retainer
ring

Patch
adhesive

Drug-coated
titanium array

ZP Drug-Coated Patch
(b)

ZP-Patch

Magnified section of drug-coated
microprojection array

Individually coated
microprojection

(c)

FIGURE 10.1   (a) ZP-Patch applicator, (b) drug-coated patch, and (c) microprojection array. (With kind permission from Springer Science+Business Media: *Pharm. Res.*, Parathyroid Hormone PTH(1-34) Formulation That Enables Uniform Coating on a Novel Transdermal Microprojection Delivery System, 27:303–313, 2010, M. Ameri, S. C. Fan, and Y. F. Maa, Figure 5 [67]. Copyright 2009.) **(See color insert.)**

is being developed for treatment of osteoporosis in postmenopausal women. The bioavailability was found to be approximately 40% in human studies, though it was affected by the application site, apparently due to different levels of metabolism. Bioavailability was <25% in the hairless guinea pig model (67). The ZP-PTH patch has completed Phase II clinical trials and was found to be safe and efficacious. It will now enter Phase III clinical trials once a partner is identified. TransPharma Medical is using radio-frequency (RF) ablation technology to also develop a hPTH(1-34) product for osteoporosis, in collaboration with Eli Lilly. In early studies, systemic delivery of therapeutic doses from a stable printed dry-form patch were reported. Two biomarkers (ionized calcium and phosphorus) were also analyzed to confirm that the bioactivity was maintained after delivery. The product is currently in Phase 2B clinical trials. The handheld device (ViaDerm™) used for RF ablation of skin is intended for home use and uses a battery-operated electronic control unit, a disposable low-cost microelectrode array, and a drug-containing patch. The device uses a

1 cm² microelectrode array in a typical configuration, but this is available in three sizes depending on the dose of the drug that needs to be delivered. The device is gently placed on the skin with light pressure when it creates microchannels within 1 second, as indicated by a beeping tone and a green LED to signify successful creation of the microchannels. In studies on 136 volunteers, the device was reported to demonstrate an excellent safety and tolerability profile. The microchannels created were of uniform depth irrespective of skin type, age group, or gender (68). More scientific details of the radio-frequency ablation process and the published literature are discussed in Section 3.3.3. The printed patch used in some of these studies can be developed by spotting precise small droplets of the drug solution on a backing liner followed by drying (69). Other technologies have been explored for delivery of PTH. Feasibility of delivery of PTH(1-34) has been investigated by other technologies such as iontophoresis and electroporation (70). In an early study, iontophoretic delivery of parathyroid hormone was investigated in swine. Doses in excess of 25 µg (therapeutic dose) were administered following 4 hours of iontophoresis. Peak plasma levels equivalent to those achieved following subcutaneous injection were achieved within 90 minutes of iontophoresis and fell rapidly when current was turned off (71).

10.5.2.3 Products in Preclinical Development

Several of the companies mentioned as having a product under clinical development also have several preclinical studies in progress. Altea has a partnership with KAI pharmaceutics to deliver some of their proprietary peptides, and with Hospira to develop a transdermal patch of enoxaparin sodium, a LMWH, which is currently available as a Lovenox® injection (65). One advantage of delivery via microporated skin may be that transport of hydrophilic molecules via aqueous microchannels is unlikely to form a skin depot of the type formed by passive diffusion of lipophilic drug molecules into the stratum corneum. For a drug that is known to have problems related to formation of skin depot (e.g., fentanyl), it may be possible to choose a salt form of the drug (e.g., fentanyl citrate) and deliver via microporated skin to avoid these problems. Altea is also developing a fentanyl citrate patch, and this is one of the rationales behind the development of this patch. Altea has also done preclinical studies with influenza and tetanus antigens, DNA vaccines, erythropoietin, and apomorphine.

3M (St. Paul, Minnesota) is developing a solid microstructured transdermal system (sMTS) as well as a hollow microstructured transdermal system (hMTS) using biocompatible polymeric microneedles. These disposable microneedles can be integrated with a traditional glass reservoir and are designed for self-administration. Using the hMTS device, 3M has reported the delivery of human growth hormone (hGH; 22 kDa) with bioavailability similar to that achieved by syringe administration. A volume of 0.9 ml was infused over 15 minutes in this study using hollow microneedles which resulted in a lower T_{max} and higher C_{max} as compared to the injection control. A proof of concept hMTS device is shown in Figure 10.2 with 18 900-µm-long hollow microneedles arranged in a 1 cm² array. Typically, these hMTS arrays have microneedles ranging from 500 to 900 µm in length and may achieve a depth of penetration of around 275 to 650 µm. Volumes of up to 1 ml can be administered over 10 to 30 minutes. Using hMTS, a study in human subjects found good tolerability when 0.75 ml to 1 ml volumes were administered to the upper arm or

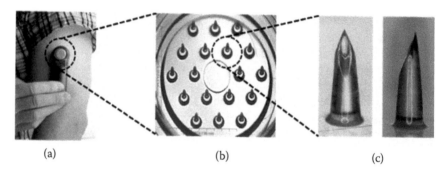

(a)   (b)   (c)

FIGURE 10.2   3M's hollow microstructured transdermal system (hMTS) integrated device for drug delivery. (a) Subject demonstrating the use of wearable hMTS integrated device. It consists of an applicator, drug reservoir, and delivery system, all in a single, compact device; (b) hMTS array consisting of 18 "mini-hypodermic" structures distributed over 1 cm^2 area; (c) front and side views of single polymeric mini-hypodermic needle (approximately 900 μm in length). (Reproduced with permission from 3M, St. Paul, Minnesota.) **(See color insert.)**

thigh of healthy adult subjects. These microneedle arrays can be designed in different shapes and sizes. In one configuration of sMTS, pyramidal polycarbonate projections of 250 μm (1000 microstructures in a patch) were inserted with an applicator into swine and hairless guinea pig skin and found to penetrate to a depth of around 120 μm. Similar to delivery of hGH by hMTS, delivery of naloxone HCl by sMTS also reported similar C_{max} and bioavailability as subcutaneous injection. In release studies with protein-coated sMTS, up to 150 μg of protein was coated on the arrays and 60% to 70% delivery was reported within 5 minutes. Preclinical studies were also reported with tetanus toxoid and ovalbumin (65,72–76). It seems that the company has mostly done preclinical studies at this point with specific drug molecules but is seeking partners for clinical studies with both proteins and vaccines.

Theraject (Fremont, California) is developing microneedle array technology (TheraJect MAT™) consisting of rapidly dissolving microneedles (0.2 to 1.5 mm in length) made of generally recognized as safe (GRAS) inert materials. Preclinical studies have been done with lidocaine and PTH. Theraject has now licensed its cosmetic business to another company and is focused on vaccine delivery (60). TransDerm (Santa Cruz, California) is investigating the delivery of siRNA by a protrusion array device (PAD) that has PVA microneedles (800 μm) that dissolve upon insertion to form a gel-like plug that then releases the drug. Each microneedle is preloaded with 10 to 20 nL of the formulation, and the polymer imbibes the formulation. Effective silencing of reporter gene by siRNA and delivery of reporter plasmids to show expression have been demonstrated. The success of this delivery technology in these early studies may eventually help in the translation of the siRNA agents developed for the treatment of skin disorders (60,77). Apogee Technology (Norwood, Massachusetts) is developing microneedles coated with PCPP, a polyphosphazene as an immunoadjuvant compound. This offers potential advantages because alum, the most common adjuvant used in vaccines, induces adverse effects if administered intradermally (78). Pantec Biosolutions (Leichtenstein) has done preclinical work with triptorelin, an LHRH superagonist. Corium (Menlo Park, California) is also developing a microneedle patch (MicroCor™), and their technology is protected by the over 50 patents they acquired from Procter & Gamble (79). Some of the other preclinical work from published literature is summarized in Chapter 3. Delivery of insulin by microporation approaches is discussed in Chapter 7. Applications to delivery of vaccines, a major advantage of microneedles, is discussed in Chapter 9.

10.6 COMMERCIAL DEVELOPMENT OF IONTOPHORETIC DEVICES

10.6.1 INTRODUCTION

Several iontophoretic devices are already on the market (see Section 10.6.4 and Table 10.2). These iontophoretic devices can be developed as single-use or reusable systems. For prototypes of single-use systems, electronics and other components are thin and fully integrated into the system, so that the systems are flexible and comfortable to wear. For reusable systems, replaceable drug pad couples with the

TABLE 10.2
Some of the Iontophoresis Products on the Market or in Development

Manufacturer	Iontophoresis Device	General Notes (Electrode/ Drug Reservoir Composition)
Wescor Inc.	Sweat Inducer	Pilogel® Discs Gel Reservoir
Life Tech Inc.	Iontophor® II Model 6111PM/ DX, Microphor® Model 6121	Meditrode electrodes; cotton/ rayon reservoir
Vyteris	Lidosite®	
Travanti Pharma (a subsidiary of Teikoku Pharma USA)	WEDD Technology Incorporated into IontoPatch®	IontoPatch® family includes: IontoPatch® 80, IontoPatch STAT®, and IontoPatch® SP
Empi	Empi Action Patch™, Hybresis™	
ActivaTek™	Trivarion®, ActivaPatch®	ActivaStim™ electrodes, ActivaDose™ controller
Dharma Therapeutics Inc. (a subsidiary of Transcu Group Ltd.)	Iontophoretic drug delivery system (IDDS)	
Isis Biopolymer	IsisIQ™	In development
Nupathe®	Zelrix Patch (SmartRelief Technology)	In development
Chattanooga Group®—a DJO® Company (formerly Iomed)	Chattanooga Ionto™ Iontophoresis System, Iontocaine®/Numby Stuff®, Companion 80™	IOGEL® disposable electrodes, TransQ FLEX® disposable electrodes
Dyna Dental Systems University Medical Pharmaceuticals Corporation	hyG Ionic™ Toothbrush WrinkleFree Eyes™	

reusable controller. An on-demand system has a button to administer a bolus dose while the system continuously administers a baseline level of the drug. An LED display indicates whether the system is in baseline or bolus delivery mode. The size of these systems will vary from 5 to 50 cm², depending on scientific and marketing considerations. Because only small currents are employed, the systems are expected to be safe for use on patients with pacemakers. These systems can deliver drug for periods from hours to days and can provide flexible dosing patterns. A reusable controller can potentially be used which can be connected and removed from the disposable housing. The controller can monitor and control the power supplied during use, thus permitting safer and more reliable operations. It can also have the ability to detect the number of times the patch has been used and record the date and time of use, and its microprocessor can detect when drug supply is exhausted. Once the drug is exhausted, the controller can potentially be rendered unusable to avoid abuse (80,81). The utilization of such a reusable controller will allow costs to be kept down because the more expensive part of the device can be reused. In one concept, the controllers and reservoirs can be designed to fit together with a male/female interlock system. A peel-away covering from the adhesive on the controller side of the reservoir is removed, and the reservoir is pressed against the controller to form the system assembly. Before use, a peel-away covering from adhesive on the user side of the reservoir is removed and the patch pressed against the user.

Based on the principles of reverse iontophoresis (see Section 4.8), a glucose-monitoring device (GlucoWatch™) was developed by the former Cygnus, Inc. (Redwood City, California). The device consisted of collection reservoirs, iontophoresis electrodes, and sensor, and was useful for continuous glucose monitoring, such as in neonates or subjects requiring frequent testing (82). Glucose is extracted from the body due to electroosmosis induced by reverse iontophoresis and by using an electrochemical reaction linked with a sensor and a control module, blood glucose levels can be continuously monitored and displayed at the push of a button. The system could sound an alert or alarm in the event of hypo- or hyperglycemia. Cygnus was later purchased by insulin pump manufacturer Animas Corp (West Chester, Pennsylvania), and a second-generation device, GlucoWatch Biographer G2 was marketed but was later discontinued in 2007. The device measured glucose every 10 minutes and was helpful in showing patterns and trends in blood sugar levels. However, the device needed a 3.5-hour warm-up time, and could not be worn while bathing. Further, it had to be calibrated by a blood sugar reading, and the wearing site had to be rotated.

Compounds of interest for iontophoretic delivery in the patent literature include antimigraine drugs (83) and new derivatives of glucagon-like peptide 1 and insulinotropin (84). Insulin analogs have also been developed for potential use in iontophoretic delivery. These are prepared by substituting at least two of the residues in human insulin by Glu and Asp at selected positions. These analogs have reduced tendency to associate and will thus overcome one of the obstacles to iontophoretic delivery of insulin (85). Other compounds of interest for iontophoretic delivery were discussed in Chapter 4. Similar to human studies where pharmacokinetics of drug delivery following application of passive patches were studied (86), several human studies have been done for investigation of iontophoretic delivery. Clinically significant doses of tacrine have been delivered to human subjects by iontophoresis (87). Similarly, clinically significant doses of fentanyl have been administered to humans with no adverse effects related to the delivery mode, except for erythema at the location of the dispersive pad which resolved without treatment within 24 hours (88). A discussion of iontophoresis of fentanyl can be found in Section 7.4, and comments about its commercial development are discussed later in this chapter. Iontophoretic delivery of leuprolide, an LHRH analogue, and a calcitonin analog to human volunteers has also been reported and is discussed in Chapter 8, which is dedicated to the delivery of polypeptides and proteins. Octreotide, a bioactive peptide (MW: 1000) has been used in a placebo controlled dose escalating study with three groups of eight healthy male volunteers. Steady-state levels were achieved quickly and increased with increasing magnitude of current; plasma levels then declined rapidly once the current was terminated (89). Another study in four ethnic groups also reported that a slight erythema observed at the active electrode disappeared within 24 hours (90). Iontophoretic delivery of alniditan, a 5 HT_{1D} agonist useful for treatment of migraine, has also been investigated in eight healthy volunteers. The molecule (MW 302) has pK_a's of 8.3 and 11.5, and is ionized over a wide pH range which hinders its passive transdermal permeation. An *in vitro* study was first performed on freshly excised hairless rat skin to optimize the factors affecting delivery. A dose of 0.5 mg of alniditan was successfully delivered within less than 1 hour (91). Iontophoretic delivery of sumatriptan and commercial development of a product for migraine is discussed in Section 10.6.4. In another study performed in two human subjects, significant amounts of salbutamol were absorbed 2 hours after current application, and plasma levels declined when the current was terminated (92). Electrotransport of metoclopramide, an anti-emetic, across the skin of healthy male subjects was reported to rapidly achieve and sustain reproducible blood levels (93).

10.6.2 IONTOPHORESIS DEVICES

The inherent disadvantages of an iontophoretic patch over the marketed passive patches would include the complexity of the delivery system and cost considerations. While wearable electric patches continue to be developed for systemic delivery, external palm or walkman-sized iontophoretic devices have been used for several decades for topical delivery of drugs. Generally, iontophoresis devices used in the physical therapy market have not been approved for any specific drug and have not undergone the series of clinical trials of safety and efficacy. This is because they were marketed prior to the passing of the medical devices amendments in 1976. They have generally been used for delivery of corticosteroids such as dexamethasone sodium phosphate to treat local tissue inflammation. These companies market external palm-sized devices that are hooked up to electrodes for topical delivery of drugs. In addition to the drug electrode, a second electrode is applied to the patient's skin to complete the circuit. This second electrode is referred to as a ground, dispersive, or indifferent electrode. The power source is a direct current generator, with a current output of 0 to 4 mA. These are designed as constant current devices (i.e., the voltage adjusts based on the skin resistance of the patient so that the current stays constant at the desired setting). The dosage is expressed as follows:

$$\text{Dosage (mA*min)} = \text{Current (mA)} \times \text{time (min)}$$

The maximum dosage (mA*min) delivered by these devices varies, with a maximum of 80 to 160 mA*min between various manufacturers. It should be noted that the use of mA*min units to express dosage can be confusing (see Chapter 4).

The device from Chattanooga Group (Figure 10.3) requires the setting of dose and current, by which it calculates the time required for the selected dose to be delivered, and adjusts it if the current setting is changed during the treatment period. Current ramp-up and -down within 30 seconds and automatic shut-off features are provided. Current can be set at the desired setting in 0.1 mA increments between 0.5 mA and 4.0 mA (94). It is a dual-channel system, where the user can set the dosage and current levels for each channel independently. This can allow delivery at two different body sites at same or different delivery rates, or two different drugs, or treatment of two different patients, simultaneously (95). Another iontophoresis device (Iontophor®) available from Life-Tech International (Stafford, Texas) has a current range of 0 to 4 mA in 0.02 mA increments. It maintains constant current against impedance as high as 40 kΩ, enabling application to all areas including scar tissue (96). Additional manufacturers of iontophoresis devices include Dagan Corporation (Minneapolis, Minnesota) and Ionto-Comed GmbH (Germany). An iontophoretic toothbrush is also commercially available (hyG Toothbrush, Dyna Dental Systems, Phoenix, Arizona) and is claimed

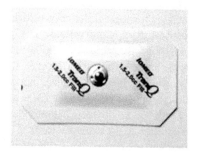

FIGURE 10.3 Iontophoresis device and electrodes. (Reproduced with permission from Chattanooga Group [DJO Global], Vista, California, http://www.chattgroup.com.) (**See color insert.**)

to remove the positively charged plaque from negatively charged teeth by attracting the plaque ions to the negatively charged bristles of the toothbrush. Iontophoresis devices are sometimes even used in beauty salons such as the Dermaculture® facial clinics (Beverly Hills, California). A wrinkle-free eye lift system based on iontophoresis is available from University Medical Pharmaceuticals Corporation (Irvine, California). Hisamitsu (Japan) and several other companies are also developing iontophoretic delivery systems. Recently, electrically charged Teflon disks (surface potential 500 to 3000 V) were used in the laboratory to enable iontophoretic delivery of salicylic acid and propofol (97). In addition to the devices discussed in this section, devices for iontophoresis of pilocarpine for diagnosis of cystic fibrosis are on the market, and devices for glucose monitoring based on the principle of reverse iontophoresis were on the market until recently (see Chapter 4). Several of these companies making specialized delivery systems often work on a contract basis for larger pharmaceutical companies. As a result, the number of pharmaceutical and device companies exploring iontophoresis and other technologies is much larger than the number of companies with in-house capabilities for large-scale manufacture. The collaboration will typically begin with an evaluation of technical feasibility and dermal tolerance of the delivery system for the drug. If initial studies are promising, then custom systems can be designed for exclusive marketing by the pharmaceutical partner.

Integrated iontophoretic patches have now become available from Travanti Pharma (St. Paul, Minnesota) (IontoPatch®), which is a subsidiary of Teikoku Pharma USA (San Jose, California), Chattanooga Group which is a part of DJO Global (Vista, California) (Companion 80™), and Empi (St. Paul, Minnesota) (Hybresis), also a part of DJO Global now. The IontoPatch is a single-use wearable electronic disposable drug delivery (WEDD®) system with a self-contained battery that is available in 4-hour or 14-hour average patient wear time configurations for the physical medicine and rehabilitation market. The Companion 80™ patch is designed to deliver an 80 mA/minute dose over 24 hours and has reserve battery capacity to compensate for patients with higher skin resistance. The Hybresis system contains a miniaturized controller that connects directly to the patch and provides a hybresis (hybrid) mode where a 3-minute skin conductivity enhancement can be carried out to cause a rapid decrease of skin resistance. This allows for reduced variability and shorter wear time when the patch is activated following this treatment (98).

Iontophoresis devices are generally not marketed for use with a specific drug and are approved under the FDA's premarket notification or 510k medical device approval pathway. These devices need a prescription to be purchased. Chattanooga Group (formerly Iomed) received approval for delivery of lidocaine HCl 2% and epinephrine 1:100,000 (Iontocaine® or Numby Stuff) from the FDA in 1995 under a new drug application (NDA) for local dermal anesthesia. This was the first FDA-approved drug labeled for use with an iontophoresis device (99). Another company that received approval for a prefilled iontophoresis drug/device system in 2004 was Vyteris (Fair Lawn, New Jersey). The LidoSite® product from Vyteris has 10% lidocaine hydrochloride with 0.1% epinephrine, and a 10-minute application can be used for superficial dermatological procedures such as venipuncture, IV cannulation, and laser ablation of superficial skin lesions (100). A fentanyl iontophoresis device that was marketed and later withdrawn is discussed in Section 10.6.4. There are hundreds of patents on iontophoresis between these companies and some individuals, with the majority belonging to the former Alza and to Vyteris (a Becton Dickinson spin off), who did much of the early work in this field.

10.6.3 Electrode Design

Historically, iontophoretic devices were typically a crude assembly of paper towels, lint clothes, orthopedic felt, gauze, or other suitable means to apply current to the site of application (101). These early electrodes did not provide uniform current density under the skin, resulting in the possibility of skin burns at sites that would have a higher current density. Also, these electrodes did not provide any mechanism to prevent electrolysis of water or remove generated extraneous ions. The resulting pH changes could make the skin susceptible to irritation or burns, and the mobile hydrogen and hydroxyl ions generated would lower the efficiency of drug delivery. To avoid these problems, electrodes are now carefully designed, and many different types are commercially available. In general, the electrode design consists of a backing to which a carbon layer is added to evenly distribute the electric current over the drug delivery electrode. The drug reservoir consists of a hydrophilic layer to hold the drug solution during iontophoresis. Composition of the hydrophilic material varies between manufacturers, but typically consists of fiber-type materials or hydrogel polymer matrix. The fiber-type material varies from polyester fleece to rayon, and the natural fibers include wool and cotton. Several manufacturers utilize an additional outer layer that is in contact with the skin. This last layer serves to stabilize the underlying layers of the electrode and act as a wicking layer to absorb the applied drug solution. Some other electrodes utilize a hydrogel formulation as the drug reservoir. A hydrogel formulation (see Section 10.7) can provide a conductive base (if an appropriate solution is used), ease of application, uniform current distribution at the treatment site, and allows for replacement of a drug-loaded hydrogel pad, while reusing the device. Several hydrogel systems have been investigated for their use in iontophoretic delivery of peptides and proteins of different molecular weights (102). Electrochemical reactions taking place at the electrodes during iontophoresis could create several problems, as discussed in Chapter 4. The commercially available iontophoresis electrodes have used different mechanisms to avoid these changes. The Trans-Q electrodes (see Figure 10.3) from the Chattanooga group (formerly Iomed, Inc.) use silver-silver chloride electrodes to stabilize drug pH without buffers. This becomes feasible because these electrodes actually participate in the electrochemistry, thereby preventing electrolysis of water (see Section 4.6.2). Trans-Q electrodes are available in two sizes, small (1.5 to 2 cc) and medium (2.5 to 3 cc) fill, for smaller or larger treatment sites. The smaller electrode is also available in a FLEX configuration to surround contoured treatment areas. They are also available as an IOGEL® version with a GelSponge® gel formulation and silver-silver chloride electrodes that conduct more efficiently than carbon electrodes. The IOGEL electrodes are available in small, medium, and large sizes. Meditrode electrodes (Life-Tech, Stafford, Texas) are available in six different sizes spread over five different shapes to allow application over a variety of surfaces such as digits, elbows, ankles, shoulders and temporomandibular joints. Possible local silver toxicity due to iontophoretic devices should be considered (103) but may be possible to avoid by designing appropriate electrodes. Alternatively, buffers can be used to maintain pH during iontophoresis. However, even if buffers can maintain pH, hydrogen and hydroxyl ions being generated will compete with the drug for carrying the current, thereby decreasing the efficiency of delivery as a function of time (34). The buffer ions will also provide extraneous competitive ions to the drug, resulting in lower drug delivery efficiency. Some manufacturers include an ion-exchange resin in their electrode design, and the resin has been reported to stabilize pH under both cathodic and anodic conditions.

10.6.4   IONTOPHORESIS-BASED TRANSDERMAL PRODUCTS ON MARKET OR IN CLINICAL DEVELOPMENT

Isis Biopolymer (Providence, Rhode Island) is developing a fully programmable iontophoretic patch (IsisIQ™) that consists of a battery-powered electrode layer, a drug-infused hydrogel layer, and a selectively permeable membrane (104). Using polymer thick film electrodes and a hydrogel (Figure 10.4), a small, thin, and pliable Band-Aid-type patch can be made. The selective membrane can be switched to facilitate or prevent transport, thereby avoiding passive diffusion during the current off time, even at elevated skin temperatures. The programmable patch has been investigated for delivery of methylphenidate, cefazolin, ibuprofen, and L-Dopa (105). Isis recently launched a cosmeceutical delivery patch that has been licensed to University Medical Pharmaceuticals for antiaging/wrinkles treatment by delivering ingredients like hyaluronic acid. However, the company continues its efforts to use its technology for bringing products to market to deliver pharmaceuticals and biologics (65).

Another product being developed for marketing as an iontophoresis patch is sumatriptan for the treatment of acute migraine. Migraine is a neurological disorder that affects about 28 million people in the United States alone. The currently marketed oral formulations have limitations due to nausea and vomiting that often accompany

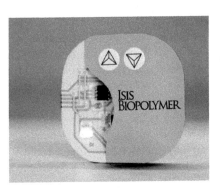 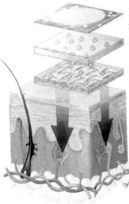

Polymer thick film electrode
Hydrogel/drug

Selective barrier membrane

FIGURE 10.4 IsisIQ™ patch with polymer thick-film electrodes that transport drug from a hydrogel across a selective barrier membrane. (Reproduced with permission from Isis Biopolymer, Providence, Rhode Island, http://www.isisbiopolymers.com.) **(See color insert.)**

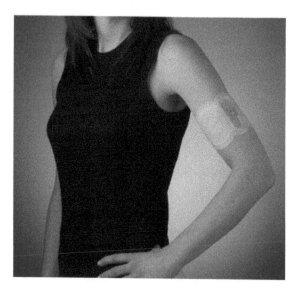

FIGURE 10.5 Developmental Zelrix patch (NuPathe Inc., Conshohocken, Pennsylvania) as worn by a patient. (Reprinted from *Neurotherapeutics*, M. W. Pierce, Transdermal delivery of sumatriptan for the treatment of acute migraine, 7:159–163, 2010 [106]. Copyright 2010, with permission from Elsevier.) **(See color insert.)**

a migraine attack. Also, the oral as well as an available nasal formulation have low bioavailability (106). Subcutaneous injections have the usual drawbacks of parenteral therapy. Transdermal delivery would be attractive, but iontophoresis is required to achieve good flux values (107). A Phase I study in eight human subjects reported a linear relationship between the total applied current and sumatriptan delivery. This study included oral and subcutaneous controls and reported favorable profiles from the iontophoresis patches, with plasma levels that were maintained above the target level ($\geq$10 ng/ml) for greater than 7 hours (108). The patch (Figure 10.5) was better tolerated than subcutaneous formulation, perhaps due to the lower C_{max} associated with the patch (106). The patch is being developed by NuPathe Inc. (Conshohocken, Pennsylvania) using their SmartRelief technology and has now completed Phase III clinical trials in 530 patients at 38 investigative sites. The patch was reported to be well tolerated during use (109).

An iontophoresis device that was successfully developed after years of research but then had to be withdrawn soon after marketing is the IONSYS™ product from the former Alza Corporation (part of Johnson & Johnson). This fentanyl iontophoretic transdermal system was an on-demand delivery system intended to allow a patient to manage acute pain by self-titrating the level of fentanyl administered according to his or her need. With this system, the onset for control of pain was almost instantaneous. Currently, transdermal fentanyl can only be used for chronic cancer pain management. The development of an iontophoretic patch allowed treatment of acute pain (e.g., postoperative pain because individualization and titration of dose becomes feasible by changing electronic parameters). For details of published literature on transdermal/iontophoretic delivery of fentanyl, see Section 7.4.2. The IONSYS product was designed for hospital patients to self-administer fentanyl for up to 24 hours or a maximum of 80 doses, whichever came first. The patch was powered by a 3-volt lithium battery, and upon activation of the dose button by pressing it twice within 3 seconds, a 40-mcg dose of fentanyl was delivered over a 10-minute period during which a red light remained on in the patch. This product was withdrawn from Europe because some defective systems were liable to be accidentally triggered due to corrosion, creating a risk of overdose (European Medicines Agency Recommendation Ref. EMEA/CHMP/613852/2008). The product was also approved by the FDA at one point, though it was never marketed in the United States. This book is on skin applications of enhancement techniques, but it should be noted that Acient Inc. (Salt Lake City, Utah) is developing an iontophoresis product for the eye. They are developing the Visulex® scleral-lens-shaped application device for delivery to the back of the eye for conditions such as age-related macular degeneration and diabetic retinopathy (110). Similarly, Iomed (Chattanooga Group) was developing an ocular iontophoresis device, OcuPhor™ (111), but its developmental status is currently not known.

10.7 FORMULATION CONSIDERATIONS

Drug in adhesive patches generally have a simple design with the drug dissolved in the pressure-sensitive adhesive at near saturation concentrations and sandwiched between a backing layer and the release liner. The adhesives used are generally acrylics, silicones, or polyisobutylene. However, looking at other type of patches as well as other types of topical formulations, a much wider variety of polymers and other formulation components have been utilized in transdermal delivery systems. Polymers such as cellulose derivatives, chitosan, carageenan, polyacrylates, polyvinylalcohol, and silicones have been widely used in transdermal delivery systems as matrices in patches or as gelating agents, crystallization inhibitors, or penetration enhancers (112). Film-forming polymeric solutions are also being investigated as alternatives to patches or semisolid topical formulations. The polymeric solution forms an invisible film by solvent evaporation when applied to the skin. Polymers from different groups such as acrylates, cellulose derivatives, polyvinylpyrrolidones, and silicones have been used in these solutions (113,114). Formulations for coating microneedles were discussed in Chapter 3.

Self-enhancing hydrophilic pressure-sensitive adhesives, without the addition of chemical enhancers and crystallization inhibitors, are also being investigated (115). Aveva (Miramar, Florida), a Nitto Denko company, successfully commercialized a gel matrix technology that can be removed gently from the skin. A hydrogel matrix system was also investigated for a buprenorphine patch (36). Hydrogel formulations are widely used in medical devices and can provide an electroconductive base; they will be especially desirable for clinical use in iontophoresis. One advantage of a hydrogel formulation would be the ease of application because a hydrogel pad can adapt to the contours of the body. A hydrogel formulation might also be helpful in reducing skin hydration during the period of medication. For iontophoresis-mediated drug delivery, a hydrogel formulation would be preferable as it can be designed as a unit dose-type drug-loaded conductive hydrogel patch that permits the daily or weekly dosage replacement while reusing the same iontophoretic device continuously. It would also be possible to control the release rates of drug from a hydrogel to some extent by changing the characteristics of the hydrogel formulation during synthesis. Furthermore, this formulation may increase skin compliance of a transdermal patch, in view of the fact that a hydrogel can absorb sweat gland secretions which, under long-term occlusion, may become irritating. The use of p-HEMA films for the iontophoretic release of propranolol HCl has demonstrated the feasibility of using electric current to control and predict release rates of drugs from polymer membranes (116,117). In a study done by the author, two types of hydrogels—polyacrylamide and p-HEMA—were synthesized (118), and these, along with carbopol 934 gel, were used to investigate the iontophoretic release and transdermal delivery of three model peptides—insulin, calcitonin, and vasopressin (102). The swelling behavior of polyacrylamide hydrogel as a function of its monomer and cross-linker concentration was studied, and the results were used to synthesize hydrogel with minimal swelling. Release of a drug from a hydrogel matrix by passive diffusion was reported to follow the Higuchi equation for matrix release. Accordingly, the cumulative amount of drug released is proportional to the square root of time, t, as per the Higuchi equation:

$$Q/A = (2\ C_0\ C_s\ D\ t)^{1/2}$$

where Q is the amount of drug released during time t, A is the surface area, D is the diffusion coefficient (cm^2/sec), C_s is the solubility of the drug in the hydrogel, and C_0 is the initial drug loading in the hydrogel. For drugs dissolved in the polymer matrix, Baker and Lonsdale proposed the following equation:

$$Q/A = 2\ C_0\ (D\ t/\pi)^{1/2}$$

Thus, a Q versus $t^{1/2}$ relationship should be followed for the release of peptide/protein molecules from the hydrogel formulation, under passive diffusion conditions. This Q versus $t^{1/2}$ linearity was observed experimentally for the release of vasopressin and insulin from hydrogels. However, under iontophoretic transport, the release profile changed from the Q versus $t^{1/2}$ relationship to a Q versus t relationship. The release kinetics of peptide/protein from hydrogel under iontophoretic transport was investigated using insulin under varying durations of iontophoresis application. The kinetic profiles suggest that insulin is released at a constant rate during iontophoresis application, and the release rate drops as soon as the current is turned off. Thus, the release of the peptide/protein from the hydrogel matrix could be modulated by iontophoresis application (102). Cross-linked hyaluronic acid hydrogels have also been investigated to make an electrically responsive and pulsatile release system for negatively charged macromolecules. The solute is released when the electric field is switched off due to the swelling of the gel, and release stops when the electric field is applied again. Hyaluronic acid is a naturally occurring polysaccharide and polyelectrolyte whose cross-linked hydrogels lose water if ionic strength is high. Application of an electric field causes partial protonation of the ionized polyelectrolyte network, causing rapid deswelling of the hydrogel (119). Iontophoretic release of antimycotic agents from hydrogels through an artificial membrane using rotary disk cells has been described in the literature (120). Gels have also been used for *in vitro* transdermal iontophoretic delivery of cromolyn sodium across hairless guinea pig skin. Gels of ionic polymers decreased the flux of cromolyn sodium, but nonionic polymers such as hydroxypropyl cellulose and polyvinyl alcohol did not affect the flux. The latter may thus be used for iontophoretic delivery via a transdermal patch system (121). Poloxamer gels have also been used, and iontophoretic release of protonated lidocaine from poloxamer 407 gels was controlled by the current density (122). Recently, iontophoretic release from a poly(acrylic acid) hydrogel and from polypyrrole/poly(acrylic acid) blend films was described. It was reported that the desired drug release can be obtained by an optimum selection of cross-linking density and electric field strength. The size of the drug molecule and its interaction with the polymeric matrix is also important (123).

10.8   PATCH DESIGN AND REGULATORY ISSUES

A discussion of the design of passive patches is beyond the scope of this book, but a few comments are being provided to lay the foundation to understand the design of active technology-based transdermal patches. When comparing formulations, comparisons should be done not at the same concentration in different vehicles but rather at saturated concentrations in the respective vehicles so that the thermodynamic activity is the same. The drug should be in the unionized form for optimal passive permeation. For ionizable drugs that are weakly acidic, pH of the dissolving buffer should be at least two units below the pK_a for most of the drug to be in the unionized form. For weakly basic drugs, pH should be maintained above the pK_a. However, in reality, marketed patches will typically have the drug dissolved in a pressure-sensitive adhesive. On the other hand, when designing iontophoretic patches, it is desired that the drug be charged and the situation will reverse. Drug crystallization in patches is a potential problem, and the manufacturer has to carry out carefully planned studies to ensure that the drug will not crystallize in the adhesive over time. Recently, a rotigotine patch was introduced in the market for Parkinson's disease, but the patch had to be recalled due to the formation of "snowflake-like" crystals (124). For regulatory approval of a patch using a drug that is already on the market in another dosage form, an NDA is still needed, whereas toxicity and safety studies may not need to be repeated. However, studies to establish clinical safety and efficacy are needed. If the drug is already available in a patch form, then the manufacturer of the generic patch can go through the ANDA route with data on CMC (chemistry, manufacturing, and control), *in vitro* release testing, bioequivalence, skin irritation, and cutaneous toxicity. However, the generic patch should use a similar release mechanism as the approved patch. For NDA applications, sensitization studies are also needed. If chemical enhancer is being used, its irritation and toxicological potential needs to be investigated, along with its mechanism of action and fate once absorbed in the body (125).

Design of an iontophoretic patch will be discussed as an example in the remainder of this section. Design of microporation patches is discussed in Chapter 3. Several optimized dose-efficient patches for iontophoretic delivery were discussed in the patents and published literature. In a patent granted to the author, a two-chamber system separated by a permselective membrane was proposed. The upper chamber had a pH-control mechanism by ion-exchange resins, and the lower chamber could house a replaceable drug-loaded hydrogel device. The permselective membrane prevented the transport of drug into the upper chamber. The patch could be used for the delivery of conventional drugs or polypeptides such as vasopressin, calcitonin, and insulin (118). In a somewhat similar design, the upper chamber contains halide ions, and a silver electrode is used. Due to problems related to use of such systems, it was suggested that the upper chamber may also be formulated with ion-exchange resins or ion-selective membranes in order to prevent ions in the upper compartment from entering the lower drug reservoir (89). Ion-exchange materials that can be used include poly(acrylic acids), styrene and sulfonated divinylbenzene copolymers, polyamines, and a ferro-cyanide salt of a styrene divinyl benzene quaternary amine. The use of these materials, as well as the patch design, is described for fentanyl and other drugs (126).

Another patch described in the literature divides the two compartments by an ion-exchange membrane. The chloride counterions cannot pass through the membrane and thus cannot compete for the current through the skin. The lower compartment contains only the drug formulation. Thus, the only ions involved are the drug ions and the ions from the skin, thereby maximizing the efficiency of the patch (127). In addition to providing increased dose efficiency, these designs separate the electrode from the peptide. Special membranes such as those made from polysulfone, cellulose, or fluororesin and treated with a cationic surfactant have low adsorptivity for peptides and have been used as an iontophoresis interface for their delivery (128). The manufacture of an iontophoretic drug delivery patch is described in the patent literature. A well is created on a first laminate, filled with drug, and then a second laminate is placed over the well and sealed to form a web from which individual patches are cut. The process is performed in an inert atmosphere which increases stability and shelf life of the drug loaded into the iontophoresis patch (129). A flexible, conductive pathway and a flexible, nonconductive polymer film can be mated to form a flexible electronic circuit. This circuit can be disposable, as it can be manufactured cheaply and can be coupled to other patch components without penetration (130). Optimization of the power supply to the patch may be beneficial to ensure improved safety.

For the palm-sized iontophoresis devices currently on the market for topical clinical applications, ramp up of current is often built in to allow the patient to adjust to the feeling of current, thus minimizing the tingling sensation as the current is turned on. A microprocessor controls the electrode output current to rise exponentially from about zero to the selected treatment level. The current stays at the treatment level for the desired length of time and then falls exponentially to about zero (131). A reversal of the polarity of electrodes during treatment has also been suggested to avoid tissue damage or eliminate undesirable skin sensations, in addition to allowing the use of delivery of multiple drugs with different polarities. Such an approach has been claimed to eliminate the need for buffering agents (132,133). Possible degradation of the drug in direct contact with the electrodes can be prevented by isolating the drug from the electrode with proper patch design. Cyclic voltammetry studies need to be done to see how the drug may react to electrode surfaces or may degrade when the current is applied. We have done such studies with methotrexate where the voltage was increased linearly and then decreased back to the starting point. Similarities in the scan showed that methotrexate was not oxidized or reduced under the electric field (134).

Commercialization of a wearable iontophoretic dosage form will require consideration of regulatory path, clinical protocols to prove safety and efficacy, and problems relating to product development. It is important to clearly understand the optimal dosing schedule and the feasibility of achieving it before proceeding too far along the developmental track (135). Unlike the iontophoresis electrodes for topical application which can be filled with virtually any drug, the design of a wearable patch will be specific for a drug. Thus, each drug will have its own iontophoresis system with a unique formulation and perhaps a unique design.

The formulation must be carefully optimized considering all the factors that affect iontophoretic delivery. Excipients used should have poor current efficiencies so that they do not compete for the current with the drug. The selection of the battery technology will be a critical part of a wearable system, because miniaturized electronics must be used. The wafer form of primary cells such as lithium or mercury will last only a day or two, while rechargeable cells may have unacceptable dimensions. A split battery technology in which air is consumed to energize the battery may be promising for development of a wearable patch. In this system, the cathode is not consumed, while the electrolyte and anode are a part of the drug reservoir that can be changed daily. With each change, a fresh battery is created. The system is small, light, inexpensive, and efficient (127). Isis Biopolymers (Providence, Rhode Island) (see Section 10.6) is using polymer film electrodes to develop an ultrathin Band-Aid-type iontophoresis patch. An iontophoretic system could also be powered by an assembly of thermocouples (thermopile), which converts thermal energy into electrical energy. A thermopile has been demonstrated to deliver an ionic compound across human cadaver skin, with an efficiency of 3 µg/µAmp hour, which is adequate to achieve therapeutic levels of a drug (136). For development of iontophoretic patches for all drugs, especially peptides, there will be several stability challenges. If the peptide does not have adequate shelf-life in solution, it may have to be marketed as a dry powder to be reconstituted before use. The best way to achieve this objective would be via the use of hydrogels (see Section 10.7). The dry hydrogel can be activated by the user just before use (89). The patent literature describes several means to accomplish this objective. A burstable pouch can be used and ruptured just before use to provide the hydrating liquid. Alternatively, pulling a tab just before use can release the hydrating liquid from an aqueous reservoir which, in turn, can hydrate the dry hydrogel (137). Another patent describes a variation where the end tab of a strip is pulled to cause progressive unsealing of the compartment to slowly hydrate a matrix (138). Polymers that are electronic conductors as a dry film may be very useful if they can be commercially developed for iontophoresis. Such polymers can be directly coated on the electrodes and serve as drug reservoirs (139). The normal development and quality control considerations for a passive patch may be applicable to an iontophoretic patch in addition to several tests that will be unique for the iontophoretic patch. The typical quality control, formulation, developmental, and scale-up considerations for a passive patch were described in the literature (140,141).

In vitro release studies may be required to assure batch-to-batch uniformity of product. This is similar to the studies required to assure batch-to-batch uniformity for topical and transdermal products. These studies may not predict the *in vivo* delivery and are only intended to detect any problems or variations in the product during routine production. For instance, crystallization of drug from solution, its nonhomogeneous distribution in polymer matrix, or electronic malfunctions are likely to be detected by such studies. Synthetic membranes can be used for such studies. A discussion about the various synthetic membranes used in iontophoresis research can be found in Section 2.2. In fact, skin should be avoided for such studies, as high variability of data (see Section 2.3) resulting due to the skin factor may mask problems related to product performance. By the same logic, a rate-determining membrane is not required as the intention is to perform quality control on the patch, not the external release membrane. The membrane should be hydrophilic so that it allows the passage of current, but it should not allow excessive movement of water. In addition to *in vitro* release studies, current profiles from the controller may be monitored by instrumentation such as oscilloscopes to detect any electronic malfunctions. A rotary disk cell approach to study iontophoretic release of drug from hydrogel vehicles has been described in the literature. A normal dissolution tester is modified to provide current to the rotating disk via the shaft and release across an artificial membrane into the vessel is monitored (120).

Regulatory approval for an iontophoretic system for a specific drug will require an NDA submission, which should include a description of patch electronics in addition to information about the drug. The U.S. regulatory pathways may differ depending on whether a prefilled iontophoretic drug delivery system or a reusable controller with a replaceable patch configuration is being developed. The most important factor to determine whether a drug delivery device is a medical device or a medicinal product is to determine if the device forms an integral product with the drug. Based on European commission guidance document, single-use, disposable iontophoresis devices incorporating a drug may be regulated as a medicinal product rather than a medical device (142). Stability studies should be performed not just for the drug but for other components as well, both alone and in combination. Possibility and implications of electronic malfunction and battery leakage should be addressed. The adhesive must adhere well to the skin or electric contact will be lost. During iontophoresis, electroporation, or sonophoresis, the excipients may migrate into skin in addition to the drug. This possibility needs to be addressed. Such codelivery of excipients may be desired in some situations such as delivery of aprotinin to minimize degradation of peptides in skin (118), but in most instances, it will not be desirable. Rotation of application site may be required similar to that done for the currently marketed passive transdermal patches. If the patch/device is placed back on the same area, it may generate more flux during the second application period. This may result due to the saturation of skin binding sites during the first application, so that the second application drives the drug directly into the vasculature or may even cause desorption of the reservoir created during the first application. Alternatively, the barrier function of the skin may be compromised by the first application (143). It is in the best interest of the handful of companies who have manufacturing expertise in the active transdermal delivery technologies to get together and propose common testing to regulatory authorities to self-regulate their production.

10.9   DOSE AND BIOAVAILABILITY

Most transdermal products have been developed for drugs that are already on the market in other dosage forms. Therefore, bioavailability is one key consideration for regulatory approval. Bioavailability studies are performed on passive patches in comparison with immediate release oral tablets, and steady-state plasma concentrations are monitored (144). For topical/dermatologicals, bioavailability has been compared by a variety of methods, including tape stripping, microdialysis, skin biopsy, suction blister, follicle removal, and confocal Raman spectroscopy (145). Tape stripping was proposed by the FDA in 1998 as a method to evaluate dermatopharmacokinetics in a draft guidance, but in 2002 the guidance was withdrawn when two investigators testing the same product reached different conclusions on which formulation was better (146). Efforts are underway to harmonize tape-stripping procedures across laboratories (see Section 2.4). Some special considerations may be involved when active patches based on skin enhancement technologies are used. For example, the total dose delivered from an electric patch could potentially be much higher than from a passive patch. This is because iontophoresis uses an electric field to deliver the drug, while a passive patch delivers drug by diffusion based on a concentration gradient. Using iontophoresis, more than 75% of the drug can be administered by proper optimization of formulation, electrochemistry, and device construction (127). However, typically the percentage of dose delivered in laboratory investigations is very low. During delivery of LHRH, only 0.7% of the dose was delivered systemically after a 5-hour active treatment period. In mass balance studies, 92% of the dose could be accounted for, with majority of the amount not even entering the skin (143). Attention to electrode design during commercialization can greatly improve the dose delivered. If the drug forms a depot, then ability to manipulate the drug once it has entered the skin will also affect the dose delivered to the circulation over a defined period of time. Though iontophoresis will result in a faster input of drug, it has not been shown *in vivo* that iontophoresis can actually deliver a true bolus dose (7). Because the amount of drug delivered by iontophoresis depends on several factors, traditional dose-response studies are more difficult to perform (147). Results may often be expressed as apparent bioavailability, which represents an underestimate of the absolute bioavailability that would result if the patches are used until all the drug is exhausted. The apparent bioavailability of octreotide in plasma of rabbits under optimal conditions (5 mg/ml at 150 μA/cm^2 for 8 hours) was about 8% (148). The use of the term *bioavailability* for active delivery devices can be confusing or misleading, and the results should be interpreted only after considering the definition of bioavailability which was used. For example, the following definition has been proposed for iontophoretic devices:

$$\text{Bioavailability} = \frac{\text{Iontophoretic response/loaded mass}}{\text{Reference response/loaded mass}}$$

This definition relates to the pharmacodynamic bioavailability as the final objective to create a response in the patient (127). Other definitions can also be used, and pharmacokinetic bioavailability needs to be calculated, but the perspective on definition and objectives should not be lost. The amount of current carried by any drug ion is a product of three major variables: the current efficiency, the electrode area, and the current density. The current efficiency, in turn, depends on what fraction of the total current is being carried by the drug ion. This will be a factor of several variables, such as the concentration of extraneous ions in the formulation. Because smaller devices would be preferred by the patient and there is an upper limit to current density, methods to increase current efficiency such as by patch design will be critical to successful commercialization of this technology. The possible exception can be when the drug is relatively inexpensive and is not a narcotic (34,147). A current density of 0.2 mA/cm^2 applied over 24 hours is well tolerated and has been taken to clinical trials. Though the upper limit is generally considered to be 0.5 mA/cm^2, this current density can only be used for short periods of time.

The incorporation of rhGH and DDAVP in dissolving microneedles was discussed in Section 3.5. In this study, the absolute bioavailability (BA) of rhGH was calculated by the following equation and was found to be about 73% for dextran-dissolving microneedles (149):

$$BA\ (\%) = (AUC_{patch}/AUC_{i.v.}) \times (Dose_{i.v.}/\ Dose_{patch}) \times 100$$

In another study that used radiofrequency ablation (see Section 3.3.3) to create microchannels in the skin, Levin et al. (150) measured the bioavailability of hGH, a 22-kDa protein, relative to subcutaneous (SC) administration and reported values of 75% in rats or 33% in guinea pigs. A dose-dependent increase in C_{max} and AUC was observed in both rats and guinea pigs up to a dose of 300 μg per 1.4 cm^2.

REFERENCES

1. A. K. Banga. *Therapeutic peptides and proteins: Formulation, processing, and delivery systems,* Taylor & Francis, London, 2006.
2. F. K. Akomeah. Topical dermatological drug delivery: quo vadis? *Curr. Drug Deliv.,* 7:283–296 (2010).
3. G. L. Flynn and B. Stewart. Percutaneous drug penetration: Choosing candidates for transdermal development, *Drug Dev. Res.,* 13:169–185 (1988).
4. V. Sachdeva and A. K. Banga. Iontophoresis: Promises and challenges, *TransDermal,* 2:17–21 (2010).
5. S. Mutalik and N. Udupa. Glibenclamide transdermal patches: physicochemical, pharmacodynamic, and pharmacokinetic evaluations, *J. Pharm. Sci.,* 93:1577–1594 (2004).
6. R. Gannu, Y. V. Vishnu, V. Kishan, and Y. M. Rao. Development of nitrendipine transdermal patches: *In vitro* and *ex vivo* characterization, *Curr. Drug Deliv.,* 4:69–76 (2007).
7. R. H. Guy. Current status and future prospects of transdermal drug delivery, *Pharm. Res.,* 13:1765–1769 (1996).
8. A. M. Wokovich, S. Prodduturi, W. H. Doub, A. S. Hussain, and L. F. Buhse. Transdermal drug delivery system (TDDS) adhesion as a critical safety, efficacy and quality attribute, *Eur. J. Pharm. Biopharm.,* 64:1–8 (2006).
9. M. Kakhi, S. Prodduturi, A. Wokovich, W. Doub, L. Buhse, and N. Sadrieh. Risk analysis of transdermal drug delivery systems, *Pharm. Engg.,* 27:60–72 (2007).

10. J. Vanakoski and T. Seppala. Heat exposure and drugs. A review of the effects of hyperthermia on pharmacokinetics, *Clin. Pharmacokinet.*, 34:311–322 (1998).

11. F. Akomeah, T. Nazir, G. P. Martin, and M. B. Brown. Effect of heat on the percutaneous absorption and skin retention of three model penetrants, *Eur. J. Pharm Sci.*, 21:337–345 (2004).

12. T. S. Shomaker, J. Zhang, and M. A. Ashburn. A pilot study assessing the impact of heat on the transdermal delivery of testosterone, *J. Clin. Pharmacol.*, 41:677–682 (2001).

13. T. S. Shomaker, J. Zhang, and M. A. Ashburn. Assessing the impact of heat on the systemic delivery of fentanyl through the transdermal fentanyl delivery system, *Pain Med.*, 1:225–230 (2000).

14. M. D. Reeves and C. J. Ginifer. Fatal intravenous misuse of transdermal fentanyl, *Med. J. Aust.*, 177:552–553 (2002).

15. J. J. Kuhlman, Jr., R. McCaulley, T. J. Valouch, and G. S. Behonick. Fentanyl use, misuse, and abuse: A summary of 23 postmortem cases, *J. Anal. Toxicol.*, 27:499–504 (2003).

16. N. Montalto, C. C. Brackett, and T. Sobol. Use of transdermal nicotine systems in a possible suicide attempt, *J. Am. Board Fam. Pract.*, 7:417–420 (1994).

17. Zars Pharma, http://www.zars.com (accessed October 8, 2010).

18. R. Paus. Immunology of the hair follicle. In J. D. Bos (ed), *Skin immune system: Cutaneous immunology and clinical immunodermatology*, CRC Press, Boca Raton, Florida, 1997, pp. 377–398.

19. P. C. Panus, J. Campbell, S. B. Kulkarni, R. T. Herrick, W. R. Ravis, and A. K. Banga. Transdermal iontophoretic delivery of ketoprofen through human cadaver skin and in humans, *J. Control. Release*, 44:113–121 (1997).

20. G. Rao, R. H. Guy, P. Glikfeld, W. R. Lacourse, L. Leung, J. Tamada, R. O. Potts, and N. Azimi. Reverse iontophoresis: Noninvasive glucose monitoring *in vivo* in humans, *Pharm. Res.*, 12:1869–1873 (1995).

21. P. W. Ledger. Skin biological issues in electrically enhanced transdermal delivery, *Adv. Drug Del. Rev.*, 9:289–307 (1992).

22. N. A. Monteiro-Riviere. Altered epidermal morphology secondary to lidocaine iontophoresis: *In vivo* and *in vitro* studies in porcine skin, *Fundam. Appl. Toxicol.*, 15:174–185 (1990).

23. S. Reinauer, A. Neusser, G. Schauf, and E. Holzle. Iontophoresis with alternating current and direct current offset (AC DC iontophoresis)—A new approach for the treatment of hyperhidrosis, *Br. J. Dermatol.*, 129:166–169 (1993).

24. J. M. Rattenbury and E. Worthy. Is the sweat test safe? Some instances of burns received during pilocarpine iontophoresis, *Ann. Clin. Biochem.*, 33:456–458 (1996).

25. M. J. Pikal and S. Shah. Transport mechanisms in iontophoresis. III. An experimental study of the contributions of electroosmotic flow and permeability change in transport of low and high molecular weight solutes, *Pharm. Res.*, 7:222–229 (1990).

26. A. Kim, P. G. Green, G. Rao, and R. H. Guy. Convective solvent flow across the skin during iontophoresis, *Pharm. Res.*, 10:1315–1320 (1993).

27. R. Y. Lin, Y. C. Ou, and W. Y. Chen. The role of electroosmotic flow on *in vitro* transdermal iontophoresis, *J. Control. Release*, 43:23–33 (1997).

28. S. M. Sims, W. I. Higuchi, and V. Srinivasan. Skin alteration and convective solvent flow effects during iontophoresis II. Monovalent anion and cation transport across human skin, *Pharm. Res.*, 9:1402–1409 (1992).

29. I. W. H. M. CraaneVanHinsberg, J. C. Verhoef, F. Spies, J. A. Bouwstra, G. S. Gooris, H. E. Junginger, and H. E. Bodde. Electroperturbation of the human skin barrier *in vitro*. II. Effects on stratum corneum lipid ordering and ultrastructure, *Microsc. Res. Tech.*, 37:200–213 (1997).

30. P. C. Kasha and A. K. Banga. A review of patent literature for iontophoretic delivery and devices, *Recent Pat. Drug Deliv. Forml.*, 2:41–50 (2008).

31. J. L. Haynes. Failsafe iontophoresis drug delivery system. Assignee: Becton Dickinson. 954,176 [U.S. Pat. 5,306,235]. 1994. New Jersey, USA.

32. S. Liden. "Sensitivity to electricity"—A new environmental epidemic, *Allergy*, 51:519–524 (1996).

33. L. L. Anderson, M. L. Welch, and W. J. Grabski. Allergic contact dermatitis and reactivation phenomenon from iontophoresis of 5-fluorouracil, *J. Am. Acad. Dermatol.*, 36:478–479 (1997).

34. B. H. Sage. Iontophoresis. In J. Swarbrick and J. C. Boylan (eds), *Encyclopedia of pharmaceutical technology*, Marcel Dekker, New York, 1993, pp. 217–247.

35. M. K. Robinson, C. Cohen, A. B. de Fraissinette, M. Ponec, E. Whittle, and J. H. Fentem. Non-animal testing strategies for assessment of the skin corrosion and skin irritation potential of ingredients and finished products, *Food Chem. Toxicol.*, 40:573–592 (2002).

36. I. Park, D. Kim, J. Song, C. H. In, S. W. Jeong, S. H. Lee, B. Min, D. Lee, and S. O. Kim. Buprederm, a new transdermal delivery system of buprenorphine: Pharmacokinetic, efficacy and skin irritancy studies, *Pharm Res.*, 25:1052–1062 (2008).

37. O. deSilva, D. A. Basketter, M. D. Barratt, E. Corsini, M. T. D. Cronin, P. K. Das, J. Degwert, A. Enk, J. L. Garrigue, C. Hauser, I. Kimber, J. P. Lepoittevin, J. Peguet, and M. Ponec. Alternative methods for skin sensitisation testing—The report and recommendations of ECVAM Workshop 19, *ATLA*, 24:683–705 (1996).

38. P. Karande, A. Jain, and S. Mitragotri. Discovery of transdermal penetration enhancers by high-throughput screening, *Nat. Biotechnol.*, 22:192–197 (2004).

39. F. Dreher, P. Walde, P. L. Luisi, and P. Elsner. Human skin irritation studies of a lecithin microemulsion gel and of lecithin liposomes, *Skin Pharmacol.*, 9:124–129 (1996).

40. H. Kalluri and A. K. Banga. Formation and closure of microchannels in skin following microporation, *Pharm. Res.*, [Epub] (2010).

41. Y. N. Kalia and R. H. Guy. The electrical characteristics of human skin *in vivo*, *Pharm. Res.*, 12:1605–1613 (1995).

42. Y. N. Kalia, L. B. Nonato, and R. H. Guy. The effect of iontophoresis on skin barrier integrity: Non-invasive evaluation by impedance spectroscopy and transepidermal water loss, *Pharm. Res.*, 13:957–960 (1996).

43. K. Kontturi, L. Murtomaki, J. Hirvonen, P. Paronen, and A. Urtti. Electrochemical characterization of human skin by impedance spectroscopy: The effect of penetration enhancers, *Pharm. Res.*, 10:381–385 (1993).

44. Y. N. Kalia and R. H. Guy. Interaction between penetration enhancers and iontophoresis: Effect on human skin impedance *in vivo*, *J. Control. Release*, 44:33–42 (1997).

45. M. J. Clancy, J. Corish, and O. I. Corrigan. A comparison of the effects of electrical current and penetration enhancers on the properties of human skin using spectroscopic (FTIR) and calorimetric (DSC) methods, *Int. J. Pharm.*, 105:47–56 (1994).

46. N. G. Turner, Y. N. Kalia, and R. H. Guy. The effect of current on skin barrier function *in vivo*: Recovery kinetics post-iontophoresis, *Pharm. Res.*, 14:1252–1257 (1997).

47. M. F. Lu, D. Lee, and G. S. Rao. Percutaneous absorption enhancement of leuprolide, *Pharm. Res.*, 9:1575–1579 (1992).

48. J. Y. Fang, P. C. Wu, Y. B. Huang, and Y. H. Tsai. *In vivo* percutaneous absorption of capsaicin, nonivamide and sodium nonivamide acetate from ointment bases: Skin erythema test and non-invasive surface recovery technique in humans, *Int. J. Pharm.*, 131:143–151 (1996).

49. S. Thysman, A. Jadoul, T. Leroy, D. Vanneste, and V. Preat. Laser doppler evaluation of skin reaction in volunteers after histamine iontophoresis, *J. Cont. Rel.*, 36:215–219 (1995).

50. S. Ollmar, M. Nyren, I. Nicander, and L. Emtestam. Electrical impedance compared with other non-invasive bioengineering techniques and visual scoring for detection of irritation in human skin, *Br. J. Dermatol.*, 130:29–36 (1994).

51. S. Ollmar and L. Emtestam. Electrical impedance applied to noninvasive detection of irritation in skin, *Contact Dermatitis*, 27:37–42 (1992).

52. R. vanderGeest, D. A. R. Elshove, M. Danhof, A. P. M. Lavrijsen, and H. E. Bodde. Non-invasive assessment of skin barrier integrity and skin irritation following iontophoretic current application in humans, *J. Control. Release*, 41:205–213 (1996).

53. E. Camel, M. O'Connell, B. Sage, M. Gross, and H. Maibach. The effect of saline iontophoresis on skin integrity in human volunteers. 1. Methodology and reproducibility, *Fund. Appl. Toxicol.*, 32:168–178 (1996).

54. S. Thysman, D. Vanneste, and V. Preat. Noninvasive investigation of human skin after *in vivo* iontophoresis, *Skin Pharmacol.*, 8:229–236 (1995).

55. M. N. Berliner. Skin microcirculation during tapwater iontophoresis in humans: Cathode stimulates more than anode, *Microvascular Res.*, 54:74–80 (1997).

56. S. Y. Oh and R. H. Guy. Effects of iontophoresis on the electrical properties of human skin *in vivo*, *Int. J. Pharm.*, 124:137–142 (1995).

57. I. Kobayashi, K. Hosaka, T. Ueno, H. Maruo, M. Kamiyama, C. Konno, and M. Gemba. Relationship between amount of beta-blockers permeating through the stratum corneum and skin irritation after application of beta-blocker adhesive patches to guinea pig skin, *Biol. Pharm. Bull.*, 20:421–427 (1997).

58. S. Chesnoy, J. Doucet, D. Durand, and G. Couarraze. Effect of iontophoresis in combination with ionic enhancers on the lipid structure of the stratum corneum: An x-ray diffraction study, *Pharm. Res.*, 13:1581–1584 (1996).

59. C. Y. Jin, M. H. Han, S. S. Lee, and Y. H. Choi. Mass producible and biocompatible microneedle patch and functional verification of its usefulness for transdermal drug delivery, *Biomed. Microdevices*, 11:1195–1203 (2009).

60. M. R. Prausnitz, J. C. Birchall, and Co-Organizers. First International Conference on Microneedles, May 23–25, Atlanta, Georgia (2010).

61. M. Ameri, X. Wang, and Y. F. Maa. Effect of irradiation on parathyroid hormone PTH(1-34) coated on a novel transdermal microprojection delivery system to produce a sterile product-adhesive compatibility, *J. Pharm. Sci.*, [Epub] (2009).

62. A. J. Harvey, S. A. Kaestner, D. E. Sutter, N. G. Harvey, J. A. Mikszta, and R. J. Pettis. Microneedle-based intradermal delivery enables rapid lymphatic uptake and distribution of protein drugs, *Pharm. Res.*, [E-pub] (2010).

63. T. R. Singh, M. J. Garland, C. M. Cassidy, K. Migalska, Y. K. Demir, S. Abdelghany, E. Ryan, A. D. Woolfson, and R. F. Donnelly. Microporation techniques for enhanced delivery of therapeutic agents, *Recent Pat. Drug Deliv. Formul.*, 4:1–17 (2010).

64. Norwood Abbey, http://www.norwoodabbey.com.au (accessed July 21, 2010).

65. Transdermal delivery—Making a comeback!, *Drug Del. Technol.*, 10:24–28 (2010).

66. M. Ameri, P. E. Daddona, and Y. F. Maa. Demonstrated solid-state stability of parathyroid hormone PTH(1-34) coated on a novel transdermal microprojection delivery system, *Pharm. Res.*, 26:2454–2463 (2009).

67. M. Ameri, S. C. Fan, and Y. F. Maa. Parathyroid hormone PTH(1-34) formulation that enables uniform coating on a novel transdermal microprojection delivery system, *Pharm. Res.*, 27:303–313 (2010).

68. TransPharmaMedical Ltd., http://www.transpharma-medical.com (accessed July 24, 2010).

69. M. Stern and G. Levin. Transdermal delivery system for dried particulate or lyophilized medications. TransPharma Medical Ltd. [US 7,335,377 B2]. 2008.

70. S. Chang, G. A. Hofmann, L. Zhang, L. J. Deftos, and A. K. Banga. The effect of electroporation on iontophoretic transdermal delivery of calcium regulating hormones. *J. Control. Rel.*, 66:127–133 (2000).

71. K. Boericke, M. A. O'Connell, C. R. Bock, P. Green, and J. A. Down. Iontophoretic delivery of human parathyroid hormone (1-34) in swine. *Proc. Int. Symp. Control. Rel. Bioact. Mater.,* 23:200–201 (1996).

72. K. Hansen and B. Haldin. Transdermal delivery: A solid microstructured transdermal system (sMTS) for systemic delivery of salts and proteins, *Drug Del. Technol.,* 8:38–42 (2008).

73. K. Hansen, J. Simons, and T. Peterson. Transdermal delivery of non-traditional APIs for systemic delivery using solid microstructures, Presented at the Annual Meeting of the American Association of Pharmaceutical Scientists, November 16–20, Atlanta, GA (2008).

74. K. Hansen, J. Simons, and T. Peterson. Transdermal delivery of high volume liquid formulations using hollow microstructures, Presented at the Annual Meeting of the American Association of Pharmaceutical Scientists, November 16–20, 2008, Atlanta, GA (2008).

75. K. Hansen, S. Burton, and M. Tomai. A hollow microstructured transdermal system (hMTS) for needle-free delivery of biopharmaceuticals, *Drug Del. Technol.,* 9:38–44 (2009).

76. A. K. Banga. Microporation applications for enhancing drug delivery, *Expert. Opin. Drug Deliv.,* 6:343–354 (2009).

77. E. Gonzalez-Gonzalez, T. J. Speaker, R. P. Hickerson, R. Spitler, M. A. Flores, D. Leake, C. H. Contag, and R. L. Kaspar. Silencing of reporter gene expression in skin using siRNAs and expression of plasmid DNA delivered by a soluble protrusion array device (PAD), *Mol. Ther.,* 18:1667–1674 (2010).

78. A. K. Andrianov, A. Marin, and D. P. Decollibus. Microneedles with intrinsic immuno-adjuvant properties: Microfabrication, protein stability, and modulated release, *Pharm. Res.,* 28:58–65 (2010).

79. Corium Group, http://www.coriumgroup.com (accessed July 30, 2010).

80. R. J. Flower. Iontophoretic drug delivery device with disposable skin patch and reusable controller—Which detects patch compatibility and exhaustion, and the date and time of usage. Assignee: Becton Dickinson. [Pat. WO 9610440]. 1996.

81. R. J. Flower, W. A. Mcarthur, S. E. Stropkay, and M. W. Tanner. Iontophoretic drug delivery system—Has disposable housing with reservoirs and high current power source and removable reusable controller. Assignee: Becton Dickinson. [Pat. WO 9610441]. 1996.

82. N. T. Azimi, B. V. Bhayani, M. Cao, R. K. Lee, L. Leung, P. J. Plante, J. Tamada, M. J. Tierney, and P. Vijayakumar. Iontophoresis sampling appts for transdermal monitoring of target substance—Has collection reservoir comprising ionically conductive hydrogel and ionically conductive solution, and two iontophoresis electrodes in contact with collection reservoirs in contact with subject's skin. Assignee: Cygnus. [Pat. WO 9600110]. 1996.

83. F. M. J. De Beukelaar, J. L. Mesens, G. Van Reet, and G. Van Rett. Iontophoretic delivery of anti-migraine drug—particularly dihydro benzopyran-alkylamino alkyl substituted guanidine compounds. Assignee: Janssen Pharm. [Pat. WO 9505815]. 1995.

84. G. C. Andrews, G. O. Daumy, M. L. Francoeur, and E. R. Larson. New derivatives of glucagon-like peptide 1 and insulinotropin—Used for enhancing insulin action in a mammal, particularly by iontophoresis. Assignee: Pfizer and Scios. [Pat. WO 9325579]. 1993.

85. J. J. V. Brange. Transdermal insulin. Assignee: Novo Nordisk. 301,838[5,597,796]. 1997.

86. R. H. Zobrist, D. Quan, H. M. Thomas, S. Stanworth, and S. W. Sanders. Pharmacokinetics and metabolism of transdermal oxybutynin: *In vitro* and *in vivo* performance of a novel delivery system, *Pharm Res.,* 20:103–109 (2003).

87. T. Kankkunen, R. Sulkava, M. Vuorio, K. Kontturi, and J. Hirvonen. Transdermal ionto-phoresis of tacrine *in vivo*, *Pharm. Res.*, 19:704–707 (2002).
88. M. A. Ashburn, J. Streisand, J. Zhang, G. Love, M. Rowin, S. Niu, J. K. Kievit, J. R. Kroep, and M. J. Mertens. The iontophoresis of fentanyl citrate in humans, *Anesthesiology*, 82:1146–1153 (1995).
89. P. Green. Iontophoretic delivery of peptide drugs, *J. Control. Release*, 41:33–48 (1996).
90. J. Singh, M. Gross, M. O'Connell, B. Sage, and H. I. Maibach. Effect of iontophoresis in different ethnic groups' skin function. *Proc. Int. Symp. Control. Rel. Bioact. Mater.*, 21:365–366 (1994).
91. A. Jadoul, J. Mesens, W. Caers, F. Beukelaar, R. Crabbe, and V. Preat. Transdermal per-meation of alniditan by iontophoresis: *In vitro* optimization and human pharmacokinetic data, *Pharm. Res.*, 13:1348–1353 (1996).
92. Y. B. Bannon, J. Corish, O. I. Corrigan, and J. G. Masterson. Iontophoretically induced transdermal delivery of salbutamol, *Drug Develop. Ind. Pharm.*, 14:2151–2166 (1988).
93. J. B. Phipps, R. Haak, S. Chao, S. K. Gupta, and J. R. Gyory. *In vivo* transdermal electrotransport in humans. *Proc. Int. Symp. Control. Rel. Bioact. Mater.*, 21:170–171 (1994).
94. Chattanooga Group, http://www.chattgroup.com (accessed September 26, 2010).
95. D. D. Maurer, T. J. Williams, and S. A. Stevens. Multiple site drug iontophoresis electronic device and method. Assignee: Empi. 649,495 [U.S. Pat. 5,254,081]. 1993. Minnesota, USA.
96. Life-Tech, http://www.life-tech.com (accessed October 8, 2010).
97. N. S. Murthy, V. A. Boguda, and K. Payasada. Electret enhances transdermal drug per-meation, *Biol. Pharm. Bull.*, 31:99–102 (2008).
98. T. M. Parkinson, M. A. Szlek, and J. D. Isaacson. Hybresis: The hybridization of tradi-tional with low-voltage iontophoresis, *Drug Del. Technol.*, 7:54–60 (2007).
99. G. A. Fischer. Iontophoretic drug delivery using the IOMED Phoresor system, *Expert Opin. Drug Del.*, 2:391–403 (2005).
100. Vyteris, http://www.vyteris.com (accessed October 8, 2010).
101. S. W. Stralka, P. L. Head, and K. Mohr. The clinical use of iontophoresis, *Phys. Ther. Prod.*, March:48–51 (1996).
102. A. K. Banga and Y. W. Chien. Hydrogel-based iontotherapeutic delivery devices for transdermal delivery of peptide/protein drugs, *Pharm. Res.*, 10:697–702 (1993).
103. M. A. Hollinger. Toxicological aspects of topical silver pharmaceuticals, *Crit. Rev. Toxicol.*, 26:255–260 (1996).
104. Isis Biopolymer, http://www.isisbiopolymer.com (accessed October 4, 2010).
105. E. A. Durand. Transdermal delivery: Intelligent sustained-release using a selective charged barrier membrane, *Drug Del. Technol.*, 9:28–31 (2009).
106. M. W. Pierce. Transdermal delivery of sumatriptan for the treatment of acute migraine, *Neurotherapeutics*, 7:159–163 (2010).
107. A. Femenia-Font, C. Balaguer-Fernandez, V. Merino, and A. Lopez-Castellano. Iontophoretic transdermal delivery of sumatriptan: Effect of current density and ionic strength, *J. Pharm. Sci.*, 94:2183–2186 (2005).
108. S. J. Siegel, C. O'Neill, L. M. Dube, P. Kaldeway, R. Morris, D. Jackson, and T. Sebree. A unique iontophoretic patch for optimal transdermal delivery of sumatriptan, *Pharm. Res.*, 24:1919–1926 (2007).
109. NuPathe, http://www.nupathe.com (accessed October 6, 2010).
110. Acient, http://www.aciont.com (accessed October 8, 2010).
111. G. A. Fischer, T. M. Parkinson, and M. A. Szlek. OcuPhor—The future of ocular drug delivery, *Drug Del. Technol.*, 2:50–52 (2002).

112. C. Valenta and B. G. Auner. The use of polymers for dermal and transdermal delivery, *Eur. J. Pharm. Biopharm.*, 58:279–289 (2004).

113. S. Zurdo, I. P. Franke, U. F. Schaefer, and C. M. Lehr. Delivery of ethinylestradiol from film forming polymeric solutions across human epidermis *in vitro* and *in vivo* in pigs, *J. Control. Release*, 118:196–203 (2007).

114. S. Zurdo, I. P. Franke, U. F. Schaefer, and C. M. Lehr. Development and characterization of film forming polymeric solutions for skin drug delivery, *Eur. J. Pharm. Biopharm.*, 65:111–121 (2007).

115. J. Zhang, Z. Liu, H. Du, Y. Zeng, L. Deng, J. Xing, and A. Dong. A novel hydrophilic adhesive matrix with self-enhancement for drug percutaneous permeation through rat skin, *Pharm. Res.*, 26:1398–1406 (2009).

116. A. D'Emanuele and J. N. Staniforth. An electrically modulated drug delivery device. I, *Pharm. Res.*, 8:913–918 (1991).

117. A. D'Emanuele and J. N. Staniforth. An electrically modulated drug delivery device. II. Effect of ionic strength, drug concentration, and temperature, *Pharm. Res.*, 9:215–219 (1992).

118. Y. W. Chien and A. K. Banga. Iontotherapeutic devices, reservoir electrode devices therefore, process and unit dose. Assignee: Rutgers University. [U.S. Pat. 5,250,022], 1–13. 1993. USA.

119. R. Tomer, D. Dimitrijevic, and A. T. Florence. Electrically controlled release of macro-molecules from cross-linked hyaluronic acid hydrogels, *J. Control. Release*, 33:405–413 (1995).

120. F. Moll and P. Knoblauch. Iontophoretic *in vitro* release of antimycotics from hydrogels, *Drug Dev. Ind. Pharm.*, 19:1143–1158 (1993).

121. S. K. Gupta, S. Kumar, S. Bolton, C. R. Behl, and A. W. Malick. Effect of chemi-cal enhancers and conducting gels on iontophoretic transdermal delivery of cromolyn sodium, *J. Control. Release*, 31:229–236 (1994).

122. W. Chen and S. G. Frank. Iontophoresis of lidocaine H$^+$ from poloxamer 407 gels, *Pharm. Res.*, 14:S-309 (1997).

123. P. Chansai, A. Sirivat, S. Niamlang, D. Chotpattananont, and K. Viravaidya-Pasuwat. Controlled transdermal iontophoresis of sulfosalicylic acid from polypyrrole/poly(acrylic acid) hydrogel, *Int. J. Pharm.*, 381:25–33 (2009).

124. K. R. Chaudhuri. Crystallisation within transdermal rotigotine patch: Is there cause for concern? *Expert. Opin. Drug Deliv.*, 5:1169–1171 (2008).

125. V. P. Shah. Transdermal drug delivery system regulatory issues. In R. H. Guy and J. Hadgraft (eds), *Transdermal drug delivery*, Marcel Dekker, New York, 2003, pp. 361–367.

126. J. R. Gyory, L. C. Moodie, J. B. Phipps, and F. Theeuwes. Electrically powered ionto-phoretic device for enhanced drug delivery—Has reservoir for drug, pref fentanyl, in electrode and ion exchange material with mobile and immobile ion species in same or other electrode. Assignee: Alza Corp. [Pat. WO 9527529]. 1995.

127. B. H. Sage, C. R. Bock, J. D. Denuzzio, and R. A. Hoke. Technological and develop-mental issues of iontophoretic transport of peptide and protein drugs. In V. H. L. Lee, M. Hashida, and Y. Mizushima (eds), *Trends and future perspectives in peptide and protein drug delivery*, Harwood Academic, Chur, Switzerland, 1995, pp. 111–134.

128. N. Higo, K. Iga, Y. Matsumoto, and S. Yanai. Iontophoresis interface useful in delivery of peptide drugs with high bioavailability—Comprises hydrophilised fluro-resin membrane with low protein adsorptivity. Assignee: Hisamitsu and Takeda. [Pat. EP 748636]. 1996.

129. B. F. J. Broberg, R. J. Clark, and P. G. Green. Forming and packaging of iontopho-retic drug delivery patches—In a continuous process in an inert atmosphere to provide increased shelf-life. Assignee: Becton Dickinson. [Pat. WO 9610398]. 1996.

130. R. P. Haak, R. M. Myers, and R. W. Plue. Iontophoretic drug delivery device—Has electrical pathway formed on one side of a flexible, non-conducting substrate. Assignee: Alza Corp. [Pat. EP 751801]. 1997.
131. L. Fabian and T. J. Williams. Iontophoresis electronic device having a ramped output current. Assignee: Empi. 984,303 [U.S. Pat. 5,431,625]. 1995. Minnesota, USA.
132. R. Tapper. Iontophoretic delivery of medicaments through skin—With electrical treatment current between electrodes periodically reversed at very low frequencies to mitigate tissue damage. Assignee: Tapper, R. [Pat. EP 776676]. 1997.
133. R. J. Flower and B. H. Sage. Iontophoretic drug delivery system—Has two electrodes and associated reservoirs with switch to reverse direction of current flow. Assignee: Becton Dickinson. [Pat. WO 9509032]. 1995.
134. V. Vemulapalli, Y. Yang, P. M. Friden, and A. K. Banga. Synergistic effect of iontophoresis and soluble microneedles for transdermal delivery of methotrexate, *J. Pharm. Pharmacol.*, 60:27–33 (2008).
135. C. Cullander and R. H. Guy. (D) Routes of delivery: Case studies (6) Transdermal delivery of peptides and proteins, *Adv. Drug Del. Rev.*, 8:291–329 (1992).
136. S. Dinh, S. E. Wouters, and J. R. Sclafani. Application of thermopiles in iontophoretic drug delivery systems, *Pharm. Res.*, 13:S-360 (1996).
137. J. R. Gyory and J. R. Perry. Iontophoretic delivery device and method of hydrating same. Assignee: Alza Corp. 892,554 [U.S. Pat. 5,310,404]. 1994. California, USA.
138. J. E. Beck, L. B. Lloyd, and T. J. Petelenz. Iontophoretic delivery device with integral hydrating assembly—Has sealed liquid storage compartment with tab pulled for progressive unsealing of compartment to release liquid onto hydratable matrix. Assignee: Iomed. [WO Pat. 9605884]. 1996.
139. L. L. Miller, G. A. Smith, A. Chang, and Q. Zhou. Electrochemically controlled release, *J. Control. Release*, 6:293–296 (1987).
140. G. A. VanBuskirk, M. A. Gonzalez, W. P. Shah, S. Barnhardt, C. Barrett, S. Berge, G. Cleary, K. Chan, G. Flynn, T. Foster, R. Gale, R. Garrison, S. Gochnour, A. Gotto, S. Govil, V. A. Gray, J. Hammar, S. Harder, C. Hoiberg, A. Hussain, C. Karp, H. Llanos, J. Mantelle, P. Noonan, D. Swanson, and H. Zerbe. Scale-up of adhesive transdermal drug delivery systems, *Pharm. Res.*, 14:848–852 (1997).
141. K. J. Godbey. Development of a novel transdermal drug delivery backing film with a low moisture vapor transmission rate, *Pharm. Technol.*, 21:98–107 (1997).
142. M. E. Donawa. Drug delivery devices: Regulatory and quality requirements for medical devices, Part I, *Pharm. Technol.*, 21:110–114 (1997).
143. M. C. Heit, N. A. Monteiroriviere, F. L. Jayes, and J. E. Riviere. Transdermal iontophoretic delivery of luteinizing hormone releasing hormone (LHRH): Effect of repeated administration, *Pharm. Res.*, 11:1000–1003 (1994).
144. Y. S. Krishnaiah, S. M. Al Saidan, D. V. Chandrasekhar, and V. Satyanarayana. Bioavailability of nerodilol-based transdermal therapeutic system of nicorandil in human volunteers, *J. Control. Release*, 106:111–122 (2005).
145. C. Herkenne, I. Alberti, A. Naik, Y. N. Kalia, F. X. Mathy, V. Preat, and R. H. Guy. *In vivo* methods for the assessment of topical drug bioavailability, *Pharm. Res.*, 25:87–103 (2008).
146. L. M. Russell and R. H. Guy. Measurement and prediction of the rate and extent of drug delivery into and through the skin, *Expert. Opin. Drug Deliv.*, 6:355–369 (2009).
147. B. H. Sage and J. E. Riviere. Model systems in iontophoresis—Transport efficacy, *Adv. Drug Del. Rev.*, 9:265–287 (1992).
148. D. T. W. Lau, J. W. Sharkey, L. Petryk, F. A. Mancuso, Z. L. Yu, and F. L. S. Tse. Effect of current magnitude and drug concentration on iontophoretic delivery of octreotide acetate (Sandostatin(R)) in the rabbit, *Pharm. Res.*, 11:1742–1746 (1994).

149. K. Fukushima, A. Ise, H. Morita, R. Hasegawa, Y. Ito, N. Sugioka, and K. Takada. Two-layered dissolving microneedles for percutaneous delivery of peptide/protein drugs in rats, *Pharm. Res.*, [Epub] (2010).

150. G. Levin, A. Gershonowitz, H. Sacks, M. Stern, A. Sherman, S. Rudaev, I. Zivin, and M. Phillip. Transdermal delivery of human growth hormone through RF-microchannels, *Pharm. Res.*, 22:550–555 (2005).

Index

Milton Keynes UK
Ingram Content Group UK Ltd.
UKHW051014071024
449327UK00012B/243

9 780367 382780